Conference Proceedings of the Society for Experimental Mechanics Series

Series Editor

Kristin B. Zimmerman, Ph.D.
Society for Experimental Mechanics
Bethel, CT, USA

More information about this series at http://www.springer.com/series/8922

Cesar Sciammarella • John Considine • Paul Gloeckner
Editors

Experimental and Applied Mechanics, Volume 4

Proceedings of the 2015 Annual Conference on Experimental and Applied Mechanics

Springer

Editors
Cesar Sciammarella
Illinois Institute of Technology
Chicago, USA

John Considine
USDA FS Forest Products Library
Edmonton, Alberta, Canada

Paul Gloeckner
Cummins, Inc.
Columbus, Indiana, USA

ISSN 2191-5644 ISSN 2191-5652 (electronic)
Conference Proceedings of the Society for Experimental Mechanics Series
ISBN 978-3-319-22448-0 ISBN 978-3-319-22449-7 (eBook)
DOI 10.1007/978-3-319-22449-7

Library of Congress Control Number: 2015951227

Springer Cham Heidelberg New York Dordrecht London

Printed on acid-free paper

Springer International Publishing AG Switzerland is part of Springer Science+Business Media (www.springer.com)

Preface

Experimental and Applied Mechanics represents one of nine volumes of technical papers presented at the 2015 SEM Annual Conference and Exposition on Experimental and Applied Mechanics organized by the Society for Experimental Mechanics and held in Costa Mesa, CA, June 8–11, 2015. The complete Proceedings also includes volumes on *Dynamic Behavior of Materials; Challenges in Mechanics of Time-Dependent Materials; Advancement of Optical Methods in Experimental Mechanics; MEMS and Nanotechnology; Mechanics of Biological Systems and Materials; Mechanics of Composite & Multifunctional Materials; Fracture, Fatigue, Failure and Damage Evolution; and Residual Stress, Thermomechanics & Infrared Imaging, Hybrid Techniques and Inverse Problems.*

Each collection presents early findings from experimental and computational investigations of an important topic within the field of Experimental Mechanics. This volume includes papers on Quantitative Visualizations, Advanced Methods for Frontier Applications, Non-homogeneous Parameters Identification, Teaching Experimental Mechanics in the Twenty-First Century, and the Mechanics of Interfaces.

Chicago, USA — Cesar Sciammarella
Edmonton, Alberta, Canada — John Considine
Columbus, Indiana, USA — Paul Gloeckner

Contents

Chapter 1
Experimental Characterization of Shape Fixity in Transversely Curved Unimorph Shape Memory Polymer Actuators

Jason T. Cantrell, Jordan R. Van Hall, Andie J. Young, and Peter G. Ifju

Abstract Shape memory polymers (SMP) have the potential to be utilized as a lightweight, solid state actuator in modern reconfigurable structures including as deployment systems for satellite solar panels or morphing aircraft wings. This study is predominantly focused on the use of Veriflex-S®, a thermally activated shape memory polymer, and bi-directional carbon-fiber-reinforced polymer (CFRP) in a unimorph actuator configuration. One of the deficiencies of shape memory polymer structures is the lack of recovery after a single or multiple thermo-mechanical cycle(s). The novel concept of incorporating transverse curvature in the CFRP substrate, similar to that of an extendable tape measurer, has demonstrated the ability to improve shape recovery by increasing the bending stiffness of the unimorph composite actuator to compensate for the lack of recovery of the SMP. Another deficiency of SMP actuators is the lack of retention of the stored or deformed shape over time. This paper is concerned with the effect of transverse curvature on the shape fixity when the unimorph composite is mechanically deformed into a stored configuration. A set of experiments was designed to determine the influence of transverse curvature, the relative widths of SMP and CFRP substrate, and SMP thickness on actuator fixity after a thermo-mechanical cycle. The performance of SMP unimorph actuators with varying levels of transverse curvature in the substrate were evaluated versus SMP unimorph actuators with a flat substrate. The ability of the unimorph actuator to minimize shape retention loss was evaluated by determining the deflection angle over time after being released from the stored configuration. Experimental results indicate that in most circumstances transverse curvature does not negatively affect the ability of the unimorph actuator to retain its shape fixity. Cases of 100 % or greater retention of shape fixity were documented which could lead to the expanded use of transversely curved shape memory polymer unimorph actuators in reconfigurable structures.

Keywords Shape memory polymer • Unimorph • Transverse curvature • Composite • Shape fixity

Nomenclature

c	Polymer width
CF	Carbon fiber
CFRP	Carbon-fiber-reinforced polymer
COV	Coefficient of variation
CTE	Coefficient of thermal expansion
L	Actuator length
MAV	Micro air vehicle
s	Substrate width
SMP	Shape memory polymer
t	Polymer thickness
T_g	Glass transition temperature
UCA	Unimorph composite actuator
ρ	Substrate radius of curvature

J.T. Cantrell (✉) • J.R. Van Hall • A.J. Young • P.G. Ifju
Department of Mechanical and Aerospace Engineering, University of Florida, 231 MAE-A, Gainesville, FL 32611, USA
e-mail: jasontcantrell@gmail.com

C. Sciammarella et al. (eds.), *Experimental and Applied Mechanics, Volume 4*,
Conference Proceedings of the Society for Experimental Mechanics Series, DOI 10.1007/978-3-319-22449-7_1

1.1 Introduction

Shape memory polymers (SMPs) are a subset of smart materials with the ability to change their shape from a programmed state to a temporary state when activated via an external stimulus [1, 2]. Veriflex-S®, the SMP provided by our sponsors for this study, uses a thermal external stimulus to allow for reconfiguration from the original/programmed state (flat beams in this study) to a deformed/stored shape (a U-shaped configuration for this study) and then return to a deployed shape (ideally the same as the original/programmed shape). Veriflex-S® and its higher glass transition temperature (T_g) counterpart Veriflex-E® have been used in applications including: deployment of satellite solar panels, active disassembly for electronics recycling, and morphing of aircraft structures [3–6]. Veriflex®, like other SMPs has two distinct classifications of stiffness and material behavior: the high modulus "glassy state" and the low modulus "rubber state" [7–10]. When the SMP is below the T_g of 62 °C it is stiff, brittle, and has a high elastic modulus (~1.1 GPa). When Veriflex® is heated above T_g the modulus drops by several orders of magnitude (~1 MPa) and becomes pliable. Veriflex® can take and hold temporary shapes indefinitely if heated above its T_g, mechanically deformed, and held in place as it cools below T_g. The sample can return to its unconstrained original/programmed state with the reapplication of heat. Ideally the sample would return to 100 % of the original shape seen before the thermo-mechanical cycle; however, in reality Veriflex® can achieve a final shape that is only close to the original shape. A typical thermo-mechanical cycle is illustrated in Fig. 1.1 which shows the ideal versus actual stress-strain-temperature behavior for a SMP. The actual and idealized cycles match until the load is released in the stored position. Ideally, the SMP retains its current configuration; however, some relaxation occurs during this step after the load is released. This is shown as shape retention or fixity loss in Fig. 1.1. Once reheated and then allowed to cool the SMP will try to return to its original/programmed shape. Unfortunately, as previously stated the SMP is often unable to recover fully and experiences some residual deformation or shape recovery loss. Researchers have studied both fixity and recoverability of SMPs extensively to determine their practicality in reconfigurable structures [9, 11–15].

Despite the lack of shape recovery and fixity in certain settings, SMPs are still seen as beneficial materials for adaptive structures. The current authors have successfully utilized SMPs in the past to provide a feasible solution for the folding of micro air vehicle (MAVs) wings and as a multipurpose morphing actuator with nearly 100 % recoverability [5, 15]. This was accomplished with the incorporation of transverse curvature (similar to an extendable tape measurer) into the carbon-fiber-reinforced polymer (CFRP) substrate of a unimorph composite actuator. A unimorph composite actuator (UCA) is described as an element capable of bi-stable configuration when supplied with an external stimulus (heat in this instance) consisting of one active layer (SMP) to which the stimulus is applied and one inactive layer (CFRP) that supports the active layer. In previous studies a simple flat CFRP beam with SMP adhered to its surface was compared against a transversely curved CFRP beam to determine the effect of transverse curvature on the recovery and residual deformation of UCAs. This study details the effects of transverse curvature on the shape retention loss (fixity) and deflection angles after the UCA has been mechanically deformed into the U-shaped stored configuration, cooled, then released from its storage container. The three types of UCAs tested are shown in Fig. 1.2 in the stored configuration. After testing it was determined that adding transverse

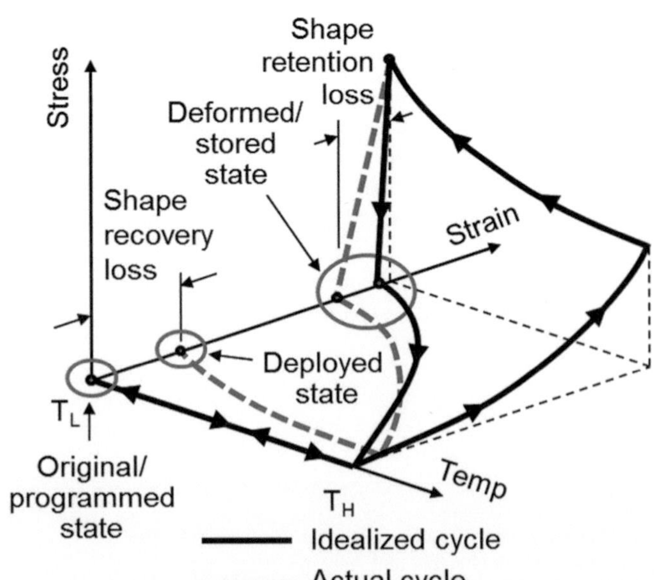

Fig. 1.1 Schematic of the stress-strain-temperature behavior of a SMP during a thermo-mechanical cycle

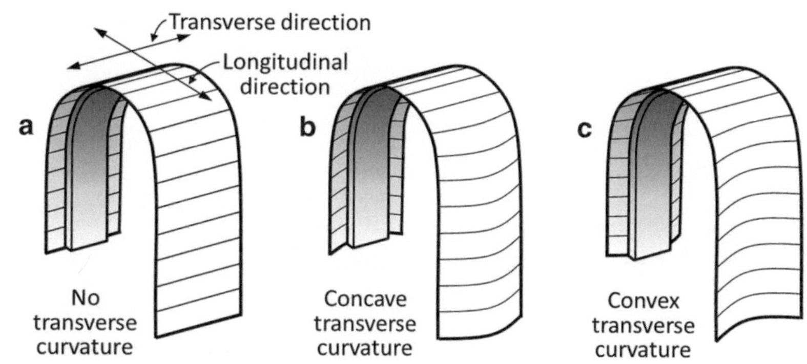

Fig. 1.2 Comparison of a UCA (**a**) without curvature (flat substrate), (**b**) with concave transverse curvature, and (**c**) with convex transverse curvature in the stored configuration

curvature to the CFRP substrate does not adversely affect the shape fixity of the UCA. The details of the UCA fixity analysis and the experimental procedures followed are described in the ensuing sections.

1.2 Fabrication Methods, Experimental Set-Ups, and Procedures

1.2.1 Materials

The individual materials used to form the UCAs are Veriflex-S® shape memory polymer and [±45°] oriented, plain weave, bi-directional carbon-fiber-reinforced polymer. The properties for the CFRP or simply carbon fiber (CF) were gathered in the [±45°] orientation as this was the direction of actuation. The materials were characterized to establish the modulus of elasticity, coefficient of thermal expansion (CTE), and Poisson's ratio of both materials below their glass transition temperatures. These values are provided below in Table 1.1.

1.2.2 Unimorph Composite Actuator Fabrication

Each unimorph composite actuator is comprised of a carbon-fiber-reinforce polymer substrate bonded to a thin sheet of SMP. Both unimorphs incorporating transverse curvature in the substrate and flat CF unimorphs followed the same fabrication procedures. A single layer of [±45°] oriented, plain weave, bi-directional CF, pre-impregnated with epoxy, was cut and placed on a Teflon® covered flat plate or curved tooling board. The [±45°] configuration was used versus a [0°/90°] configuration because it allowed for the CF to be rolled to a smaller diameter in the stored state and was more stable during storage. The entire assembly was covered in an additional layer of Teflon® film, wrapped with breather fabric, vacuum bagged, and cured in a convection oven at 130 °C for 4 h. After curing, the CF was cut to the proper dimensions, and then a Veriflex-S® shape memory polymer panel was bonded to it using Araldite® 2011 two-part epoxy at room temperature to create a unimorph composite beam.

1.2.3 Camera Set-up

The main objective of this investigation was to determine the deflection angle and shape fixity of the UCA samples. This was done using a Point Grey® 5-megapixel grayscale camera to capture images of the side profile of the samples. The cameras were set up perpendicular to the samples to capture the fixity angle over time after the specimens are released from the U-shaped fixity sample holder. Figure 1.3 shows the experimental set-up to measure the fixity angle of the UCA specimens. All of the samples had an initial angle of 0° due to being cooled in the sample holder. Immediately after being released from the holder data collection began to determine the fixity angle over time. Images were captured via VIC Snap™ 2009 and processed via MATLAB® to determine the deflection angles.

Table 1.1 Material properties for [±45°] oriented, plain weave, bi-directional carbon fiber and Veriflex-S® shape memory polymer

Property	CFRP [±45°]		Veriflex-S®	
	Value	COV (%)	Value	COV (%)
Coefficient of thermal expansion (α)	$2.5 \pm 0.2\ E^{-6}\ K^{-1}$	8	$160 \pm 15\ E^{-6}\ K^{-1}$	9.6
Young's modulus (E)	11.8 ± 0.3 GPa	2.8	1.1 ± 0.05 GPa	4.5
Poisson's ratio (ν)	0.79 ± 0.02	2.5	0.45 ± 0.01	2.2
Glass transition temperature[a] (T_g)	121 °C	–	62 °C	–

[a]Manufacturer published values

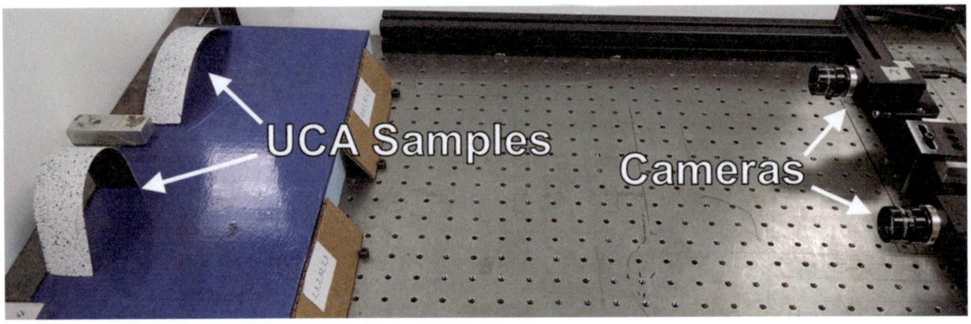

Fig. 1.3 Experimental set-up used to photograph the UCAs after cooling

1.2.4 Environmental Chamber Set-up

The UCAs were placed in a Sun Systems® Model EC12 environmental chamber used to regulate the temperature to the desired point above the SMPs glass transition temperature. Each sample was placed in the pre-heated chamber for an hour to allow it to reach steady state. The temperature was monitored via a thermocouple inside of the chamber and confirmed via a Fluke® 561 series infrared thermometer. Actuator samples were placed, unconstrained, on a 12 mm thick Teflon® plate to allow for full expansion under elevated temperature conditions.

1.2.5 UCA Sample Holder Set-up

Once the actuators were removed from the environmental chamber they were folded into a U-shaped configuration and stored in the retainer shown in Fig. 1.4 to ensure equivalent loading conditions for all actuators. This apparatus consisted of a 63.5 mm diameter by 76 mm half-round block mounted to an aluminum block of the same thickness and width. The UCAs are secured on the top and bottom by a 6 mm plate and 13 mm diameter rod respectively. The entire structure is secured to an optical table with ¼–20 bolts during the hour cool down.

1.2.6 Procedure to Measure UCA Fixity

Step-by-step procedure to measure shape fixity of the UCA.

The procedure for measuring the deflection angle is enumerated below.

Step 1. Place the undeformed UCA in the environmental chamber for 1 h at 85 °C.
Step 2. Bend the UCA beam into the U-shaped stored configuration and place it within the holder to cool to room temperature (22 °C) for 1 h.
Step 3. Remove UCA from the fixity holder and place it in the camera set-up.
Step 4. Monitor via camera while the UCA approaches its final fixity position.

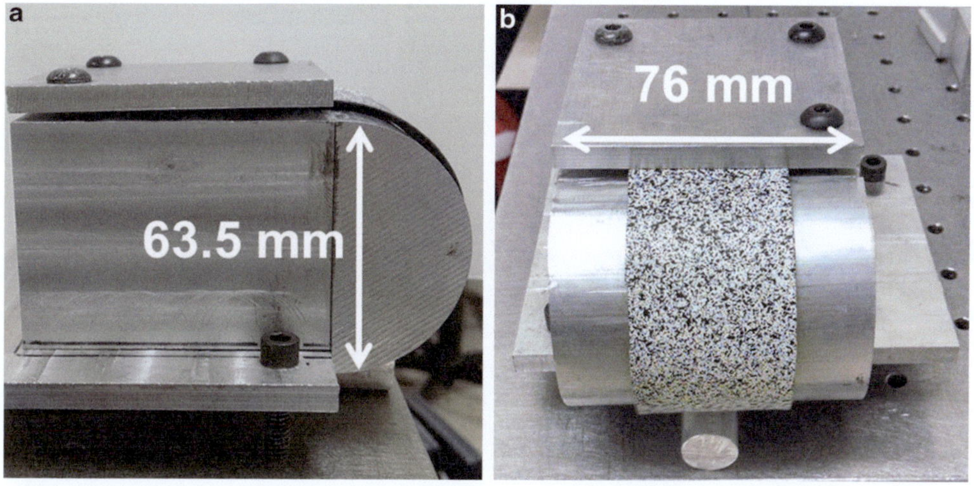

Fig. 1.4 (**a**) Side view of the sample container and (**b**) front view of the sample container holding a UCA

1.3 Unimorph Composite Actuator Results

The main focus of the experiments conducted was the deflection angle (°) as it was used to determine the shape fixity or shape retention loss of the stored UCA specimens over time. Ideally, a programmed UCA would be heated above T_g, mechanically deformed into the stored configuration, cooled below T_g, and would be capable of holding the desired stored shape indefinitely without shape retention loss and have deflection angle less than or equal to zero degrees. Initial samples consisted of 200 mm long by 38 mm wide flat (infinite radius of curvature) and 63.5 mm radius of curvature CF substrates. These samples consisted of a 12.7 mm wide and 1.6 mm thick strip of flat SMP adhered via Araldite® 2011 epoxy to the center of the CF beam. Figure 1.5 shows a mock-up of these variables on a concave UCA.

Post-processing of the images was required to determine the deflection angle for each UCA over time. Post-processing was accomplished via MATLAB® where each image was evaluated with an edge detection algorithm to find the boundaries of the UCA. The image was then filtered to eliminate any white noise the edge detection procedure incorrectly interpreted as a boundary of the UCA. Finally, the image was evaluated on both sides to find the angle of the outer edges of the UCA. This was accomplished by plotting lines along the outer edges of the UCA and finding the angle of each side with respect to a vertical line at the midplane. Positive deflection was recognized as angles returning the UCA back to the original/programmed state and negative deflection as angles further establishing the UCA in a stored configuration. A deflection angle of zero would correspond to UCAs that retained the U-shaped stored configuration without any relaxation or contraction. This process is illustrated in Fig. 1.6a while the deflection angles are shown in Fig. 1.6b.

Data for the flat, concave, and convex UCA samples were collected at various intervals to establish a pattern of relaxation or contraction in the stored configuration. Initial testing determined that the deflection angle of the UCAs displayed asymptotic behavior and thus images over several time scales were needed. The deflection angle versus time data is shown in Fig. 1.7 for the first hour, first 4 h, and for 7 days. The experiments were terminated after 7 days as general trends could be established and a majority of the UCA relaxation had already occurred. The data clearly shows all three UCAs have positive deflection angles after 7 days but only the concave sample has displayed acceptable fixity behavior. The concave sample initially has a deflection angle of −4.6° after being removed from the fixity holder and after 60 min had contracted slightly to an angle of −5.4°. The concave sample had a significantly lower deflection angle compared to the convex sample's initial deflection angle of 22.5° which expanded to 45.5° after 60 min. After 4 h the concave sample experiences some relaxation and has an angle of −3.6° while the flat and convex samples continue to have increasingly large deflection angles of 20.1° and 61.1° respectively. Finally, after 7 days the concave sample had a deflection angle of 7.1° while the flat and convex samples had angles of 55.0° and 121.4° respectively. The concave UCA displays a nearly flat growth rate with movement of less than a degree per day. The flat sample's deflection angle continued to increase at a rate of 3° per day while the convex sample expands at 5° per day. The results indicate that the concave configuration should be used for any further testing due to the minimal deflection angles seen in the fixity configuration under comparable conditions. The flat UCA does not perform as well as the concave configuration but will be used as a baseline to compare to the concave UCA. The convex configuration performed incredibly poorly and was not used in further testing due to its excessive shape retention loss.

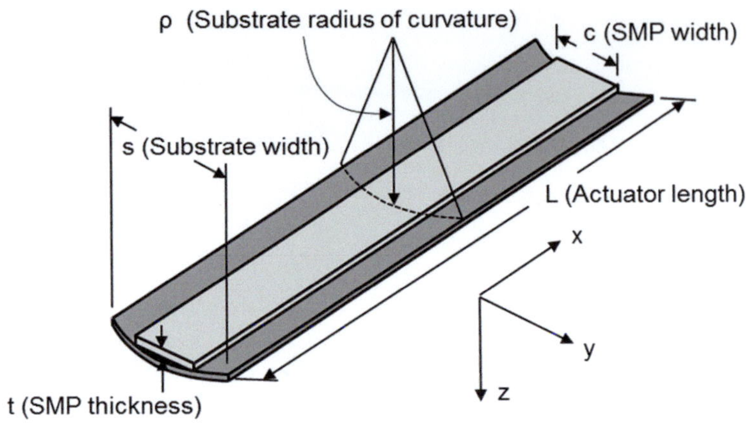

Fig. 1.5 Illustration of the variables of interest for this study on an original/programmed UCA

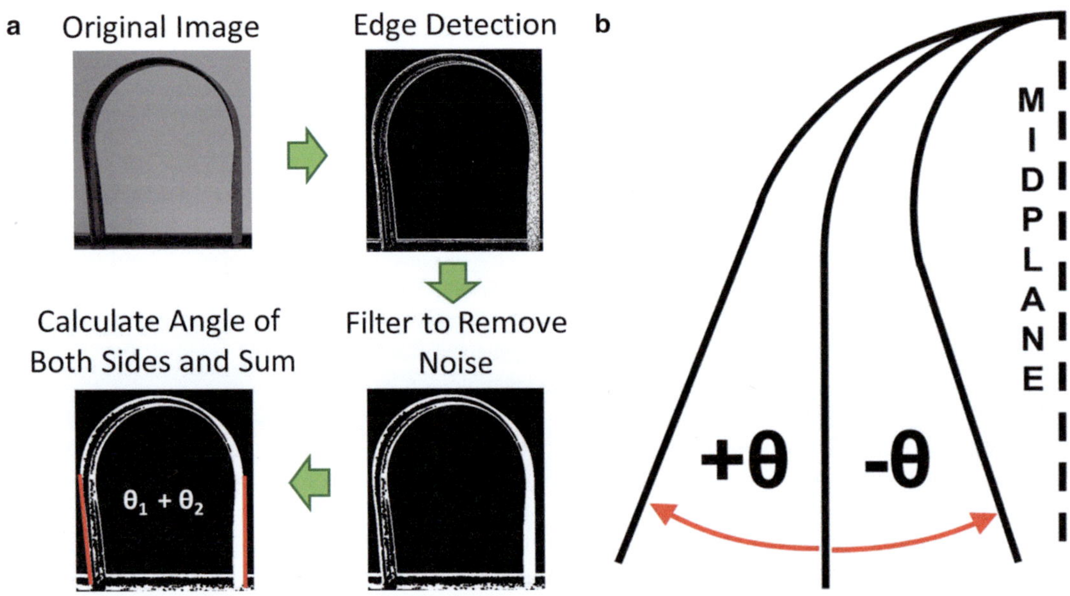

Fig. 1.6 (**a**) Illustration of the post-processing of the images and calculation of the deflection angles for a UCA. (**b**) Diagram illustrating the definition of positive and negative deflection angles for this study

The preliminary tests showed that the concept of applying concave transverse curvature to a unimorph substrate did not substantially hinder the shape fixity of the UCA. The tests also show that a UCA with concave transverse curvature tends to have less shape retention loss than flat UCAs with all other parameters identical. Further investigation was needed to determine the interactions between the assorted variables present in the construction of the UCAs and shape fixity. As shown previously in Fig. 1.5 there were five variables: actuator length, substrate radius of curvature, substrate width, SMP width, and SMP thickness. Fixing the SMP width and actuator length were required to conserve material due to the limited amount of Veriflex-S® available. After eliminating these variables, substrate radius of curvature, SMP thickness, and substrate width were left as the variables that would be altered during the expansion of the design space. Three transverse curvatures were investigated and given designations; **flat** (infinite radius of curvature), **low transverse curvature** (127 mm radius of curvature), and **high transverse curvature** (63.5 mm radius of curvature). Two SMP thicknesses were investigated; **thin** (0.8 mm) and **thick** (1.6 mm). Finally, three substrate widths were tested; **narrow** (25.4 mm), **intermediate** (38.1 mm), and **wide** (50.8 mm). This led to a total of 18 different combinations of UCAs. Table 1.2 shows the variations of curvature, thickness, and width chosen for investigation and their respective designations for convenience of the reader for the remainder of the paper. All of the samples were tested in the same manner as the original three samples and the maximum deflection angles calculated for each UCA. The data is shown below in Fig. 1.8 for the 18 samples investigated. The flat and

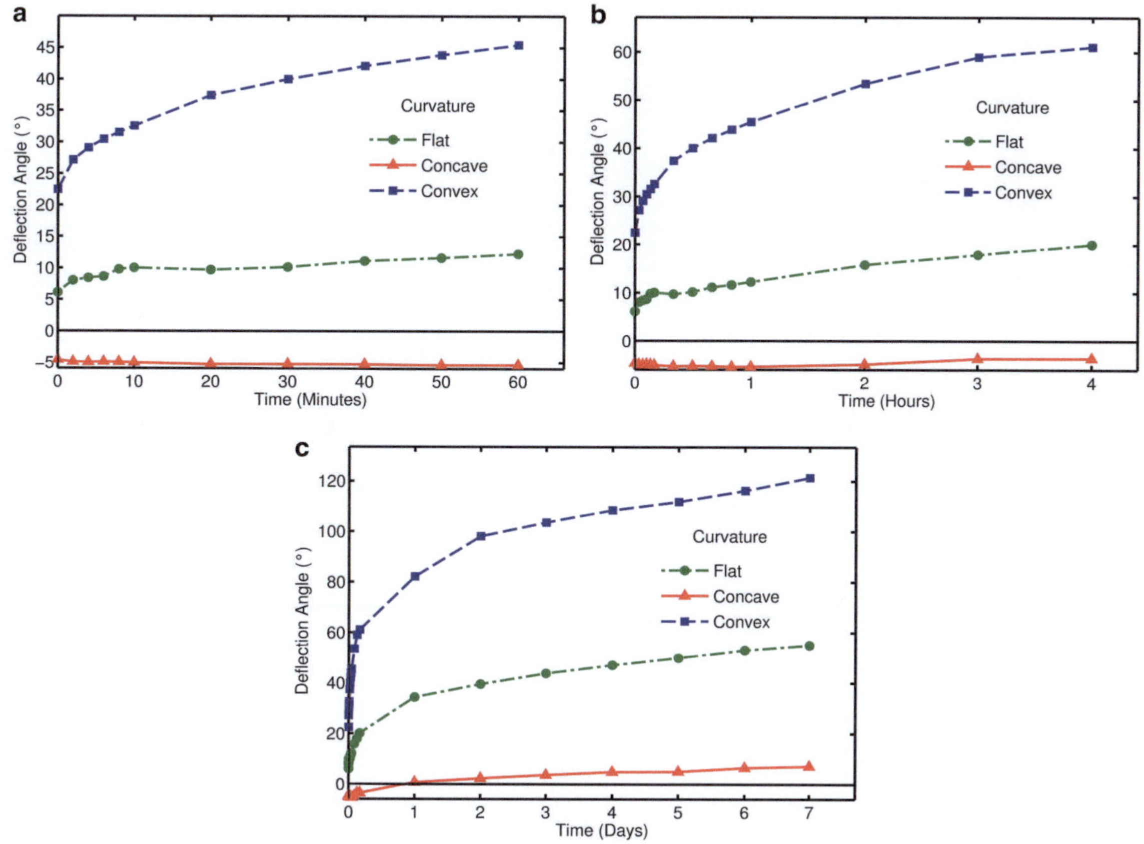

Fig. 1.7 Deflection angle versus time for (**a**) the first 60 min, (**b**) 4 h, and (**c**) 7 days for the three UCAs

Table 1.2 Radius of curvature (ρ), SMP thickness (t) and substrate width (s) possibilities and their respective designations for the UCAs studied

Nomenclature and supporting values			
Transverse curvature	Flat	Low	High
Radius of curvature (ρ)	∞	127 mm	62.5 mm
SMP thickness (t)	*Thin*	*Thick*	–
Thickness values	0.8 mm	1.6 mm	–
Substrate width (s)	*Narrow*	*Intermediate*	*Wide*
Width values	25.4 mm	38.1 mm	50.8 mm

high radius of curvature samples with thick SMP and intermediate substrate width are equal in dimension to the flat and concave samples tested previously. Analysis of the data shows a majority of the UCAs behave much differently than the samples presented previously. The UCAs from the previous data set experience relaxation and positive deflection angles while the other 16 samples contracted further into the fixed position. While the data seems to be unsystematic, trends can be seen when evaluating the design space with respect to SMP thickness and substrate width. Decreasing SMP thickness tends to decrease the deflection angle and allow the UCA to contract further into the stored configuration. This trait is pronounced in the narrow substrate widths as the UCAs with transverse curvature see a 12–14° decrease in deflection angle with all other parameters held constant. Additionally, as the substrate width narrows in the thin SMP samples the deflection angle decreases. This effect is diminished in the thicker SMP as most of the samples are in the 25–35° range. Definite trends with respect to substrate curvature cannot be seen with the data available as each substrate width and corresponding SMP thickness data set appears to perform differently. However, a negative deflection angle is a favorable result as it shows that transverse curvature in the substrate does not negatively affect shape fixity and cause an excessive amount of shape retention loss.

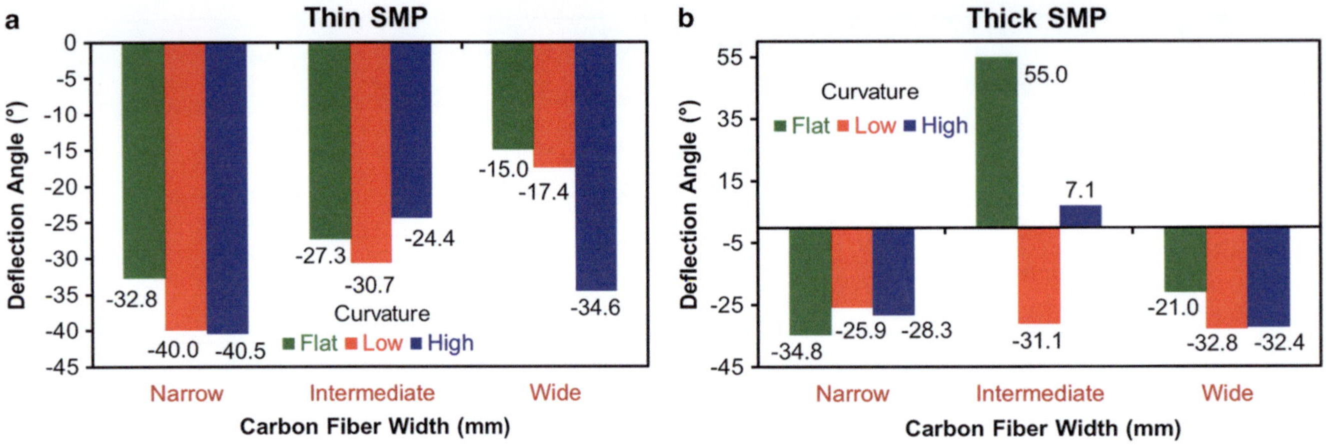

Fig. 1.8 Deflection angles keeping thickness constant for the (**a**) thin SMP and (**b**) thick SMP at 7 days

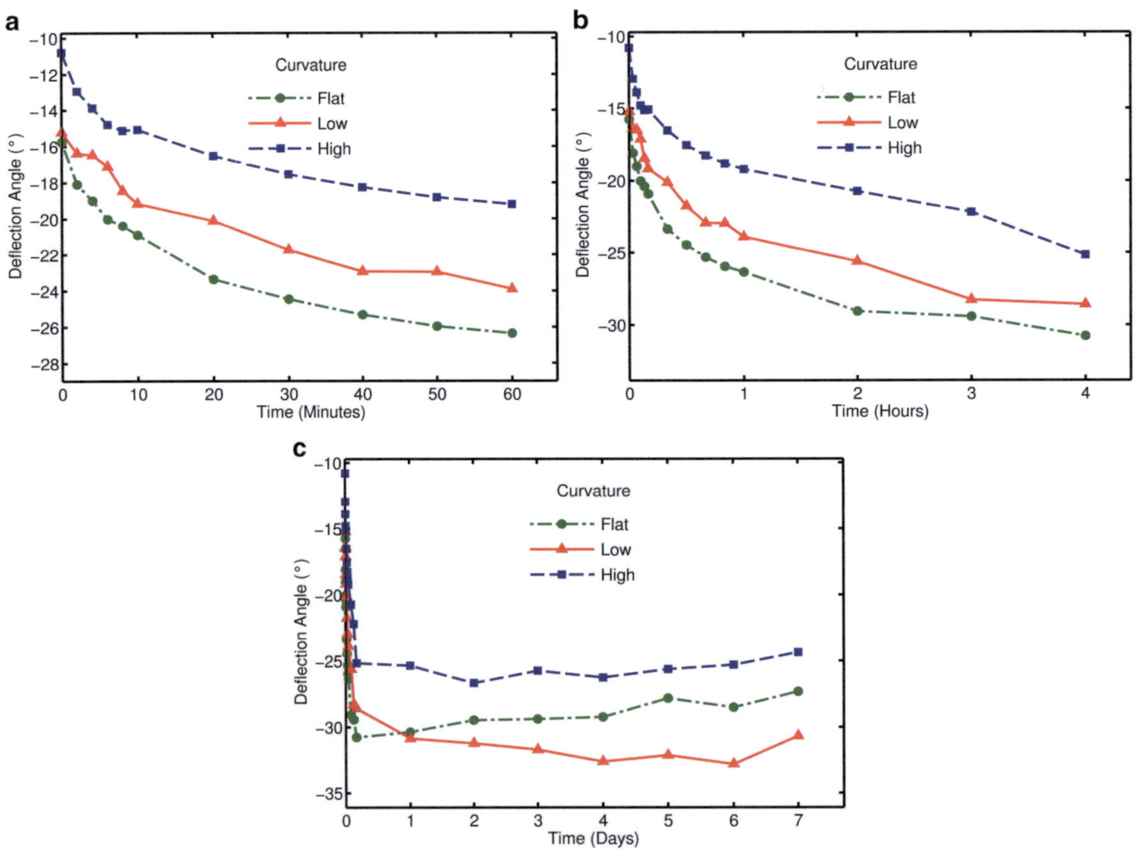

Fig. 1.9 Deflection angle versus time for (**a**) the first 60 min, (**b**) 4 h, and (**c**) 7 days of the thin SMP, intermediate substrate, UCAs with flat, low, and high transverse curvature respectively

The development of deflection angle versus time for the thin SMP, intermediate substrate UCAs, with flat, low, and high transverse curvatures is shown in Fig. 1.9. Analysis of this data shows that the UCAs contract rapidly over the first 4 h and then reach an approximate equilibrium. All three samples contract approximately 10° over the first 60 min and a total of 15° after the 4 h. After a day all of the samples maintain nearly constant deflection angles though the seventh day although this value does drift between images suggesting UCA misalignment during data capture that resulted measurement error or sensitivity to the room temperature. Further testing will be done to determine the UCAs sensitivity to variations in room temperature. Misalignment errors attributed to an uncertainty of approximately $\pm 0.5^\circ$ in the data presented. The results indicate that most of the UCAs reach equilibrium after the first day with the exception of the flat and high curvature UCAs with thick SMP and intermediate substrate width. Further evaluation of these results will be required to determine why these samples behave much differently than the rest of the design space presented here.

1.4 Conclusion

A Veriflex-S® shape memory polymer and carbon-fiber-reinforced polymer unimorph composite actuator was evaluated in a series of tests to determine the effect of transverse curvature, carbon-fiber-reinforced polymer substrate width, and shape memory polymer thickness on shape fixity. Images of the actuators were captured over time after being cooled and released from a storage container. These images were then processed to measure the deflection angle with respect to the original fixity position. In the research conducted, a unimorph composite actuator with a 63.5 mm radius of curvature was able to retain nearly an order of magnitude lower deflection angle than a flat unimorph composite actuator keeping all other variables constant. Increasing the design space yielded mixed results as decreasing shape memory polymer thickness lead to a negative deflection angle which may not be advantageous depending upon the unimorph composite actuator application. Decreasing the substrate width also contributes to an increasingly negative deflection angle. Additional testing of the unimorph composite actuators will be required to determine the effect of transverse curvature on shape fixity as results were inconclusive during testing. However, specimens with negative deflection angles and transverse curvature provide validation that introducing the transverse curvature into the substrate does not adversely affect the shape fixity of the unimorph composite actuator. The majority of the specimens tested display encouraging results which could facilitate the expanded use of shape memory polymers on a variety reconfigurable structures including micro air vehicle wings and satellite solar arrays.

Future research will include developing a model for both the fixity and recovery testing cycles of the unimorph composite. This model will be used to further evaluate the design space presented in this paper and determine the trade-offs between fixity and recoverability to develop an ideal unimorph composite actuator.

References

1. V.A. Beloshenko, V.N. Varyukhin, Y.V. Voznyak, The shape memory effect in polymers. Russ. Chem. Rev. **74**, 265–283 (2005). doi:10.1070/RC2005v074n03ABEH000876
2. J. Leng, H. Lu, Y. Liu et al., Shape-memory polymers—a class of novel smart materials. MRS Bull. **34**(11), 848–855 (2009)
3. X. Lan, Y. Liu, H. Lv et al., Fiber reinforced shape-memory polymer composite and its application in a deployable hinge. Smart Mater. Struct. **18**, 024002 (2009). doi:10.1088/0964-1726/18/2/024002
4. J. Carrell, D. Tate, S. Wang, H.-C. Zhang, Shape memory polymer snap-fits for active disassembly. J. Clean. Prod. **19**, 2066–2074 (2011). doi:10.1016/j.jclepro.2011.06.027
5. J.T. Cantrell, P.G. Ifju, *Novel Design Concepts for Micro Air Vehicles and Unimorph Composite Actuators Incorporating Transverse Curvature, Shape Memory Polymers for Aerospace Applications: Novel Synthesis, Modeling, Characterization and Design* (Lancaster, DEStech, Forthcoming 2016), pp. 517-548
6. J. Joo, B. Smyers, R. Beblo, et al., Load-bearing multi-functional structure with direct thermal harvesting for thermally activated reconfigurable wing design. In: *International Conference on Composite Materials*, Society of Composite Materials, Jeju, 2011, pp. 1–6
7. A.J.W. McClung, G.P. Tandon, J.W. Baur, Strain rate- and temperature-dependent tensile properties of an epoxy-based, thermosetting, shape memory polymer (Veriflex-E). Mech. Time-Depend. Mater. **16**, 205–221 (2011). doi:10.1007/s11043-011-9148-7
8. A.J.W. McClung, G.P. Tandon, K.E. Goecke, J.W. Baur, Non-contact technique for characterizing full-field surface deformation of shape memory polymers at elevated and room temperatures. Polym. Test. **30**, 140–149 (2011). doi:10.1016/j.polymertesting.2010.11.010
9. A.J.W. McClung, G.P. Tandon, J.W. Baur, Fatigue cycling of shape memory polymer resin, in *Mechanics of Time-Dependent Materials and Processes in Conventional and Multifunctional Materials*, ed. by T. Proulx (Springer, New York, 2011), pp. 119–127
10. J. Klesa, V. Placet, E. Foltête, E. Collet, Experimental evaluation of the rheological properties of Veriflex® shape memory polymer. In: *ESOMAT 2009—8th European Symposium on Martensitic Transformations*, Prague. doi: 10.1051/esomat/200904006
11. G.P. Tandon, K. Goecke, K. Cable, J. Baur, Durability assessment of styrene- and epoxy-based shape-memory polymer resins. J. Intell. Mater. Syst. Struct. **20**, 2127–2143 (2009). doi:10.1177/1045389X09348255
12. M.N.H. Nahid, M.A.A. Wahab, K. Lian, Degradation of shape memory polymer due to water and diesel fuels, in *Mechanics of Time-Dependent Materials and Processes in Conventional and Multifunctional Materials, Society of Experimental Mechanics*, ed. by T. Proulx (Springer, New York, 2011), pp. 37–48

13. A.J.W. McClung, G.P. Tandon, J.W. Baur, Deformation rate-, hold time-, and cycle-dependent shape-memory performance of Veriflex-E resin. Mech. Time-Depend. Mater. **17**, 39–52 (2011). doi:10.1007/s11043-011-9157-6
14. J.T. Fulcher, H.E. Karaca, G.P. Tandon, Y.C. Lu, Thermomechanical and shape memory properties of thermosetting shape memory polymer under compressive loadings. J. Appl. Polym. Sci. **129**(3), 1096–1103 (2013). doi:10.1002/app.38791
15. J.T. Cantrell, P.G. Ifju, Unimorph shape memory polymer actuators incorporating transverse curvature in the substrate. In: *Proceedings of the 2014 Annual Conference on Experimental Applied Mechanics* (Springer, Greenville, 2014) pp. 1–10

Chapter 2
Full-Field Three-Dimensional Characterization of Non-repetitive Motions by Single-Shot Multiplexed Digital Holography

Morteza Khaleghi, Jérémie Guignard, Cosme Furlong, and John J. Rosowski

Abstract Typically, in 3D displacement measurements, the sample is repeatedly loaded at least three times and the displacement fields are obtained from multiple sensitivity vectors. However, for studying transient and non-repetitive phenomena, including, but not limited to, displacement fields of temporally-varying biological tissues, repeating the experiment is not an option. Therefore, to measure 3D displacements in such applications, all the measurements have to be done concomitantly. In this paper, single-frame, multiplexed holography is used for simultaneous quantification of 3D displacement fields. In our approach, the hologram of an object of interest is recorded in an off-axis configuration, with three simultaneous incoherently-superimposed pairs of reference and object beams. Three different spatial carrier frequencies are realized by small differences in the angle of illumination of each reference wave with respect to the CCD sensor. Therefore, the reconstructed image corresponding to each illumination direction is reconstructed at a particular position on the image plane. Because of the differences in the position of each reference beam and wavelength of each pair of beams, the reconstruction distance and magnification of each sensitivity vector are different. Therefore, we developed and implemented registration algorithms to accurately translate individual views into a single global coordinate system. Representative results highlighting the 3D measuring capabilities of our holographic system are presented.

Keywords 3D displacement measurements • Digital holography • High-speed measurements • Image registration • Multiplexed holography

2.1 Introduction

Typical 3D displacement measurement techniques require repeating the measurements from at least three sensitivity directions. Repetitive measurements are not applicable to temporally-varying objects such as biological samples. Therefore a novel, unique method based on multiplexed holography is developed to enable simultaneous acquisition of 3D displacement fields of objects and, in turn, to minimize the measurement time. In this method, the hologram of an object of interest is recorded with three simultaneous incoherently-superimposed pairs of reference and object beams, such that the modulation image corresponding to each illumination direction is reconstructed at a particular position on the image. Registration algorithms are implemented to accurately translate individual views into a single global coordinate system. The method is illustrated by acquiring three images obtained from three sensitivity vectors in a multiplexed hologram, which are registered and, consequently, 3D data are extracted.

M. Khaleghi (✉)
Center for Holographic Studies and Laser micro-mechaTronics (CHSLT), Mechanical Engineering Department, Worcester Polytechnic Institute, Worcester, MA 01609, USA
e-mail: khaleghi29@gmail.com

J. Guignard • J.J. Rosowski
Eaton-Peabody Laboratory, Massachusetts Eye and Ear Infirmary, Boston, MA 02114, USA

Department of Otology and Laryngology, Harvard Medical School, Boston, MA 02114, USA

C. Furlong
Center for Holographic Studies and Laser micro-mechaTronics (CHSLT), Mechanical Engineering Department, Worcester Polytechnic Institute, Worcester, MA 01609, USA

Eaton-Peabody Laboratory, Massachusetts Eye and Ear Infirmary, Boston, MA 02114, USA

Department of Otology and Laryngology, Harvard Medical School, Boston, MA 02114, USA

C. Sciammarella et al. (eds.), *Experimental and Applied Mechanics, Volume 4*,
Conference Proceedings of the Society for Experimental Mechanics Series, DOI 10.1007/978-3-319-22449-7_2

2.2 Methods

2.2.1 *Off-Axis Multiplexed Lensless Digital Holography*

For 3D displacement measurements using off-axis multiplexed holography, three incoherently-superimposed pairs of object and reference beams are simultaneously recorded with the CCD sensor. The object of interest is illuminated concurrently with three object beams from different directions to provide three sensitivity vectors that are required for 3D displacement measurements [1]. The reference beams provide slightly different spatial carrier frequencies on the CCD to enable single-frame phase extraction [2]. The total intensity recorded on the CCD detector, $I(k,\ l)$, due to incoherent superposition of three pairs of reference-object beams is [3–5]

$$I(k,l)=\sum_{n=1}^{3}I_n(k,l)=\sum_{n=1}^{3}|R_n(k,l)+O_n(k,l)|^2, \tag{2.1}$$

where k and l are the coordinates of the pixels in the CCD (hologram) plane, R_n and O_n are the nth reference and object waves, respectively, with n = 1, 2, 3. (Equation 2.1) can be expanded to

$$I(k,l)=\sum_{n=1}^{3}a_n(k,l)+c_n(k,l)\exp\left[2\pi i\left(f_{n\xi}x+f_{n\eta}y\right)\right]+c_n^*(k,l)\exp\left[-2\pi i\left(f_{n\xi}x+f_{n\eta}y\right)\right], \tag{2.2}$$

where $a_n(k,l)=R_n^2(k,l)+O_n^2(k,l)$, and $c_n(k,l)=O_n(k,l)R_n(k,l)exp[i\varnothing_n(k,l)]$. The two-dimensional Fourier Transform of (2.2) is

$$\text{FT}\{I(k,l)\}=\sum_{n=1}^{3}A_n(\xi,\eta)+C_n\left(\xi-f_{n\xi0},\eta-f_{n\eta0}\right)+C_n^*\left(\xi+f_{n\xi0},\eta+f_{n\eta0}\right), \tag{2.3}$$

where uppercase notation denotes the Fourier-transformed components. The terms C_n and $C_n{}^*$ describe the spatial frequency distribution of the twin images components that mathematically are expressed as complex conjugates of one another, each containing the required phase information. As shown in Fig. 2.1b, these terms are shifted in the Fourier domain due to the off-axis angles of the reference waves and can be separated from one other by appropriate calculation and setting of carrier frequencies induced by the incident angles and positions of the reference beams. The detail of the theoretical and experimental design requirements of this method have been described in detail in our previous article [5]. Figure 2.1b

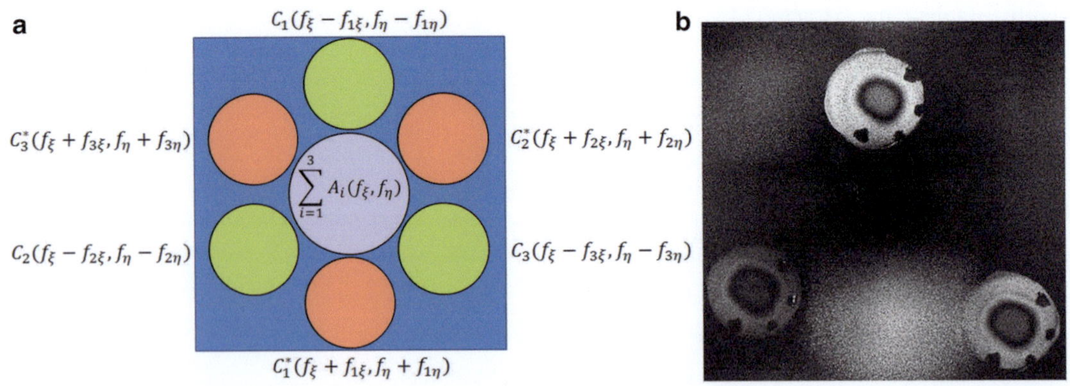

Fig. 2.1 Multiplexed off-axis holography used for simultaneous acquisition of 3D displacements: (**a**) spectrum of the recorded multiplexed off-axis hologram consisting of three incoherently-superimposed interference patterns; and (**b**) representative example of numerically-reconstructed double-exposure multiplexed hologram that is used for 3D displacement measurements. The object of interest is a 1 cm diameter latex membrane, acoustically excited by a tone of 1.5 kHz. Sound-induced concentric fringe patterns and four painted marks used to test registration algorithms are visible on each of the reconstructed images

shows the reconstructed multiplexed hologram showing the image of an acoustically-loaded latex membrane illuminated from three different angles.

Our techniques are based on single-frame lensless digital holography [6, 7] in which reconstructions of the multiplexed holograms are obtained by the Fresnel-Kirchhoff integral that is efficiently computed with the 2D FFT as

$$\Gamma(m,n) = \mathrm{FFT2}[R(k,l)h(k,l)\psi(k,l)], \tag{2.4}$$

where $\Gamma(m, n)$ is the complex reconstructed hologram at coordinates m and n in the reconstruction plane, $R(k, l)$ is the complex amplitude of the plane reference wave that is considered to be one in the numerical reconstruction, and $\psi(k, l)$ is the two-dimensional chirp function defined with

$$\psi(k,l) = \exp\left[\frac{-i\pi}{\lambda d}\left(k^2\Delta x^2 + l^2\Delta y^2\right)\right], \tag{2.5}$$

where Δx and Δy are the pixels size of the CCD sensor, λ is the laser wavelength, and d is the reconstruction distance. Additionally, the DC component can be mathematically removed from the multiplexed hologram by recognizing that the DC component in the Fourier domain is due to the low-frequency spatial variation in the hologram, and it can thus be isolated by applying a high-pass filter [8–10]. Therefore, the optical phase of the reconstructed hologram is obtained with

$$\varphi(x,y) = \mathrm{atan2}\left[\frac{\Im(\Gamma(m,n))}{\Re(\Gamma(m,n))}\right], \tag{2.6}$$

where $\Im$ and $\Re$ denote the imaginary and real parts of the complex number $\Gamma(m, n)$. Thus, in one single frame of the camera, optical phase information corresponding to all three sensitivity vectors exists. With the use of double-exposure techniques, i.e., subtracting optical phases of two reconstructed holograms corresponding to deformed and reference states of the object, 3D motion components of the objects can be obtained with the method of multiple sensitivity vectors in holographic interferometry. Considering the fact that the fringe-locus function Ω, i.e., the unwrapped optical phase difference between the two states, is the scalar product of the sensitivity vector, $\boldsymbol{K}(\boldsymbol{K}_x, \boldsymbol{K}_y, \boldsymbol{K}_z)$, with the object's displacement vector, $\boldsymbol{d}(\boldsymbol{d}_x, \boldsymbol{d}_y, \boldsymbol{d}_z)$, the 3D displacements components are obtained by

$$\begin{bmatrix} \boldsymbol{d}_x \\ \boldsymbol{d}_y \\ \boldsymbol{d}_z \end{bmatrix} = \frac{\lambda}{2\pi}\begin{bmatrix} \boldsymbol{K}_x^1 & \boldsymbol{K}_y^1 & \boldsymbol{K}_z^1 \\ \boldsymbol{K}_x^2 & \boldsymbol{K}_y^2 & \boldsymbol{K}_z^2 \\ \boldsymbol{K}_x^3 & \boldsymbol{K}_y^3 & \boldsymbol{K}_z^3 \end{bmatrix}^{-1} \times \begin{bmatrix} \Omega_1 \\ \Omega_2 \\ \Omega_3 \end{bmatrix}. \tag{2.7}$$

However, in the case of off-axis multiplexed holography, the differences in the position of each reference beam and wavelength of each pair of beams, causes the position, reconstruction distance, and magnification of each image corresponding to each sensitivity vector to differ. Therefore, for quantification of the 3D displacement, an image registration algorithm is developed and implemented to position each image into a common global coordinate system [5].

2.2.2 *Multiplexed Holographic System*

In our experimental system, incoherent superposition of the beams is implemented by using three different near-infrared external cavity tunable laser sources with wavelengths centered at 779.8, 780.2, and 780.6 nm. As shown in Fig. 2.2, Acousto-Optic Modulators (AOM) contained in each laser delivery subsystem (LD1 to LD3) are used to switch the laser on/off to enable stroboscopic measurement capabilities. A dual-channel function generator is used with one of the channels set to acoustically stimulate the TM sample with a calibrated speaker while the second channel is set to pulse mode to simultaneously drive all three AOMs. The duty cycle of the pulse signal sent to the AOMs is typically set to 2–5 % of the period of the tonal stimulus. As shown in Fig. 2.2, each laser is coupled into single-mode polarization maintaining fibers and

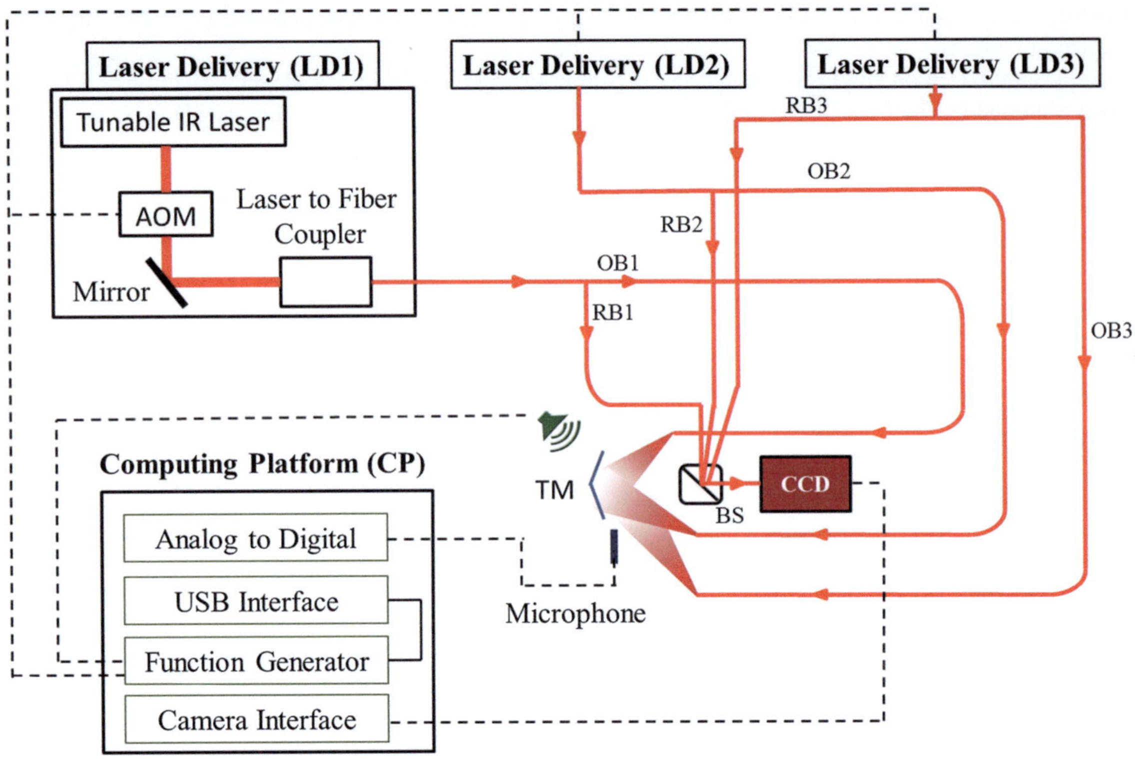

Fig. 2.2 Experimental system for simultaneous 3D displacement measurements with multiplexed holography. The *solid lines* denote the beam paths and the *dashed lines* the connections between components. AOM is acousto-optic modulator; RB and OB are reference and object beams, respectively; BS is the beam splitter; and TM is the tympanic membrane sample. Each of the three laser delivery subsystems contains similar components

then split into reference and object beams. The reference beams illuminate the CCD in an off-axis configuration by a beam splitter and the object beams concomitantly illuminate the sample from three different directions to define the sensitivity vectors for 3D displacement measurements. In Fig. 2.2, the optoelectronic components are shown for only one of the laser delivery subsystems; however, all three laser delivery subsystems contain similar components. The Computing Platform (CP) performs multiple tasks that include, synchronizing the stroboscopic illumination of the lasers with different stimulus phases, acquiring multiplexed holograms with a 5 MPix CCD camera having a pixel size of 3.45 × 3.45 μm^2, and reconstructs the holograms in real-time. A live 2D FFT is used to ensure that the components of the frequency spectrum of the hologram are separated from each other.

2.3 Results

2.3.1 3D Motion of an Artificial Sample

In order to validate the accuracy of the present method, and prior to the deployment into a medical research environment, the results of an artificial test sample obtained with our multiplexed holographic system (MHS) are compared with the ones obtained with a documented repetitive holographic interferometric method [7]. The artificial sample is a thin circular latex membrane excited with a tonal excitation of 2.2 kHz. Figure 2.3 shows the modulation and wrapped optical phase images of both twin components to illustrate how the three pairs of complex conjugated reconstructed holograms are distributed.

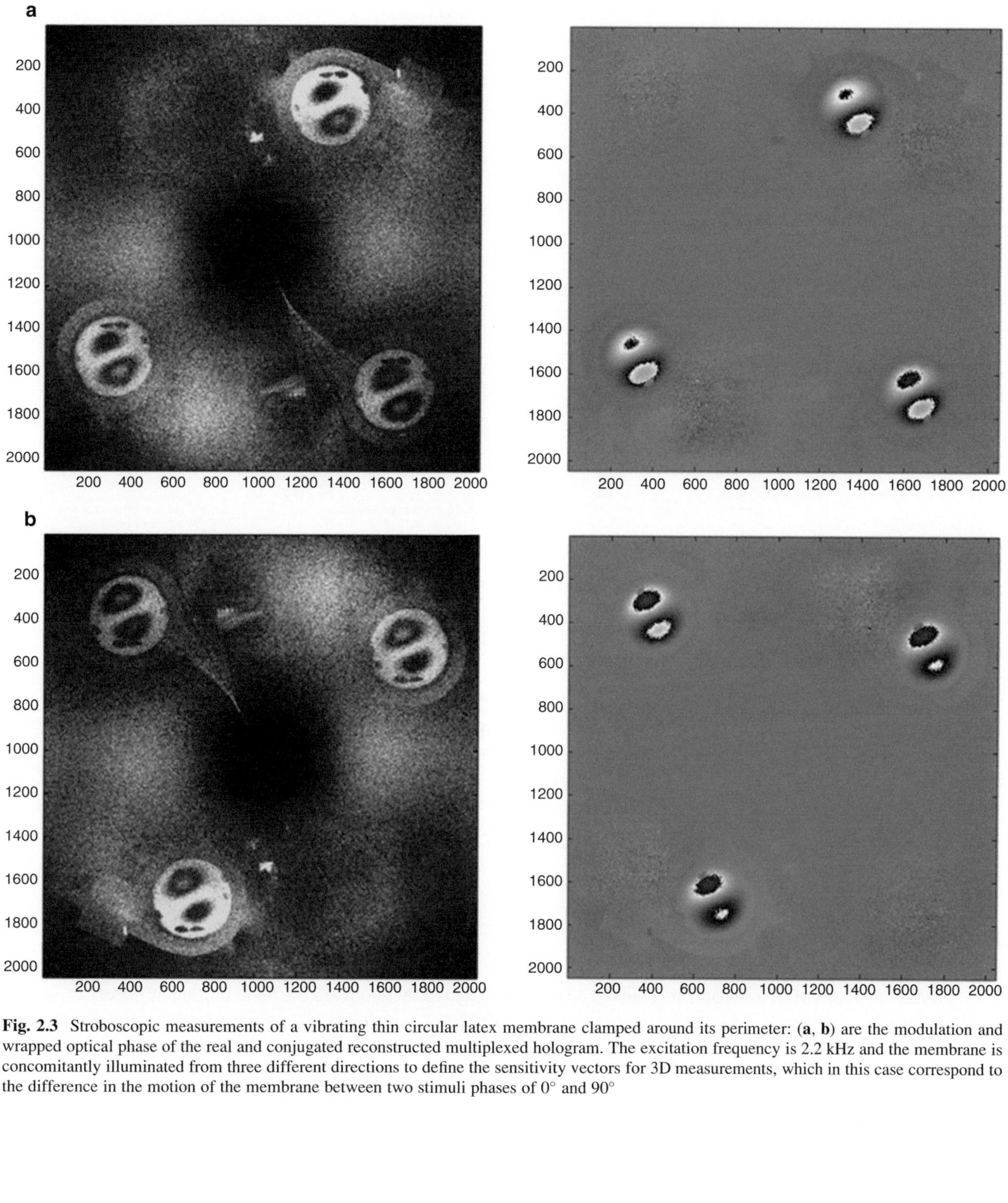

Fig. 2.3 Stroboscopic measurements of a vibrating thin circular latex membrane clamped around its perimeter: (**a**, **b**) are the modulation and wrapped optical phase of the real and conjugated reconstructed multiplexed hologram. The excitation frequency is 2.2 kHz and the membrane is concomitantly illuminated from three different directions to define the sensitivity vectors for 3D measurements, which in this case correspond to the difference in the motion of the membrane between two stimuli phases of 0° and 90°

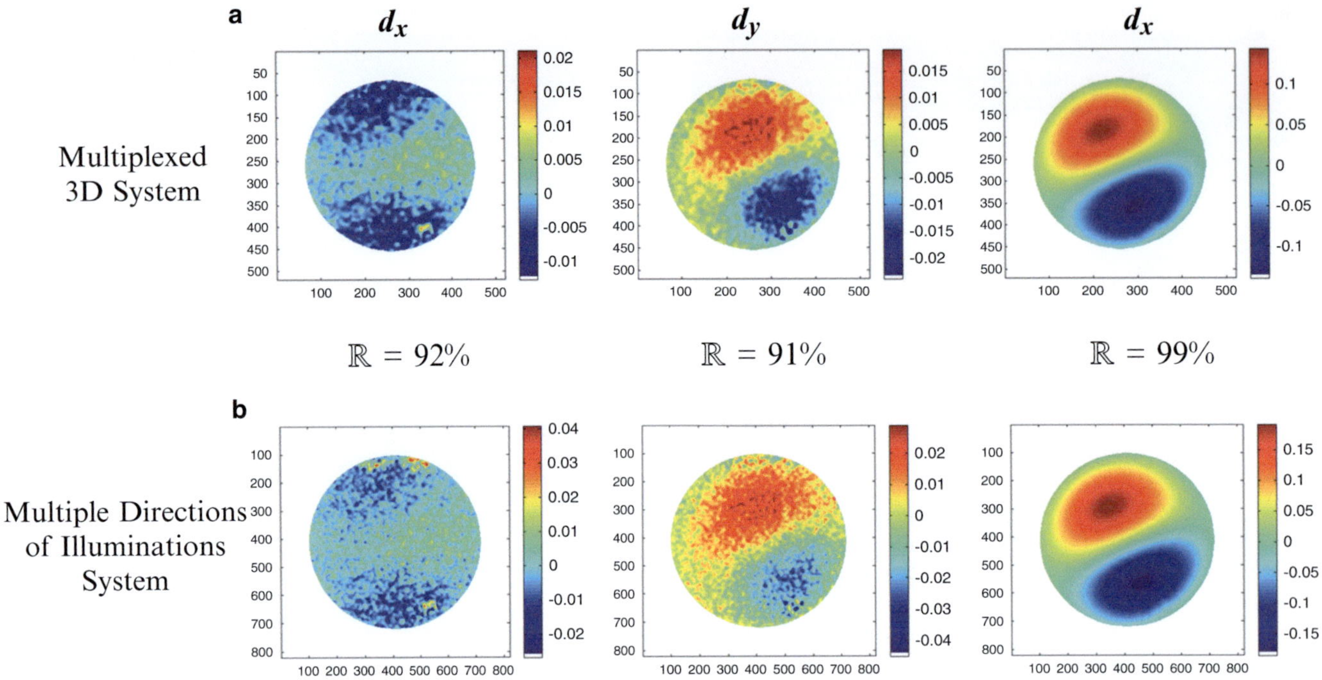

Fig. 2.4 Validation of the results of 3D displacement measurements versus a documented method: (**a**) displacement components along x, y, and z obtained using MHS; and (**b**) displacement components along x, y, and z obtained with the method of repetitive holographic interferometry. R is the Pearson correlation coefficient between the results obtained with the two methods

The registered, unwrapped, normalized phases are used to quantify 3D deformation. To test and verify the validity of the measurements obtained with MHS, the results are compared with the ones obtained with a documented repetitive holographic interferometric method based on multiple directions of illuminations [7]. The results obtained with both methods are shown in Fig. 2.4. A point-by-point correlation measure indicates a Pearson correlation coefficient of 92, 91 and 99 % along the *x*-, *y*- and *z*-axis, respectively, indicating the competence of the MHS with high-end repetitive holographic systems.

2.3.2 3D Motions of Human TM

The cadaveric human TM of a female, 46 year-old donor was prepared by removing the bony portion of the ear canal in a lightly fixed temporal bone. The middle-ear space of the sample was widely opened, which enabled assessment of the normality of the TM and ossicles. Due to semi-transparency of the mammalian TMs, the sample was coated with a thin layer of zinc oxide to increase the laser light reflection. The effect of coating on shape and deformation patterns have been studied by several researchers and found to be negligible [11, 12]. The temporal bone was held with an adjustable clamp and mounted on a post in front of the holographic system. Figure 2.5 shows strobed sound-induced 3D motion of a human TM at four different excitation frequencies corresponding to the differences in the motion of the TM between two stimuli phases of 0° and 90°.

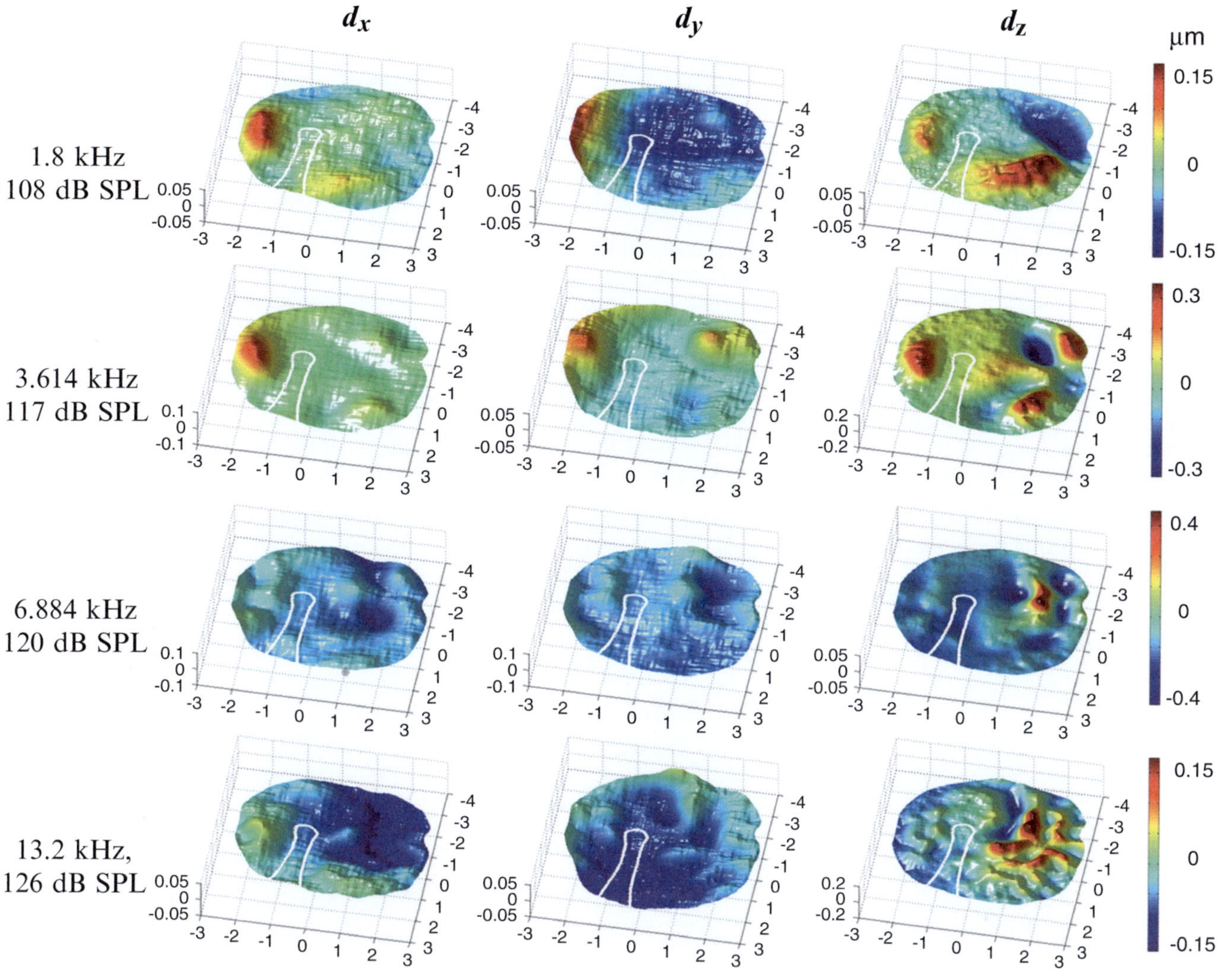

Fig. 2.5 Three-dimensional sound-induced displacements of a human TM excited at three different frequencies obtained along three orthogonal axes x, y, and z. The outline of the manubrium is shown in *white* in all the figures

2.4 Conclusions

Due to the time-varying nature of biological tissues like the TM, a novel method for 3D displacement measurements based on multiplexed holography is being developed. The method allows simultaneous holographic measurements along multiple sensitivity vectors within one single hologram. Such methodology is a critical step towards *in-vivo* measurements of 3D TM motions with a minimum acquisition time. In our approach, the hologram of an object of interest is recorded with three simultaneous incoherently-superimposed pairs of reference and object beams, such that the modulation image corresponding to each illumination direction is reconstructed at a different position of the image. The time needed for a given 3D displacement measurement is decreased at least threefold. We demonstrate that the present method is a valid alternative to repetitive holographic methods and offers promising perspectives towards faster accurate displacement measurements of biological specimens.

Acknowledgments This work was supported by the National Institute on Deafness and other Communication Disorders (NIDCD), Massachusetts Eye and Ear Infirmary (MEEI), the Swiss National Science Foundation (SNSF), and the Mechanical Engineering Department at Worcester Polytechnic Institute. We also acknowledge the support of all of the members of the CHSLT labs at WPI and Eaton-Peabody labs at MEEI, in particular Ellery Harrington, and Jeffrey Tao Cheng.

References

1. U. Schnars, C. Falldorf, J. Watson, W. Jüptner, *Digital Holography and Wavefront Sensing* (Springer, Heidelberg, 2015)
2. M. Takeda, H. Ina, S. Kobayashi, Fourier-transform method of fringe-pattern analysis for computer-based topography and interferometry. J. Opt. Soc. Am. **72**, 156–160 (1982)
3. S. Schedin, G. Pedrini, H.J. Tiziani, F. Mendoza Santoyo, Simultaneous three-dimensional dynamic deformation measurements with pulsed digital holography. Appl. Opt. **38**(34), 7056–7062 (1999)
4. P. Picart, E. Moisson, D. Mounier, Twin-sensitivity measurement by spatial multiplexing of digitally recorded holograms. Appl. Opt. **42**(11), 1947–1957 (2003)
5. M. Khaleghi, J. Guignard, C. Furlong, J.J. Rosowski, Simultaneous full-field 3D vibrometry of the human eardrum using spatial-bandwidth multiplexed holography. J. Biomed. Opt. **20**(11), 111202 (2015)
6. M. Hernández-Montes, C. Furlong, J.J. Rosowski, N. Hulli, J.T. Cheng, F. Mendoza Santoyo, M.E. Ravicz, E. Harrington, Optoelectronic holographic otoscope for measurement of nano-displacements in tympanic membranes. J. Biomed. Opt. **14**(3), 034023–03402 (2009)
7. M. Khaleghi, C. Furlong, M. Ravicz, J.T. Cheng, J.J. Rosowski, Three-dimensional vibrometry of the human eardrum with stroboscopic lensless digital holography. J. Biomed. Opt. **20**(5), 051028 (2015)
8. N. Pavillon, C.S. Seelamantula, J. Kühn, M. Unser, C. Depeursinge, Suppression of the zero-order term in off-axis digital holography through nonlinear filtering. Appl. Opt. **48**(34), H186–H195 (2009)
9. M. Khaleghi, I. Dobrev, E. Harrington, C. Furlong, J.J. Rosowski, Study of the transient response of tympanic membranes under acoustic excitation, in *Mechanics of Biological Systems and Materials*, ed. by F. Barthelat, P. Zavattieri, C.S. Korach et al., vol. 4 (Springer, New York, 2014), pp. 1–9
10. A. Alamdari, V. Krovi, Active reconfiguration for performance enhancement in articulated wheeled vehicles. In: *ASME 2014, Dynamic Systems and Control Conference*, San Antonio, 2014, p. V002T27A004
11. J.J.J. Dirckx, W.F. Decraemer, Coating techniques in optical interferometric metrology. Appl. Opt. **36**(13), 2776–2782 (1997)
12. J.J. Rosowski, J.T. Cheng, M.E. Ravicz, N. Hulli, M. Hernández-Montes, E. Harrington, C. Furlong, Computer-assisted time-averaged holograms of the motion of the surface of the mammalian tympanic membrane with sound stimuli of 0.4–25 kHz. Hear. Res. **253**(1), 83–96 (2009)

Chapter 3
Basic Foundations of Signal Analysis Models Applied to Retrieval of Displacements and Their Derivatives Encoded in Fringe Patterns

Cesar Sciammarella and L. Lamberti

Abstract All the techniques that measure displacements, whether in the range of visible optics or any other form of field methods require the presence of a carrier signal. The carrier signal is a wave form that is modulated (modified) by an input, deformation of the medium. The carrier is tagged to the medium under analysis and deforms with the medium. The wave form must be known both in the unmodulated and the modulated conditions. There are two basic mathematical models that can be utilized to decode the information contained in the carrier, phase modulation or frequency modulation, both are closely connected. Basic problems that are connected to the detection and recovery of displacement information that are common to all optical techniques will be analyzed. This paper is concentrated in the general theory common to all the methods independently of the type of signal utilized. The aspects discussed are those that have practical impact in the process of data gathering and data processing.

Keywords Fringe pattern analysis • Basic mathematical models for detection and recovery of displacements • Phase and amplitude modulation • Hilbert transform • Multiphase methods and in-quadrature signals analysis

3.1 Introduction to Signal Analysis of 2D Patterns That Contain Displacement Information

Fundamental mathematical problems in the analysis of 2-D and 3-D images containing displacement information are: (1) data retrieving; (2) data analysis; (3) data processing. One must have a robust method to retrieve signal information contained in an image against stochastic and deterministic local deviations of the recorded information. As stated in [1], the phase of a signal is a robust representation of a signal in the presence of noise. In [1], it is shown that the information contained in a 2-D signal can be recovered to a great extent only on the basis of phase information by setting the amplitude to unit value. On the contrary, if one makes the phase zero and the intensity is recovered, the signal cannot be reconstructed. This is a very important aspect of fringe pattern analysis, where following the classical approach in Optics one assumes that the displacement information is contained in the phase of the recorded signals that assume the form of a phasor. For the sake of simplicity in the text that follows a one dimensional approach will be utilized. In the case of methods that measure displacements, the general equation of a fringe system for a one dimensional signal is of the form [2, 3] and Chapter 10 of [4],

$$I(x) = I_0 + I_1 \cos \Psi(x) \tag{3.1}$$

where I_o is a background term that that ideally should be a constant value for all x's. In actual applications it is assumed to be a slow changing term as a function of x. This means that in the FT power spectrum of the signal, I_o must be a spike at the origin of coordinates of the frequency space. The term containing the displacement information is the second term of (3.1), a phase modulated (frequency modulated) sinusoidal signal restricted to the first harmonic. In the ideal model, I_1 (the amplitude of the first order harmonic of the signal) is a constant. However, in actual signals, it is also a function of x: the phase term $\Psi(x)$ that can be of the form,

C. Sciammarella (✉)
Department of Mechanical, Materials and Aerospace Engineering, Illinois Institute of Technology, 10 SW 32nd St., Chicago, IL 60616, USA
e-mail: sciammarella@iit.edu

L. Lamberti
Dipartimento di Meccanica, Matematica e Management, Politecnico di Bari, Viale Japigia 182, Bari 70126, Italy

C. Sciammarella et al. (eds.), *Experimental and Applied Mechanics, Volume 4*,
Conference Proceedings of the Society for Experimental Mechanics Series, DOI 10.1007/978-3-319-22449-7_3

$$\Psi(x) = [2\pi f_c + \phi(x)] \tag{3.2}$$

where f_c is the frequency of the carrier that is phase modulated by $\phi(x)$. The function $\phi(x)$ is the term containing the information on the optical path change caused by the displacement field. Due to the presence of $\phi(x)$, the cosine term in (3.1) is not a harmonic function but it can be under certain specific restrictions assumed to be a quasi-harmonic function, a very important fact in modeling the actual signals that contain the displacement information. In the more general signal there will be higher order terms, the importance of these higher orders will become clear later on in the development of the paper. Equation (3.2) corresponds to the case where the carrier is imaged by the optical system. There is an alternative expression for the phase, the carrier is not resolved by the optical system and only the modulation function is captured,

$$\Psi(x) = \phi(x) \tag{3.3}$$

Equations (3.1–3.3) constitute the basic models utilized in fringe pattern analysis (see Chapter 10 of [4]). The image must contain a carrier to encode information. The carrier can be a deterministic signal in the case of moiré patterns, harmonic phase analysis (HARP) or a stochastic signal. Stochastic signals are utilized by speckle methods in all the different approaches, holography, speckle interferometry, speckle photography, white light speckles, digital image correlation (DIC). The stochastic signals lead to the formation of fringes by optical means or the equivalent of fringes by numerical methods that replace the optical correlation.

The purpose of this paper is to utilize the Theory of Signal Processing to get general conclusions concerning the retrieval of displacement information and its derivatives. Therefore the concept of phase introduced in (3.1–3.3) is a fundamental tool in this paper. Although, for example, the DIC method does not overtly uses this concept, to get conclusions of general validity in the process of retrieval of displacement information and its derivatives it is necessary to cast this process in the more general context of the Theory of Signal Processing and in this theory the concept of phase is a fundamental variable.

The phase of a fringe pattern can be obtained from a recorded image by the method of multiphase recording of signals, for example the four phases signal (see Chapter 10 of [4]). From the very beginning of the fringe pattern analysis of isothetic signals (moiré pattern fringes) an alternative was proposed, the utilization of the in-quadrature method [3]. However, this alternative although at the time that it was proposed seemed to be fully equivalent to the multiple phase method proved not to be so in practice when implemented trough the FFT method that provides the in-phase and in-quadrature signals. In this paper, it will be shown both methods have a common root and properly implemented lead to the same result.

3.2 Frequency Modulation Model of Isothetic Fringes

The next step in the process of fringe pattern analysis is to introduce the concept of frequency modulation. The classical approach is followed in this sections but it is done in terms relevant to fringe analysis an essential point to understand spatial frequency modulation and its relationship to fundamental variables of Continuum Mechanics. In what follows are presented developments that are related to fringe pattern processing whether they are coming from a deterministic carrier or a stochastic carrier. The argument $\Psi(x)$ given in (3.2) can be expressed in the following form,

$$[2\pi f_c x + \phi(x)] = \left(2\pi f_c x + 2\pi \int_0^x m(x)dx\right) \tag{3.4}$$

where we have made the substitution $\phi(x) = \int_o^x \frac{d\phi(x)}{dx} dx$ and called,

$$\frac{d\phi(x)}{dx} = m(x) \tag{3.5}$$

In (3.5), m(x) is the derivative of the modulation function, or instantaneous frequency of $\Psi(x)$. Figure 3.1 illustrates an assumed cosinusoidal frequency rate of change of the modulation function. Estimates of the bandwidth of the signal that will appear in the power spectrum of the FT of $\Psi(x)$ can be made on the basis of the above assumption. Then,

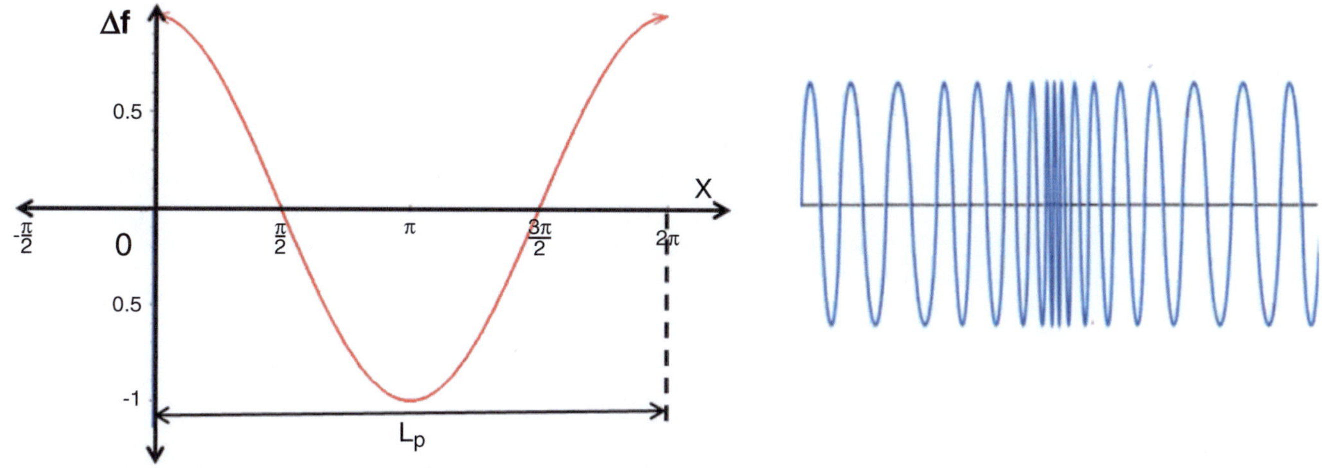

Fig. 3.1 A cosinusoidal derivative of the modulation function assumed sinusoidal. Lp is the one dimensional space where the signal is defined; Δf maximum value change of the frequency in the interval of definition

$$m(x) = \left[\frac{d\phi(x)}{dx}\right]_m \cos\frac{2\pi}{L_p}x \tag{3.6}$$

We define

$$\left[\frac{d\phi(x)}{dx}\right]_m = \Delta f \tag{3.7}$$

is the maximum value of the frequency change in the interval L_p where the signal is defined.

In Fig. 3.1, f_m is the frequency of the modulation function signal, $f_m = 1/L_p$. The frequency f_m corresponds to a given term of the expansion of the modulation function in a FT. Consequently, for this particular term we can write,

$$[2\pi f_c x + \phi(x)] = \left(2\pi f_c x + 2\pi\int_0^x \Delta f\cos\left(2\pi f_m x\right)dx\right) \tag{3.8}$$

Upon integration one obtains,

$$[2\pi f_c x + \phi(x)] = \left(2\pi f_c x + \frac{\Delta f}{f_m}\sin\left(2\pi f_m x\right)\right) \tag{3.9}$$

Then the signal containing the displacement information is of the form

$$S_m(x) = I_1(2\pi f_c x + \beta\sin\left(2\pi f_m x\right)) \tag{3.10}$$

where β is defined as the modulation index

$$\beta = \frac{\Delta f}{f_m} \tag{3.11}$$

The original sinusoidal carrier signal is not longer sinusoidal because to the carrier it is added a modulating term that transforms the signal from a sinusoid to a frequency modulated function that contains many harmonics. In order to evaluate the harmonics associated with the modulation function, (3.10) is expanded in a series of Bessel functions of the first kind and of order n of the argument β,

Fig. 3.2 FT spectrum of a frequency modulated carrier

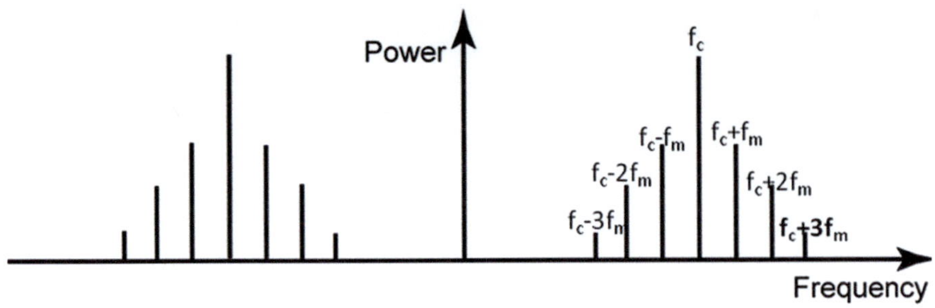

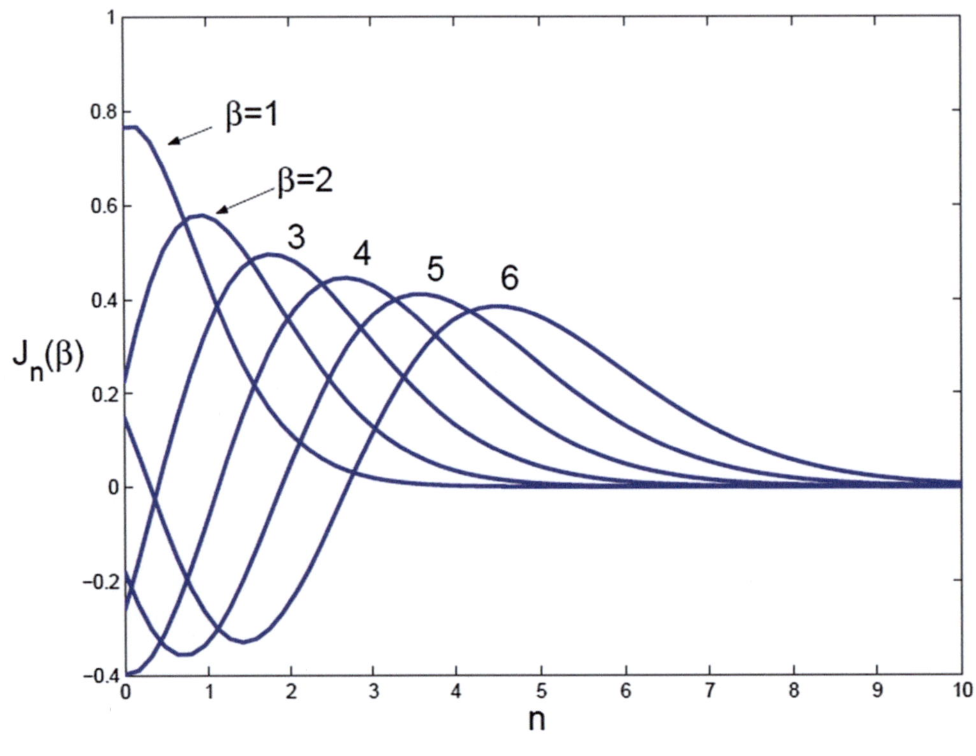

Fig. 3.3 Values of the function Jn(β) as a function of the order n

$$S_m(x) = I_1 \sum_{n=-\infty}^{n=+\infty} J_n(\beta) \cos \left[2\pi(f_c + nf_m)^m\right] x \tag{3.12}$$

Utilizing the above expansion, a discretization of the actual distribution of harmonics is obtained; that is in place of one constant I_1 that concentrates all the energy of the sinusoidal carrier, a distribution of amplitudes is obtained. If one computes the FT of (3.12), the power spectrum of the signal is made of an infinite number of components of amplitude $I_{1\beta} = I_1 J_n(\beta)$ and frequencies, $f_n = f_c \pm nf_m$ (see Fig. 3.2).

One can notice that the spectrum is concentrated around the carrier signal and although the spectrum of the signal is infinite, from the plot of the Bessel functions of the first kind of order n (see Fig. 3.3) it is possible to see that the extent of the spectrum of the modulated signal depends on the value of these arguments. From a table of Bessel function of order n is possible to see that $|J_n(\beta) \to 0|$ as n $\to \propto$. Beyond a certain value of β the amplitude of the corresponding harmonic becomes negligible as illustrated in Fig. 3.3.

The above arguments lead to the Carson's rule [5] to estimate the bandwidth of a signal,

$$BW = 2(1 + \beta)f_m \tag{3.13}$$

In [5], the estimation of bandwidth is limited to $\beta < 2$. The larger β is the larger will be the bandwidth of the signal. This is a very important fact for the process of data retrieval: small values of β lead to a narrow distribution that becomes closer to the model of (3.1) where the energy is concentrated in the I_1 constant and the signal is assumed to be a sinusoid.

3.3 Spectrum of Isothetic Signals

The next task is to relate the preceding developments to Continuum Mechanics variables. In the context of one dimensional signals, we can analyze the lines of equal projected displacement or isothetic lines (see Chapter 13 of [4]). Returning to the initial ideal model, recalling (3.1) and (3.2), S(x) is of the form,

$$S(x) = I_1 \cos\left[2\pi f_c x + \phi(x)\right] \tag{3.14}$$

For the isothetic signal, the following relationship exists between the displacement field and the modulation function $\phi(x)$ (see Chapter 13 of [4])

$$u(x) = \frac{p}{2\pi}\phi(x) \tag{3.15}$$

where $f_c = 1/p$, p is the pitch of the carrier tagged on the surface under analysis. The modulation function is proportional to the projected displacement u(x). The instantaneous frequency of the argument is (see Chapter 13 of [4]),

$$\frac{d\phi(x)}{dx} = \frac{2\pi}{p}\frac{du}{dx} \tag{3.16}$$

Taking into consideration (3.7), the above expression gives,

$$\Delta f = \frac{1}{p}\frac{du}{dx} \tag{3.17}$$

Taking into consideration (3.11), it follows,

$$\beta = \frac{\Delta f}{f_m} = \frac{1}{f_m p}\frac{du}{dx} \tag{3.18}$$

For the linear strain tensor,

$$\beta = \frac{\varepsilon(x)}{f_m p} \tag{3.19}$$

Recalling (3.13), it can be written

$$BW = 2\left(1 + \frac{\varepsilon(x)}{f_m p}\right) f_m = 2\left(f_m + \frac{\varepsilon(x)}{p}\right) \tag{3.20}$$

A very important point can be seen from Fig. 3.4: the signal is centered on the carrier frequency that carries the spectrum of the signal.

A very important conclusion can be derived from (3.20). The bandwidth of the signal as defined by Carson's rule relates the bandwidth with the frequency of the carrier: the higher is the frequency of the carrier (i.e. smaller pitch p), the widest is the spectrum [6]. The bandwidth increases also with the gradient of the displacement field and for the linearized strain tensor with the strain. Furthermore, the bandwidth depends of how fast the modulation function changes in space; this dependency is defined by L_p.

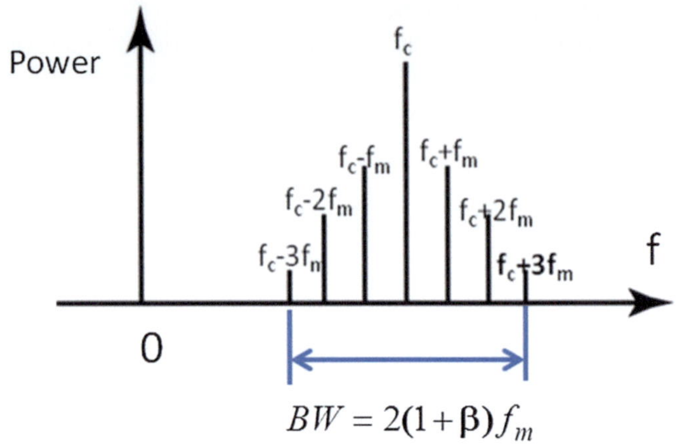

Fig. 3.4 Bandwidth of the frequency modulated signal centered on the carrier frequency

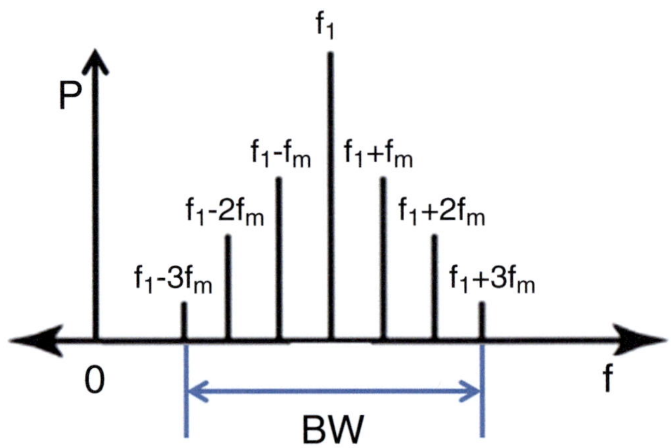

Fig. 3.5 Spectrum of the first order harmonic after removing the carrier term

For the purposes of retrieving the modulated phase, one can utilize other definitions of bandwidth that may reflect better the variables that one need to utilize for phase retrieving. If the pitch p of the carrier is large, the sensitivity will be reduced but also the bandwidth will be reduced making the signal to come closer to the quasi-sinusoidal model that will be defined later on in the paper. As we are going to see, this reduction of the bandwidth could be important for the process of retrieving information from the isothetic lines. The case we have considered corresponds to the event where both the carrier and the modulation function are captured. This case occurs when one records the modulated carrier. If the carrier tagged to a surface is not visible the signal becomes, according to (3.3),

$$S_m(x) = I_1 \cos \phi(x) \tag{3.21}$$

The signal contains only the modulation function. In this case, the carrier is removed from (3.9) and Fig. 3.4: in the frequency space, the signal is centered on the first harmonic (see Fig. 3.5).

For reasons that will be explained later on in the paper in the actual application of data retrieval operations, carrier fringes are introduced in the image (see Chapter 13 of [4]). The corresponding signal is,

$$S_{mc}(x) = I_1 \cos \left[\frac{2\pi}{p_{ca}} x + \phi(x)\right] \tag{3.22}$$

In (3.22), we have assumed that the carrier has amplitude I_1 and the pitch is p_{ca}.

The FT of the carrier is a pair of pulses at frequency $f_{ca} = 1/p_{ca}$. The frequency f_{ca} is such that,

$$f_{ca} >> f_s \tag{3.23}$$

where f_s represents the frequency content of (3.21). Again the signal in the frequency space (Fig. 3.4) becomes centered on the carrier. The presence of an added carrier has important consequences in the data processing of the signal. In the next section, we will analyze some of the effects of the presence of carrier fringes.

Summarizing this section, we started with models of the signals containing displacement information represented by (3.1–3.3). After the concept of modulation factor is introduced, the actual theoretical signal is expressed in discrete form by (3.12). The required displacement information is contained in an infinite number of harmonics. These harmonics depend on the modulation factor β, a key factor in the retrieval of displacement information.

3.4 Differences Between the Theoretical Model and the Actual Recorded Signals

We have started with a model based on phase (frequency) modulation. The real signal has a structure that can be quite different from the adopted model equations (3.1–3.3). The signal is not only frequency modulated but is also amplitude modulated.

Computing the FT of Fig. 3.6 it is possible to compare the spectra of the two patterns (see Fig. 3.7). Figure 3.7a shows the FT of the moiré pattern while Fig. 3.7b shows the FT of the disk with a carrier. Figure 3.7c shows the FT of the moiré pattern without the background tern that is shown in Fig. 3.7d. A very important conclusion can be drawn: there is an overlapping of the zero order that provides the FT of a disk without any fringes and the actual FT of the fringes that carry the displacement information. It is evident from the analysis of Fig. 3.7 that unless overlapping of the orders is overcome it is not possible to get the correct values of the displacement field, no matter which method of fringe pattern analysis is applied. It is also possible to conclude that the introduction of carrier fringes in the moiré pattern has provided a separation of the background pattern from the fringe pattern information. Increasing the frequency of the carrier it is possible to separate the information of the background from the information of the displacement signal (see Fig. 3.6c).

The analysis of Figs. 3.6 and 3.7 brings to light a very important aspect that affects all the methods that provide displacement information: the overlapping of the background terms that define the geometrical configuration of the surface under observation and the way that this surface is illuminated with the signal containing the displacement information and that has been characterized in some detail in the preceding sections. Not only the background information is mixed with the displacement information. Due to the fact that the carrier grating has been modulated orders higher than the first order are present in the image, frequency bands of these orders may overlap. It can be seen in Fig. 3.7b that the second order of the displacement signal appears in the region that contains information of the first harmonic (see the circled area in the FT pattern) but the intensity of the components is lesser that the intensity of the first order. The experimental evidence shows then the problem of order overlapping in the signal that gives displacements.

There is another important source of signal contamination, the presence of amplitude modulation. This effect can be seen in Fig. 3.8. The displacement signal is amplitude modulated by the effect produced by the intensity distribution in the image. The intensity distribution is the result of the combined effect of the utilized illumination system and the transfer function of the optical and the recording system. The amplitude modulation introduces further harmonics in the signal and thus compounds the problem of displacement information recovery.

In order to understand the effects of the above described phenomena in the recovery of the displacement signals one must resort to the theory of analytical functions. The reader should be aware that the above developments are of general validity and independent of any particular technique that may be utilized for information retrieval from moiré, holography, speckle

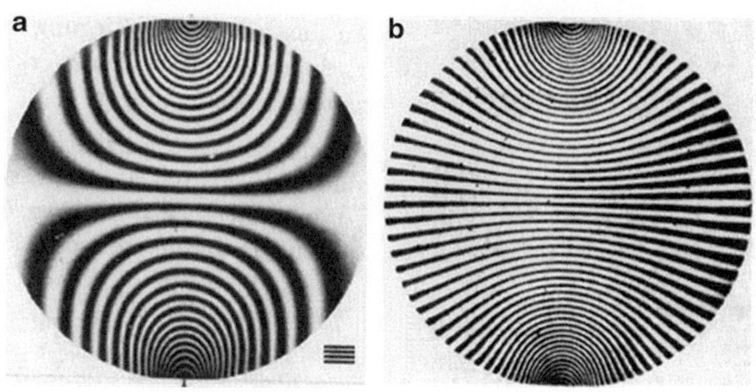

Fig. 3.6 Isothetic lines of the vertical displacements of the points of a disk under diametrical compression: (**a**) Moiré fringes; (**b**) Moiré fringes with a carrier

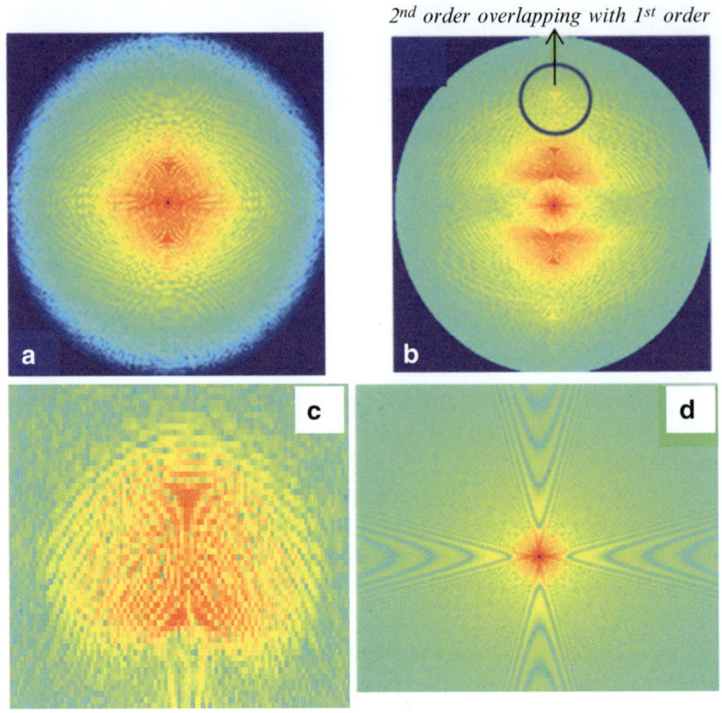

Fig. 3.7 FT of the disk moiré patterns of Fig. 3.6: (**a**) FT of the pattern without carrier; (**b**) FT of the modulated pattern; (**c**) FT of the moiré pattern without background term; (**d**) Magnified view of the FT region corresponding to the background

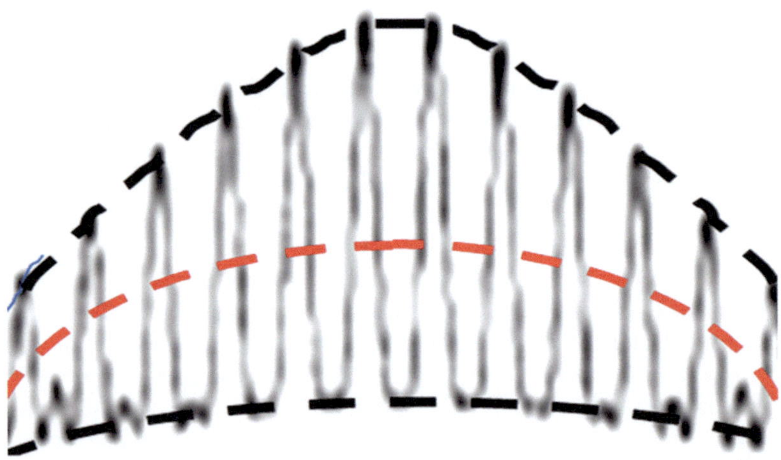

Fig. 3.8 Amplitude modulation along the vertical axis of the disk due to the effect of the optical transfer function

methods, digital image correlation (DIC) etc. One should separate techniques to process data and the mathematical models that are behind these techniques. The mathematical foundations of techniques to retrieve displacement information must satisfy the basic optical physics phenomena that underpin all of them.

3.5 Analytical Signal Theory Applied to the Methods of Retrieving Information from Isothetic Fringe Pattern

The next step is to recall some of the basic principles involved in the connection of analytic functions and signal analysis [7]. The problem to be addressed is as follows: we have an AM-FM signal produced by a linear system, a camera lens, and from this signal recorded by the lens we want to extract displacement information that is encoded in the phase of the signal.

The general problem of AM-FM signals is analyzed in great detail in [7]. We will utilize some of the basic results contained in [7] in the developments that follow. The signal is encoded in a phasor where both components depend on the coordinates,

$$\vec{I}(x) = I_1(x)e^{2\pi i\psi(x)} \tag{3.24}$$

The aim is to know under which conditions from the images of the lens system that receives this input it is possible to get expressions that in closed form will provide reliable values of Ψ(x). Should not be possible to get closed form solutions, we want to estimate the errors involved.

Resorting to solution of linear systems in [7], it is proven that a modulated frequency signal of the form given by (3.24), where Ψ(x) is a modulated frequency signal, is not an eigenfunction of the linear system. This means that one cannot get a closed form solution for this kind of inputs. However, it is shown that it is possible to get approximate quasi-eigenfunction if certain conditions are satisfied; these conditions strongly depend among other variables mainly on the gradient $\nabla\Psi(x)$.

In order to complete the outline of the mathematical model, it is necessary to introduce the concept of complex extension of real-valued signals as is the case with fringes present in an image. The complex extension plays a fundamental role in the process of transforming gray levels into phase information. As pointed out in [7], the instantaneous amplitude and phase components of a complex signal are uniquely defined as well as the instantaneous frequency. However, this statement is not true for a real signal, for example, levels of gray of an image. Hence, the complex extension is necessary to achieve the objective of uniqueness. In what follows some elements of the theory of analytical functions are reviewed in order to facilitate the reader understanding of the developments that follow. The model has direct implications with the process of multiphase technique to transform fringes into phases or on the signal in-quadrature process, an alternative to get phase from fringe signals.

Without going through all the formalisms of analytic functions, Fig. 3.9 provides a quick view of the meaning of the role of analytic function in handling displacements signals. Figure 3.9a represents the space where an analytic function is defined and be shows the connection between the concept of phasor and an analytic function [8]. The optical signal takes the form,

$$I_z(x) = I_p(x) + iI_q(x) \tag{3.25}$$

$$I_p(x) = \operatorname{Re} z(x) \tag{3.26}$$

$$I_q(x) = \operatorname{Im} z(x) \tag{3.27}$$

$$\left\|\vec{I}_1(x)\right\| = \sqrt{I_p^2 + I_q^2} \tag{3.28}$$

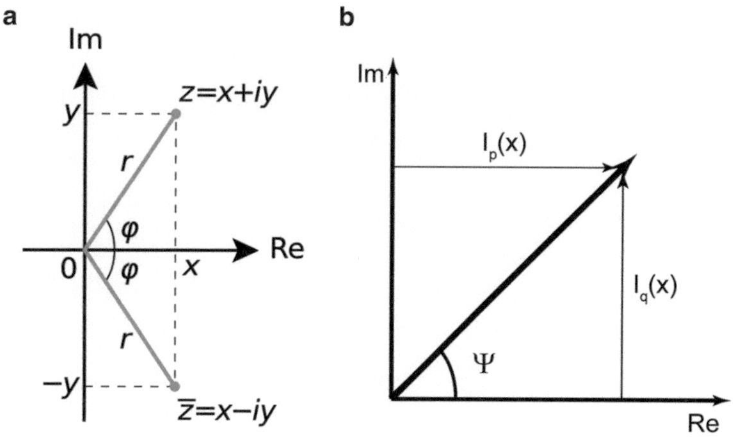

Fig. 3.9 (**a**) Complex plane; (**b**) phasor signal

$$\Psi = arctg^{I_q}/_{I_p} \tag{3.29}$$

$$\vec{I}_z(x) = I_1\, e^{i\psi(x)} = I_1[\cos\Psi(x) + i\sin\Psi(x)] \tag{3.30}$$

The preceding equations provide the mathematical basis for the analysis of displacements signals. The gray level projection on the real axis is the signal directly captured by the optical system utilized to retrieve displacement information. The phase part of the signal that corresponds to the optical path of the signal or path length converted to an angle is the component of the signal that provides the displacement information.

There is a connection between the interpretation of the phasor signal representing the propagation of the displacement optical signal and the Hilbert transform [8],

$$I_q = H\left[I_p(x)\right] = \frac{1}{\pi}PV\int_{-\infty}^{\infty}\frac{I_p(\eta)}{x-\eta}d\eta \tag{3.31}$$

where PV stands for the Cauchy principal value of the integral when this principal value exists and H is the symbol representing the Hilbert transform.

The Hilbert transform can be inverted and it follows,

$$I_p(x) = H^{-1}\left[I_q(x)\right] = \frac{1}{\pi}PV\int_{-\infty}^{\infty}\frac{I_q(\eta)}{\eta-x}d\eta \tag{3.32}$$

It is possible to see that, if one gets the real part of the signal (level of gray) by applying the Hilbert transform, one can get the in-quadrature component, and hence the phase that provides the displacement information.

Equations (3.31 and 3.32) are at the foundation of the phase stepping technique and to the in-quadrature technique to recover displacement information from fringe patterns. The defined mathematical framework is also at the basis of alternative techniques such as DIC that, in order to be valid, must assume that the above mathematical foundations apply.

3.6 Application of the Bedrosian and the Bedrosian-Nuttall Theorems to the Recovery of Signals

The question of utilizing methods that involve arctangent routines to get the phase has been partially answered in the preceding section and can be extended to all methods that provide as an answer the displacement vector directly or components of the displacement vector.

We concluded that both the amplitude of the phasor and the phase are function of x's. How to separate phase and amplitude? Answers to the above question are provided by the Bedrosian-Nuttall theorems [9–11].

The generalized Bedrosian-Nuttall [11] theorem states that for square integrable functions f(x) and g(x) it holds:

$$H[f(x)g(x)] = f(x)H[g(x)] \tag{3.33}$$

if the following conditions are satisfied:

a) The FT of f(x), F(f), vanishes for $f < -f_c$, and the FT of g(x), G(f), vanishes for $f > f_c$, where f_c is an arbitrary positive constant;
b) f(x) and g(x) are square integrable.

The meaning of the theorems is symbolized in a nut-shell [12] by Fig. 3.10. Separation is possible if the different orders spectra do not overlap in the frequency space. This is a necessary but not a sufficient condition [13].

It is necessary to analyze the conditions required to avoid the overlapping orders. The background term of the displacement signal does not appear in this analysis; a term at the zero frequency in the Hilbert transform implies a singularity. One must understand that the removal of the background term does not mean the elimination of the effect of the background in the term that contains the displacement information $I_1(x)$.

We can analyze the contributions to the amplitude of the $I_1(x)$ term

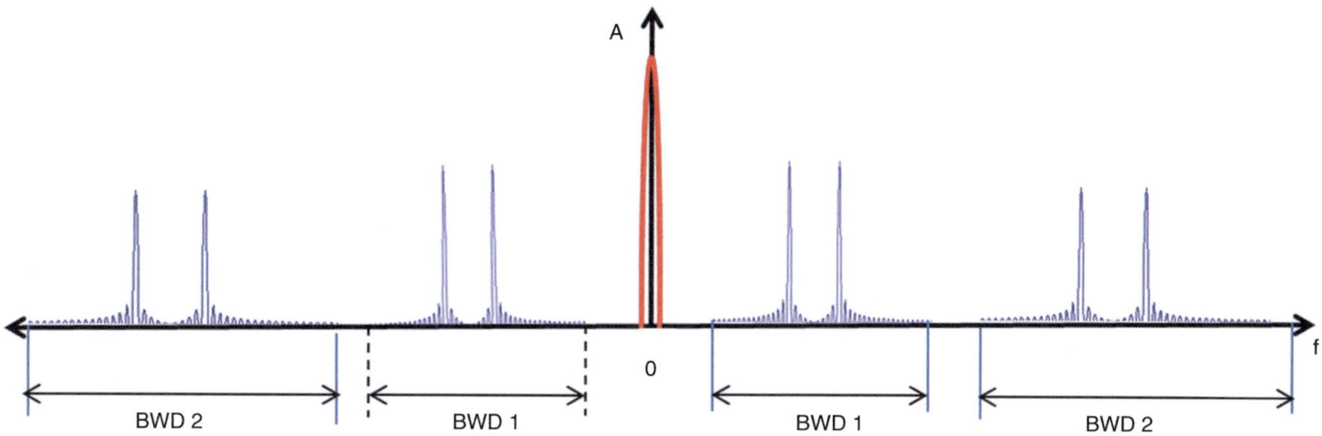

Fig. 3.10 Graphical representation of a condition for the validity of the Bedrosian-Nuttal theorems

$$I_1(x) = I_o(x) + I_A(x) + I_1 \sum_{i=-n}^{i=+n} J_n[\beta(x)] + \sum_{k=1}^{k=m} I_{khigo}(x) \tag{3.34}$$

Equation (3.34) contains all the terms that contribute to the term I(x):

- $I_o(x)$ is the contribution of zero order;
- $I_A(x)$ are the amplitude changes from factors other than the displacement field;
- $I_1 \sum_{i=-n}^{i=+n} J_n[\beta(x)]$ are the amplitude changes due to the frequency modulation;
- $\sum_{k=1}^{k=m} I_{khigo}(x)$ are the amplitude changes due to overlapping of higher orders.

It is necessary to make some assumptions concerning the terms that influence $I_1(x)$.

a) $\left\|\frac{dI_o(x)}{dx}\right\| << \left[f_m + \frac{\varepsilon(x)}{p}\right]$, the gradient of the background term must be smaller than the frequencies contained in the signal;

b) $\left\|\frac{dI_A(x)}{dx}\right\| << \left[f_m + \frac{\varepsilon(x)}{p}\right]$, the changes caused by optical effects and the modulation transfer function must satisfy the same condition.

c) $\left\|\frac{d\left[I_1 \sum_{i=-n}^{i=+n} J_n[\beta(x)] + \sum_{k=1}^{k=m} I_{ksigo}(x)\right]}{dx}\right\| << \left[f_m + \frac{\varepsilon(x)}{p}\right]$, identical conditions.

Since a frequency modulated function is not a sinusoidal function, the image produced by the lens will not be a sinusoidal function. The quadrature condition cannot directly be invoked to apply Bedrosian's theorems. However, the Bedrosian's theorems are approximately valid utilizing a quasi-eigenfunction approximation. This approximation was originally pointed out by Bedrosian-Nuttall [11] and they provided an expression of the error involved.

There is a useful interpretation of the Hilbert transform as a filter to explain the quasi-eigenfunction approximation. There is a well known property of linear systems: their response to the unit impulse symbolized by δ(x) is given by the output of the system h(x) referred also as the convolution kernel of the system. Since the Hilbert transform is a linear system, h(x) is the Hilbert transform of δ(x), that is the response of the system to the unit impulse. In symbols,

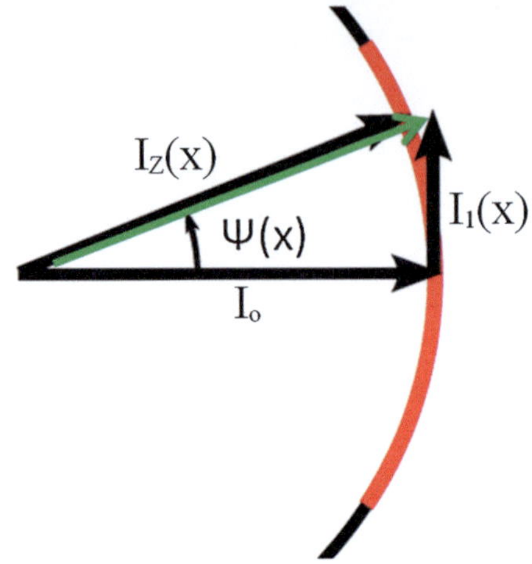

Fig. 3.11 Graphical interpretation of the quasi-harmonic approximation of frequency modulated functions

$$I_1(x)e^{i\psi(x)} \rightarrow h\left[(\psi(x)\right]I_1(x)e^{i\psi(x)} + e(x) \tag{3.35}$$

Symbolically, the above expression represents the quasi-harmonic response of the Hilbert transform; the response of the transform depends on the gradient of the modulating function and involves an error. If there is no carrier present,

$$I_1(x)e^{i\phi(x)} \rightarrow h[\phi(x)]I_1(x)e^{i\phi(x)} + e(x) \tag{3.36}$$

Hence, methods that are based on the arctangent routines are approximately valid and the phase can be recovered from the gray levels with a certain error.

It is possible to develop a simple and intuitive synthesis of all the different theoretical developments. The preceding developments have the simple intuitive graphical representation of Fig. 3.11. The modulated signal can be assumed to be generated by a rotating vector of amplitude I_o and angular position given by,

$$\Psi(x) = [2\pi f_c x + \phi(x)] \tag{3.37}$$

If the frequency change of ϕ(x) is much smaller than f_c, it is possible to make the following approximation,

$$I_o e^{i\Psi(x)} = I_o[1 + \mathrm{Im}\Psi(x)] \tag{3.38}$$

Then

$$\mathrm{Im}[I_o\Psi(x)] = I_1(x) \tag{3.39}$$

$$\tan\Psi(x) = \frac{I_{1(x)}}{I_o} \tag{3.40}$$

$$I_Z(x) = \sqrt{I_o^2 + I_1^2(x)} \tag{3.41}$$

The green vector represents an approximated vector while the actual vector is the black vector. Hence, the phase given by (3.40) contains an error.

3.7 Practical Aspects of the Application of the Quasi-Harmonic Extension and the Bedrosian Theorems

One can derive important conclusions about ways to recover the displacement information and its derivatives from optical signals. It is possible to discuss possible strategies starting from the effect of the background modulation and following with the analysis of the different terms that modulate the first harmonic of the signal

A first step is to remove the zero order term of the power spectrum of FT by utilizing a low-pass filter in the power spectrum and replace the zero order with a normalized background term to be utilized in the recovery of the signal.

The second problem to be solved is to separate the background terms from the displacement signal; there are several solutions to this problem. A direct solution is that one computes the FT of the initial image without the carrier and repeats this operation for the image with the displacement signal. In view of the linearity of FT,

$$FT\left[I_{sig}(x)\right] = FT\left[I_{image}(x)\right] - FT[I_{back}(x)] \tag{3.42}$$

By inversion of the preceding equation one gets $I_{sig}(x)$.

It was previously mentioned that by adding a carrier it is possible to separate the background from the signal. The other alternative is to utilize the multiple phase technique that provides a way to separate background from signal.

It remains to consider the effect of the overlapping of orders; this effect can be controlled by dealing with the modulation factor β. The expression of the bandwidth can be written,

$$BW = \frac{2}{p}\left[\frac{p}{L_p} + \varepsilon(x)\right] \tag{3.43}$$

In order to avoid overlapping of orders and mixing of signals, there are two ratios that control bandwidth as defined in the Carson's rule,

$$\rho = \frac{\varepsilon_x}{p} \tag{3.44}$$

$$\gamma = \frac{p}{L_p} \tag{3.45}$$

It can be seen from (3.43) that the bandwidth is inversely proportional to the carrier pitch p. Hence, it is possible to control bandwidth by changing the ratios ρ and γ. The ratio ρ depends on the gradient of the displacement field in the field of view of the image. The ratio γ depends again on the gradients of the modulation function in the field of view. Large gradients require reducing the field of view in order to avoid the overlapping of orders. If one has large fields of view the pitch of the carrier has to be increased, vice versa in small fields of view the pitch can be reduced keeping under control the overlapping of orders.

3.8 Examples of Phase Recovery from Gray-Levels

If the quasi-harmonic condition is satisfied there are alternative approaches to convert gray levels into displacements.

Two traditional forms are: (1) multiple phase method; (2) signal in-quadrature method. These methods have been considered as competing alternatives. From the preceding sections one can conclude that both alternatives are based on a common mathematical model of fringe signal analysis. Hence, both procedures should yield similar results if they are properly implemented. In what follows an example will be given that support the above introduced assumption.

The pattern shown in Fig. 3.12a is a computer generated pattern utilizing the Theory of Elasticity solution. This provides reference values that can be checked against the numerically generated values called in Fig. 3.12b "experiment" (that is, numerical experiment). Two different alternatives were utilized for the disk analysis, the four phases method and the in-quadrature method by applying the Hilbert transform. The four phases method was implemented by writing an algorithm that from a single picture and utilizing the FT produces numerically fringes with phases difference π/2, π and 3π/2, to get phase

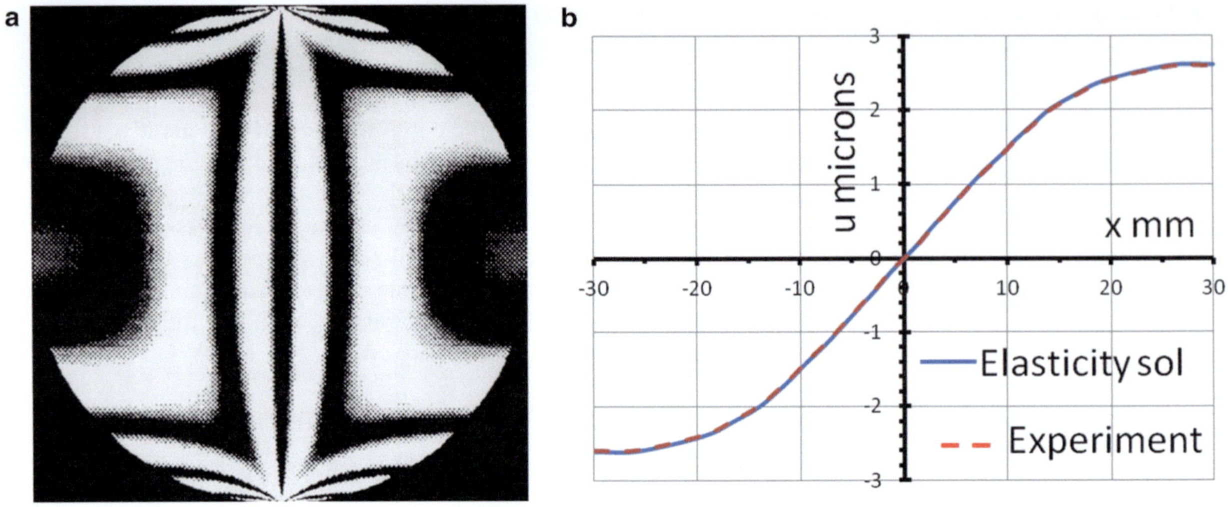

Fig. 3.12 (**a**) u(x) displacement of a disk under diametrical compression; (**b**) displacements along the diameter choosing as zero relative displacement the center of symmetry

values via the arctangent algorithm. The Hilbert transform was implemented utilizing the MATLAB Hilbert transform available in the Signal processing toolbox of Matlab R2014b.

In Fig. 3.12b, with the adopted scale of the plot it can be seen that the theoretical displacement values and those of the numerical computation cannot be visually separated in the plot. Furthermore, the two approaches, four phases and in-quadrature techniques, agree within periodic changes that oscillate around 50 nm.

3.9 Summary and Conclusions

The paper evaluated basic assumptions utilized to analyze fringe patterns as signals that contain displacements and their derivatives information or, in the case of contouring methods, metrological information.

The starting equations assume sinusoidal signals that are represented by phasors. It was pointed out in the article that when to a harmonic phasor is added a modulation function containing displacement information, the phasor is not longer a harmonic signal but a phase (frequency) modulated signal.

The effect of the signal modulation was analyzed here using a discrete series representation of the modulated signal. The power spectrum of the FT of the series contains an infinite number of harmonics; the number of significant harmonics is tied to a parameter, the modulation index β. Equations were derived connecting β to fundamental variables related to displacement fields and their corresponding gradients.

In order to compare the theoretical model of the signals that encode displacement information, actual patterns were analyzed and it was concluded that the real signals are both amplitude and frequency modulated. Furthermore, it was observed that patterns providing different optical information signals overlap in the frequency space and thus direct information recovery is not possible unless steps are taken to make the recovered signals as close as possible to the theoretical model.

A mathematical framework was introduced that makes possible to handle amplitude-phase modulation (AM-FM), and to recover phase from mixed signals. From the analytical signals model arises the quasi-harmonic assumption that is at the basis of many of the methods that retrieve displacement information from optical signals.

A numerical example was presented showing that if, the quasi-harmonic signal hypothesis is satisfied, displacement information can be accurately retrieved. It is also concluded that both the multi-phase approach and the in-quadrature signal approach give similar results when properly implemented.

For a long time in the literature of fringe analysis it was assumed that the multiphase method and the in-quadrature signals analysis where competitive alternatives and that the in-quadrature method could not practically be applied. In this paper, it was shown that both methods have a common basis and that the in-quadrature method properly implemented gives the same results than a multiple phase method. The multiple phase method provides a direct solution for the overlapping of the background term with the displacement signal while to apply the Hilbert transform it is necessary to introduce a carrier that separates the displacement information from background information.

References

1. A.V. Oppenheim, J.S. Lin, The importance of phase in signals. Proc. IEEE **69**(5), 529–541 (1981)
2. C.A. Sciammarella, Basic optical law in the interpretation of the moiré patterns applied to the analysis of strains-Part 1. Exp. Mech. **5**(5), 154–160 (1965)
3. C.A. Sciammarella, A numerical technique of data retrieval from moiré or photoelastic patterns. In: Seminar in depth pattern recognition studies. Proc. SPIE. **16**, 91–101 (1969)
4. C.A. Sciammarella, F.M. Sciammarella, *Experimental Mechanics of Solids* (Wiley, Chichester, 2012)
5. J.R. Carson, Notes on the theory of modulation. Proc. Inst. Radio Eng. (IRE) **10**(1), 57–64 (1922)
6. C.A. Sciammarella, F.M. Sciammarella, Heisenberg principle applied to the analysis of speckle interferometry fringes. Opt. Lasers Eng. **40**, 573–588 (2003)
7. J.P. Havlicek, D.S. Harding, A.C. Bovik, Multidimensional quasi-eigenfunction approximations and multicomponent AM-FM models. IEEE Trans. Image Process. **9**(2), 227–242 (2000)
8. A. Papoulis, *Systems and Transforms with Applications to Optics* (McGraw-Hill, New York, 1968)
9. E. Bedrosian, A product theorem for Hilbert transforms. Memorandum RM-3439-PM. The Rand Corporation, Santa Monica (1962)
10. E. Bedrosian, Hilbert transform of bandpass functions. Proc. IEEE **51**, 868–869 (1963)
11. A.H. Nuttall, E. Bedrosian, On the quadrature approximation to the Hilbert transform of modulated signals. Proc. IEEE **54**(10), 1458–1459 (1966)
12. C. Pachaud, T. Gerber, N. Martin, Consequences of non-respect of the Bedrosian theorem when demodulating. In: *Proceedings of The Tenth International Conference on Conditions Monitoring and Failure Prevention Techniques*, Krakow, 2013
13. Y. Xu, D. Yan, The Bedrosian identity for the Hilbert transform of products of functions. Proc. Am. Math. Soc. **134**(9), 2719–2728 (2006)

Chapter 4
Recent Optical Measurements for the Mechanical Properties of Thin Films

Chi-Jia Tong, Y.-T. Wang, Y.-C. Cheng, and M.-T. Lin

Abstract With advances in the technology behind micro-systems, the mechanical properties of sub-micron and nano-scale thin films have become an item of particular interest. We demonstrate two recent development of using optical methods to measure the mechanical properties of thin films. First, a paddle-like cantilever beam test structure with nano-scale metal films deposited on its top was successfully fabricated and calibrated using standard CMOS processes. Paddle cantilever beam deflection was obtained using a four-step phase-shifting process with a Michelson interferometer. Film strain was determined using a simple force equilibrium equation. Residual stresses were measured and observed at different thicknesses. High tensile stress forms during the early deposition stage for thin film due to grain coalescence, and a decrease in stress with an increase in film thickness. With thicknesses greater than 150 nm of Cu film, lattice relaxation associated with the surface mobility of metallic atoms changes residual stress from tension to compression. Second, it summaries the XRD measurements of the bulge tested thin film. We annealed thin films and tracked the texture transformation using X-ray diffraction while independently varying the stress in the film using a bulge test apparatus. The bulge height was measured as a function of pressure using a optical interferometer, using the bulge as the fully reflective surface, and an optically flat half-silvered mirror as a reference surface. A CCD camera was used to record interference fringe motion as the pressure was increased. Results show that applied stresses have no links on the kinetics of the texture transformation in films.

Keyword Optical methods • Thin films • Mechanical properties of thin films • Optical measurements • Interferometer

4.1 Introduction

The micro-systems technology has grown rapidly during the last decade; with advances in the technology behind micro-systems the need to enhance performance and cost efficiency has increased demand for further miniaturization. Accordingly, the mechanical properties of micron and nano-scale thin films have become an item of particular interest. Thin metal films are critical elements in a many nano- and micro-fabricated technologies. Although many studies had performed on the characterization of mechanical behavior on thin films [1–3], the results obtained from different measurement techniques were vary widely for nominally identical samples due to the difficulty with the techniques.

Here, we introduce two recent development of using optical methods to measure the mechanical properties of thin films. First, a paddle-like cantilever beam test structure with nano-scale metal films deposited on its top was successfully fabricated and calibrated using standard CMOS processes. Paddle cantilever beam deflection was obtained using a four-step phase-shifting process with a Michelson interferometer. The test structure was designed on a cantilever beam specimens with dimensions as few hundred nanometers to less than 15 nm. The experiment is designed to be operated by using the electrostatic deflection on the cantilever beam and then measure the deposited thin metal film materials on top of it. Thus allow us to study mechanical function mechanisms in thin films and film of the thickness on nanoscale regime.

Second, we introduce the X-ray diffraction (XRD) measurements of the bulge tested thin film. We annealed thin Ag films and tracked the texture transformation in-situ using synchrotron X-ray diffraction while independently varying the stress in the film using a bulge test apparatus. The bulge height, h, was measured as a function of pressure, P, using a simple Fabry-Perot optical interferometer, using the bulge as the fully reflective surface, and an optically flat half-silvered mirror as a reference surface. A CCD camera was used to record interference fringe motion as the pressure was increased. The bulge height was obtained by counting the number of fringes that passed a given point. Both methods allow us to study the mechanical properties in thin films with film thickness on micro and nanoscale regime.

C.-J. Tong • Y. Wang • Y. Cheng • M. Lin (✉)
Graduate Institute of Precision Engineering, National Chung Hsing University, 250, Kuo-Kuang Rd., Taichung 40227, Taiwan, R.O.C.
e-mail: mingtlin@nchu.edu.tw

C. Sciammarella et al. (eds.), *Experimental and Applied Mechanics, Volume 4*,
Conference Proceedings of the Society for Experimental Mechanics Series, DOI 10.1007/978-3-319-22449-7_4

4.2 Experimental

In microbeam bending, cantilevered beams of the film of interest are fabricated by micromachining. The end of the beam is deflected using an indentation apparatus which senses vertical displacement. Elastic properties of the thin film material can be determined from the measured load-displacement characteristics [1–3]. If a thin film is deposited on a thicker free-standing cantilever, plastic properties of the thin film can be probed as well [3]. The "paddle" cantilever beam approaches here was to design a "constant stress" cantilever beam that eliminate the non-uniform distribution of stresses along the cantilever. In a cantilever beam with a single localized load at the free end, the bending moment varies linearly from zero at the point of load application to a maximum at the built in end. Thus for a parallel sided beam the axial stress on the beam surface is proportional to the bending moment. Here, through the design of a cantilever section which modulus can be made proportional to the width proportional and holding the thickness constant, stress is constant from one end of the beam to the other.

4.2.1 Testing System I

The single test chip is designed (Fig. 4.1) to fit in an apparatus where one side of the paddle plate is pulled by an electrostatic field. The measured strain was converted through the capacitance measurement for the deflection of the cantilever.

The sample is shown as the structure of a 20 × 20 mm chip with paddle as shown above in plane view and cross section. The thin silicon beam (40 μm thick) has a triangular shape in plane view to provide uniform strain in a metallic film on its surface when the beam is bent with force on the paddle. Figure 4.1 shows the schematics view of the sample.

The sample fabrication procedure is using standard clean room processing. The sample is fabricated on a standard 4 in. diameter double polished silicon wafer. The detail of the fabrication process sequence is shown in Fig. 4.2. The structure can be metallized from both sides. No processing is required after the metal has been deposited. Later, a metallic conducting surface is required on both sides of paddle for stable deflection and capacitance measurements to be made.

The apparatus is custom design and the system set-up is design to measure beam deflection by Michelson Interferometer. It consists of the laser beam and the deflection electrode. Figure 4.3 shows the schematic of the measuring system. The sample chip is mounted together with He-Ne laser and the electrode shown below. A spacing of 100 μm to the window frame chip surface around the paddle structure is defined by a metallic spacer. During the experiment, the bending force was generated by underneath electrode. The deflection of the paddle beam can be measured from the interfere image using "Four Phase Shifting Technique". The system schematic is shown in Fig. 4.4. During the standard tests, the electrode is mounted below the paddle plate and the measurement is controlled by PC through IntelliWave program. The control optical include the Michelson Interferometer system, the CCD, the sample fix block and computer. Data was digitizing from interferometer image to value by IntelliWave software.

Four step phase-shifting method, which uses PZT to push the mirror to generate four images of interference fringes to calculate the data. Each image will obtain one equation, four equations list in below. Their intensities are [5].

$$I_1 = I_o + I_r + 2\sqrt{I_o I_r}\cos(\varphi) \tag{4.1}$$

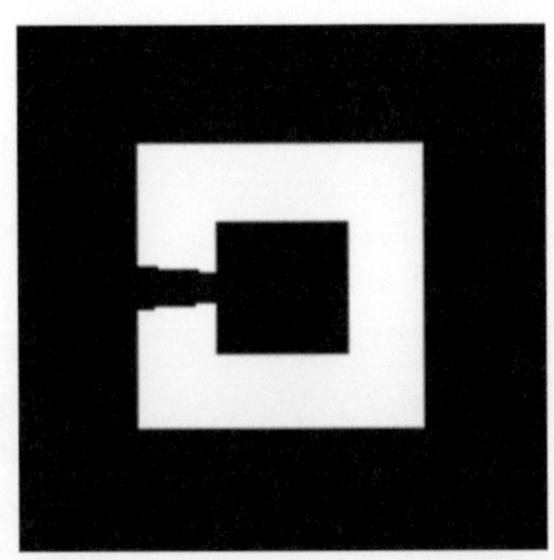

Fig. 4.1 Schematic view of the sample

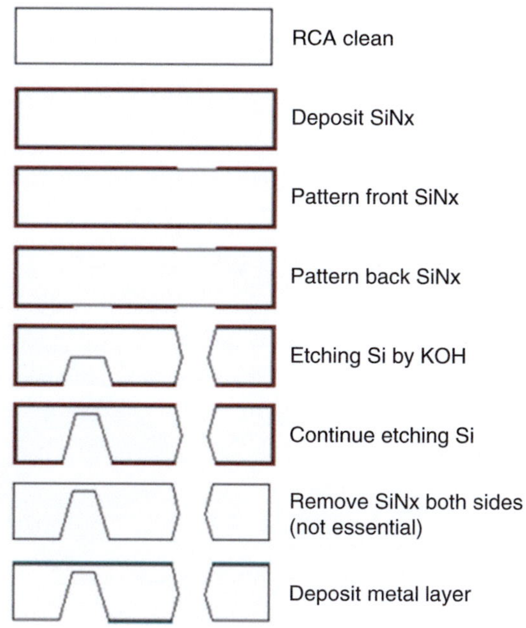

Fig. 4.2 The detail of the fabrication process sequence

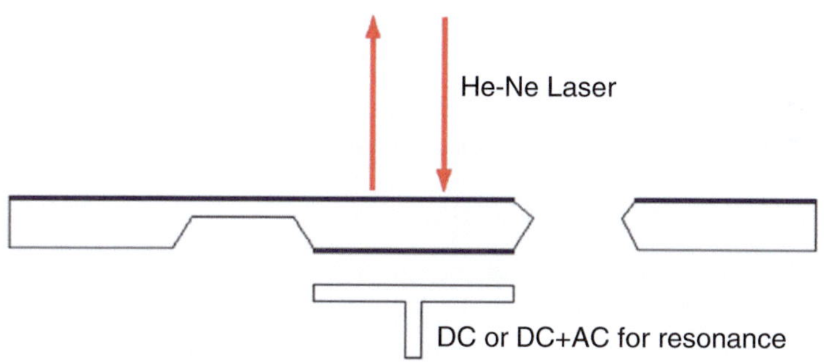

Fig. 4.3 Schematic of the measuring system

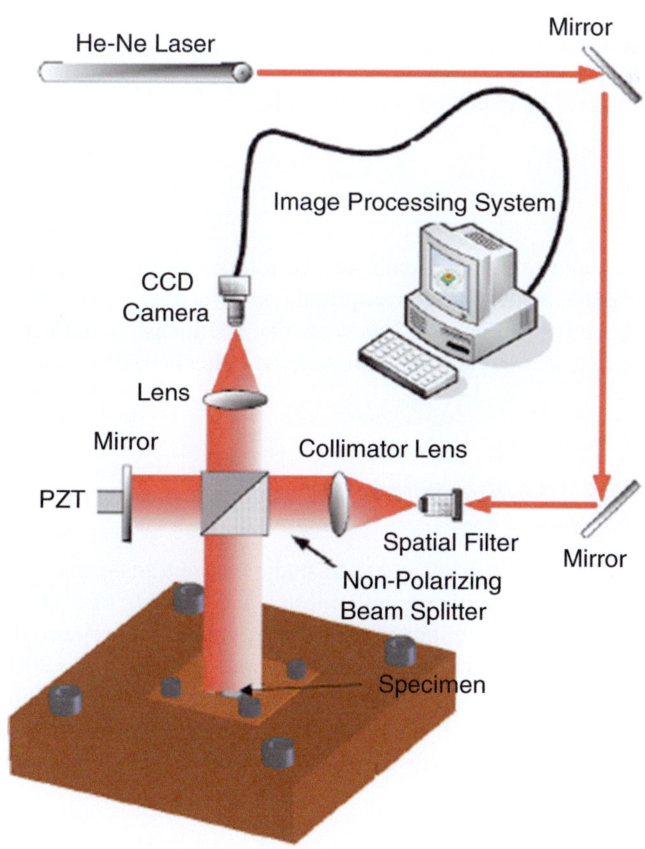

Fig. 4.4 Schematic of the system

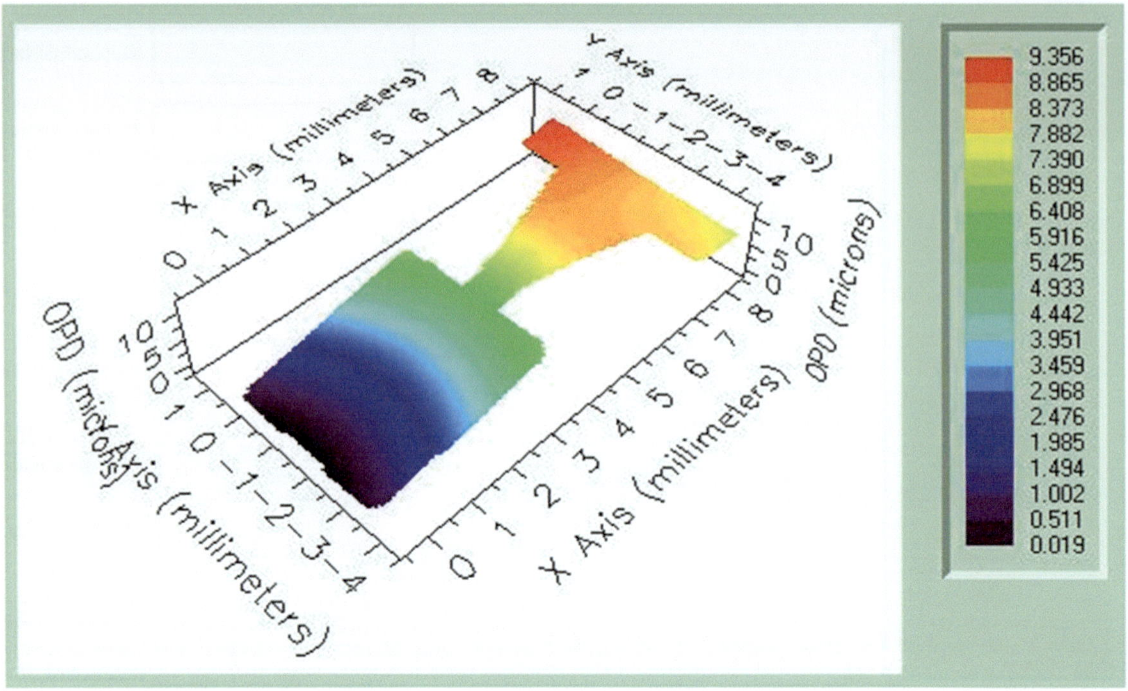

Fig. 4.5 Interference fringe images which generated in four different phase

$$I_2 = I_o + I_r + 2\sqrt{I_o I_r}\cos\left(\varphi + \frac{\pi}{2}\right) \tag{4.2}$$

$$I_3 = I_o + I_r + 2\sqrt{I_o I_r}\cos(\varphi + \pi) \tag{4.3}$$

$$I_4 = I_o + I_r + 2\sqrt{I_o I_r}\cos\left(\varphi + \frac{3}{2}\pi\right) \tag{4.4}$$

Then the expected phaseφ will be

$$\varphi = \tan^{-1}\left[\frac{I_4 - I_2}{I_1 - I_3}\right] \tag{4.5}$$

I_o and I_r are amplitudes of two interference beams respectively and φ is the phase distribution which has relation with displacement, vibration amplitude or refractive index. These interference fringes of the phase angle are 0, π/2, π, and 3π/2 respectively. Figure 4.5 presents the 3-D image of deflection of paddle cantilever beam by solving (4.1–4.5). The results present that the test structure and the optical (FSPS) method work properly.

4.2.2 Testing System II

The second experiment was to study texture transformation of thin Ag film. The film stress was varied in this study using a plane strain membrane bulge tester that consists of a sealed chamber behind a sample. The chamber, was filled with an atmosphere of Helium at pressure, P, creating a pressure differential across the sample which causes the sample to bulge inducing stresses in the film. The sample comprised of a 200 nm SiN_x membrane bridging a 3 mm × 12 mm windows in the center of 20 mm × 20 mm silicon dies. Thin silver films were deposited using e-beam evaporation onto the nitride membrane at room temperature. A 40 nm thick titanium adhesion layer was first deposited using e-beam evaporation onto the bulge test substrates, at a deposition rate of 5 Å/s. On top of the adhesion layer, silver films were deposited at a rate

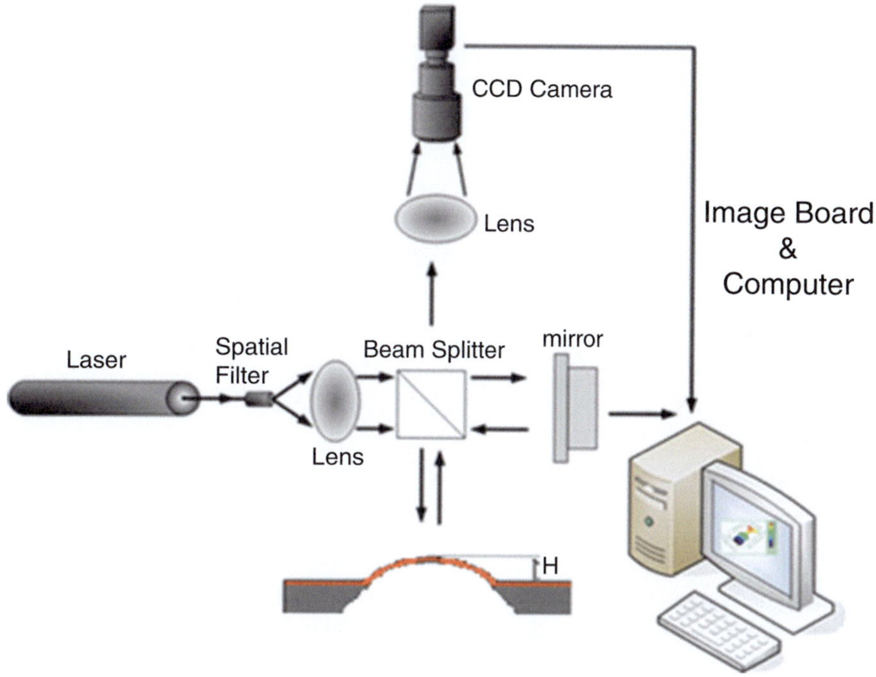

Fig. 4.6 Interferometer system setup schematic

of 40 Å/s to final thickness of 1200 nm. Both the silver and titanium layers were evaporated out of molybdenum crucibles. The chamber base pressure was on the order of 10^{-6} Torr for all depositions.

The bulge height, H, was measured ex-situ as a function of pressure by using an optical interference system as shown in Fig. 4.6. The system consists of a laser light source with a 532 nm wavelength, spatial filter, beam splitter, CCD camera, lens, and several mirrors. Measurements were taken after the membrane bulged into the desired calibration position. Reflection mirror was used to give the reference mirror in 90° segments to generate interference fringe images at different phases. A CCD camera was used to catch the phase images prior to calculating bulge height. Bulge height is obtained by counting the moving numbers of interference fringes and then multiplies the half length of the wavelength.

The bulge tester was mounted on a heating stage also made out of brass. The heating stage was connected to a water heat exchanger. Once the sample was mounted, the heat water flow was started which resulted in a fast temperature increase to a steady temperature state. The temperature of the sample was measured directly at the brass sample holder. Within 3 min, the sample temperature reached 67 °C. The sample temperature stabilized between 68 and 69 °C within another 5 min. Samples were tested at different applied stresses but always at the target temperature of 69 °C. Sample had no stress applied, sample with a stress of 140 MPa, and sample with a stress of 200 MPa. Before stress and temperature were applied to the samples, an initial diffraction cycle was performed at room temperature.

4.3 Results

Silver, gold, and copper were respectively deposited on the different paddle cantilever beams, but the thickness of every paddle cantilever beam is equal to 40 μm. Figure 4.7 show the deflections of silver, films respectively with the comparison of the paddle cantilever beam without thin metal film on its paddle plane. The tensile stress causes the position of paddle plane upper than original position, and the compressive stress is lower than the initial position. In Fig. 4.7, the curve shows the deflection decreases after deposited silver to demonstrate that the tensile residual stress in the film. The deflections of silver, gold, and copper films (50–250 nm) driven by the electrostatic force were measured using four step phase-shifting method. The results present that the test structure and the optical (FSPS) method work properly. Furthermore, silver presents that the deflection decreases after deposition to demonstrate that the tensile residual stress in the film. Oppositely, gold and copper films show compressive residual stress in them.

The volume fractions of the 100/111 texture component were measured. The 100 texture component volume fraction of the 140 and 200 MPa samples were at the same point of time as the sample with no applied stress. The times of the first

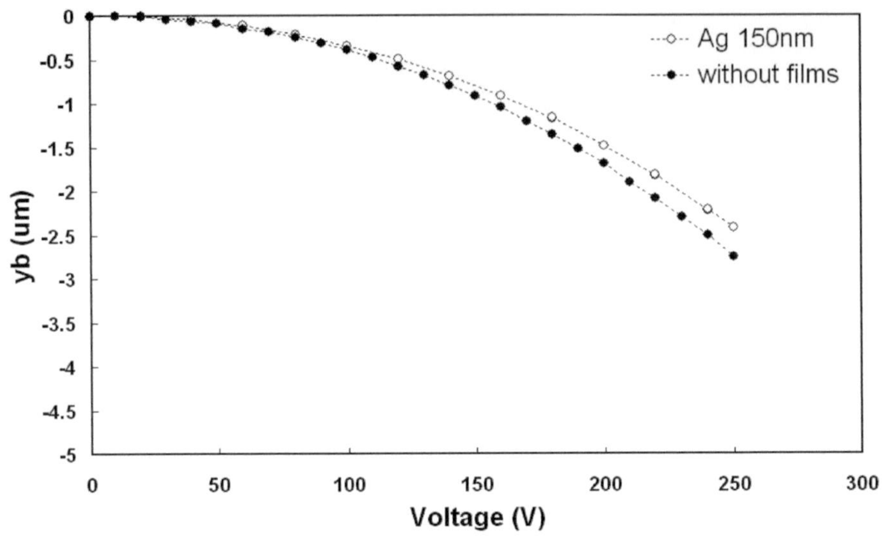

Fig. 4.7 y_b versus voltage with Ag 200 nm on paddle cantilever beam

experimental points for the samples with applied stress do not indicate the beginning times of the measurements. The sample with no additional stress applied, had initially a 100 volume fraction of 3 % which increased over 260 min to 96 %. The 140 and 200 MPa samples showed a 100 volume fraction of 40 % and 42 %, respectively, at the beginning of the X-ray experiments. Their final 100 volume fraction after 150 min and 135 min was 93 % and 91 %, respectively. Due to the high purity of the silver films, the silver films transformed already at room temperature which resulted in different texture volume fractions at the beginning of the experiments. To compare the texture transformations of all three samples, a modified Avrami fit was used to fit the transformed 100 volume fraction, f_{100}, versus time, t. A shift from the sample with no stress, to the 140 MPa and finally 200 MPa sample in all coefficients was observed.

4.4 Conclusions

An optical measurement had been performed using paddle cantilever beam specimens with dimensions as few hundred nanometers to less than 10 nm. The experiment can be operated by using the electrostatic deflection on the paddle cantilever beam and then measure the deposited thin metal film materials on top of it. Interference techniques were used to measurement on the other side of the deflected plate to measure its deflection with respect to the force. The measured strain was converted through the interference measurement for the deflection of the cantilever. The results help to predict system performances of cantilever beam position in different residual stress. In addition, we had found that thin film transformations conducted in-situ during bulge tests showed no detectable effect of applied stress. The results may suggest that variations in initial nucleus density with film thickness control the transformation kinetics.

References

1. T.P. Weihs, S. Hong, J.C. Bravman et al., Mechanical deflection of cantilever microbeams: a new technique for testing the mechanical properties of thin films. J. Mater. Res. **3**(5), 931–942 (1988)
2. W.D. Nix, Mechanical properties of thin films. Metall. Trans. A **20**, 2217–2245 (1989)
3. J.A. Schweitz, Mechanical characterization of thin films by micromechanical techniques. MRS Bull. **17**(7), 34–45 (1992)
4. D. Kingsbury, Integrated mechanical testing laboratory, Arizona State University, Laboratory notes, unpublished
5. C.J. Tong, Y.C. Cheng, M.T. Lin, K.J. Chung, J.S. Hsu, C.L. Wu, Optical micro-paddle beam deflection measurement for electrostatic mechanical testing of nano-scale thin film application to MEMS. Microsyst. Technol. **16**, 1131–1137 (2010)
6. Y.T. Wang, T.C. Hu, C.J. Tong, M.T. Lin, Novel full range vacuum pressure sensing technique using free decay of trapezoid micro-cantilever beam deflected by electrostatic force. Microsyst. Technol. **18**, 1903–1908 (2012)
7. S. Hyun, W.L. Brown, R.P. Vinci, Thickness and temperature dependence of stress relaxation in nanoscale aluminum films. Appl. Phys. Lett. **83**(21), 4411–4413 (2003)

8. R. Carel, C.V. Thompson, H.J. Frost, Computer simulation of strain energy effects vs surface and interface energy effects on grain growth in thin films. Acta Mater. **44**, 2479–2494 (1996)
9. S.P. Baker, B. Hoffmana, L. Timiana, A. Silvernaila, E.A. Ellisa, Texture transformations in Ag thin films. Acta Mater. **61**(19), 7121–7132 (2013)
10. S.P. Baker, K. Saha, J.B. Shu, Effect of thickness and Ti interlayers on stresses and texture transformations in thin Ag films during thermal cycling. Appl. Phys. Lett. **103**, 191905 (2013)
11. P. Sonnweber-Ribic, P.A. Gruber, G. Dehm, H.P. Strunk, E. Arzt, Kinetics and driving forces of abnormal grain growth in thin Cu films. Acta Mater. **60**, 2397–2406 (2012)

Chapter 5
Evaluation of the Penalized Least Squares Method for Strain Computation

Raphaël Moulart and René Rotinat

Abstract This work proposes an alternative procedure to smooth and differentiate experimental full-field displacement measurements to get strain fields. This one, the penalized least squares method, relies on the balance between the fidelity to original raw data and the smoothness of the reconstructed ones. To characterize its performance, a comparative study between this algorithm and two other commonly implemented strategies (the 'diffuse approximation' and the Savitzky-Golay filter) is achieved. The results obtained by the penalized least squares method are comparable in terms of quality of the reconstruction to those produced by the two other algorithms, while the proposed technique is the fastest as its computation time is totally independent from the asked amount of smoothing. Moreover, unlike both other considered methods, it is possible with this technique to perform the derivation to obtain strain maps before smoothing them (while the smoothing is normally applied to displacement maps before the differentiation) which can lead in some cases to a more effective reconstruction of the strain fields.

Keywords Full-field kinematic maps • Smoothing algorithms • Numerical differentiation • Penalized least squares • Performance characterization

5.1 Introduction

Most of the full-field kinematic measurement methods (such as digital image correlation, electronic speckle pattern interferometry, grid method, ...) provide displacement maps when, in solid mechanics, the useful information is normally given by strain fields (for identification of mechanical properties for instance). Consequently, the raw displacement data usually have to be numerically differentiated. However, a numerical differentiation strongly amplifies the level of noise that affects a signal. An intermediary step of smoothing is thus often necessary before proceeding to the numerical derivation.

Surprisingly, only a few smoothing techniques are currently used by the experimental mechanics community [1]:

- Local data filtering by a kernel [2]. This technique provides a fast approach but does not efficiently reconstruct the data near the boundaries of their domain.
- Global regression techniques (on a 2D polynomial function basis for instance) [3]. This approach can be efficient for very low spatial frequency signals but can induce reconstruction artefacts on the borders and parasitic oscillations within the reconstructed field when the number of parameters of the function increases.
- Global regression based on a "finite elements-like" approach [4].
- Local polynomial regression utilizing a moving weighted least squares algorithm: the so-called "diffuse approximation" [4]. This technique provides very satisfactory results (especially efficient for reconstructing the data at the boundary) but can be very time-consuming when the span of the moving least squares algorithm increases.

In order to improve the currently applied methodologies, a fast bibliographical review in statistics and data analysis literature shows that several competitive alternatives do exist. This article is aimed at adapting one of the available techniques to the processing of full-field kinematic measurements and to evaluate its performances compared to two others methods: the "diffuse approximation" (called DA later in the paper) and the Savitzky-Golay filtering approach (called SG later in the paper). SG, if not generally used in processing full-field kinematic maps, is a very popular smoothing technique since it combines the quickness of the kernel smoothing with a local polynomial basis approach such as in DA. Moreover, it can compute directly the derivatives of a noised data set [5].

R. Moulart (✉) • R. Rotinat
Mechanics Surfaces and Materials Processing, Arts et Métiers ParisTech, Rue Saint Dominique BP 508, Chalons-en-Champagne 51006, France
e-mail: Raphael.Moulart@ensam.eu

C. Sciammarella et al. (eds.), *Experimental and Applied Mechanics, Volume 4*,
Conference Proceedings of the Society for Experimental Mechanics Series, DOI 10.1007/978-3-319-22449-7_5

After having recalled the principle of the chosen algorithm, the penalized least squares (called PLS later in the paper), the technique is applied to several cases (an analytical 2D function, FE simulated displacement maps and actual experimental maps) and its performances compared to those of the two other algorithms. A final section concludes the paper and gives some prospects of the chosen method that could easily be extended to fit new requirements induced by the current evolution of full-field kinematic measurements (especially, the 3D tomographic measurements that are becoming more and more popular).

5.2 The Penalized Least-Squares Algorithm

The penalized least squares method is not new. It dates back to approximately a century [6, 7]. A modern formulation of PLS can be found in [8, 9].

The basic idea of the method is to minimize a quantity expressing the balance between the deviation of the reconstructed data from the raw noised data and the roughness of the reconstructed data. To illustrate this, let us consider a 1D data set of n observations y_i made on equally spaced abscissae. One has to find the reconstructed values z_i that minimize the quantity Q expressed in (5.1):

$$Q = \underbrace{\sum_{i=1}^{n} (y_i - \hat{y}_i)^2}_{\text{Deviation from data}} + \lambda \underbrace{\sum_{i=1}^{n-d} \left(\Delta^d \hat{y}_i\right)^2}_{\text{Roughness}}. \tag{5.1}$$

In (5.1), the term $\Delta^d z_i$ is the dth order difference of the z_i values. So, the deviation term is the square of the norm of residuals and the deviation term, the square of the norm of the dth order difference of the reconstructed values. The term λ is the smoothing parameter that gives a relative weight to the roughness term in the functional Q to be minimized: the higher value of λ, the smoother the reconstructed data.

The minimization of Q has a close-form solution. Adopting a matrix formalism, the vector of the reconstructed values $\{z\}$ can be written as:

$$\{z\} = \left([I] + \lambda [D_d]^T [D_d]\right)^{-1} \{y\}, \tag{5.2}$$

where { . } denotes a column vector, [.] a matrix; $[I]$ is the $n \times n$ identity matrix and $[D_d]$ the matrix such as $[D_d].\{z\}$ is the d th order difference of the z_i values.

As the method leads to a close-form solution, it is very fast compared to other currently used techniques.

To fit to the requirements of the processing of kinematic maps, the algorithm as to be able to deal with missing values (that can occur when regions of interest that are not rectangle-shaped are observed or when the samples present holes). This missing values result in NaNs in Matlab® or equivalent numerical analysis software. To deal with these NaNs, a weighted version of the algorithm can be set up. Each residual term is assigned by a weight w_i equal to 1 for valid points and to 0 for rejected points. This leads to the reconstructed data expressed in (5.3):

$$\{z\} = \left([W] + \lambda [D_d]^T [D_d]\right)^{-1} [W]\{y\}, \tag{5.3}$$

where $[W]$ is a diagonal matrix containing the w_i values.

To deal with 2D data sets, $[D_d]$ matrix has to be slightly adapted. A second order differentiation operator is considered with two additional lines at the beginning and the end corresponding to a reflective boundary condition. $[D_2]$ is now a square matrix and can be eigen decomposed leading to:

$$\{z\} = [U]\left([I] + \lambda [\Omega]^2\right)^{-1} [U]^T \{y\} = [U][\Gamma][U]^T \{y\}. \tag{5.4}$$

The left multiplication by $[U]^T$ and by $[U]$ corresponds to the application of a discrete cosine transform (DCT) and inverse discrete cosine transform (IDCT) [9, 10]. Equation (5.4) thus becomes:

$$\{z\} = [U][\Gamma][U]^T\{y\} = IDCT([\Gamma]\mathrm{DCT}(\{y\})). \tag{5.5}$$

Due to the properties of the discrete cosine transform, the same kind of approach can be applied to 2D equally spaced data. For a 2D grid of values, (5.5) becomes:

$$[z] = \mathrm{IDCT2}\left(\left[\Gamma^{\mathrm{II}}\right] \circ \mathrm{DCT2}([y])\right), \tag{5.6}$$

with DCT2 and IDCT2 respectively denoting the 2D discrete cosine transform and inverse discrete cosine transform, $\circ$, the Hadamard product (elementwise product) and $[\Gamma^{\mathrm{II}}]$ a square matrix depending on λ and the eigenvalues of the 2D version of $[D_2]$.

5.3 Smoothing of Kinematic Maps: Evaluation of Performances

5.3.1 *Smoothing of a Noised Map of a 2D Analytical Function*

The performances of the chosen algorithm have been first evaluated on a noised analytical function. The advantage of using an analytical function is that one has access to the analytical derivatives which allows a comparison of the reconstructed derivatives to the actual ones getting rid of every source of error on these last values.

The analytical function that has been chosen is the "peaks" function of Matlab®. Here, a (1000 × 1000) matrix has been generated. A Gaussian noise with mean zero and standard deviation one has been added to the map which leads to a typical high resolution noised displacement maps both in terms of spatial localizations and of signal to noise ratio. Views of the raw and noised functions and its analytical derivative with respect to x are shown in Fig. 5.1.

First of all, the time spent to perform the reconstruction has been looked at. The computation has been performed using Matlab® on a 64 bits computer equipped with an Intel® Core™ i7-2760QM (2.4 GHz) CPU and a 8 GB DDR3 1333 MHz RAM. The computation time against the smoothing parameters for DA, SG and PLS algorithms are shown in Fig. 5.2.

It can be seen that:

- for both DA and SG, the computation time is strongly dependent of the span of the window (h) on which the reconstruction is performed: the more h, the longer the processing;
- if the computation time remains reasonable for SG (inferior to half a minute for $h = 85$ pixels), it becomes very large for DA (about 6 h for $h = 85$ pixels) which can be prohibitive;
- the computation time is very low (about 0.78 s) and independent of the smoothing parameter λ for PLS

From these results, it is clear that PLS is far quicker than both DA and SG which makes it very challenging.

In a second time, the accuracy of the reconstruction has been investigated. For that, an approach similar to the one developed in [4] has been set up. The criterion to qualify the reconstruction is the average distance between the reconstructed derivatives and the exact ones defined as:

$$d_{\partial P} = \left\langle \sqrt{\left(\frac{\partial P^{\mathrm{rec.}}}{\partial x} - \frac{\partial P^{\mathrm{exact}}}{\partial x}\right)^2 + \left(\frac{\partial P^{\mathrm{rec.}}}{\partial y} - \frac{\partial P^{\mathrm{exact}}}{\partial y}\right)^2} \right\rangle, \tag{5.7}$$

where $\langle\ .\ \rangle$ denotes here a spatial averaging of the data all over the map (P represents the values of the "peaks" function).

The optimal results, leading to the lowest value of $d_{\partial P}$, are given in Table 5.1 for each method.

It can be seen that, for an optimal value of the smoothing parameter, PLS conducts to a lower value of this average deviation than DA and SG (0.12 for PLS to be compared to 0.22 for DA and 0.23 to SG). It is worth noting that, with PLS, the smoothing algorithm can also be applied after a point-to-point derivation of the raw map leading to satisfactory performance (which is almost impossible with the other methods).

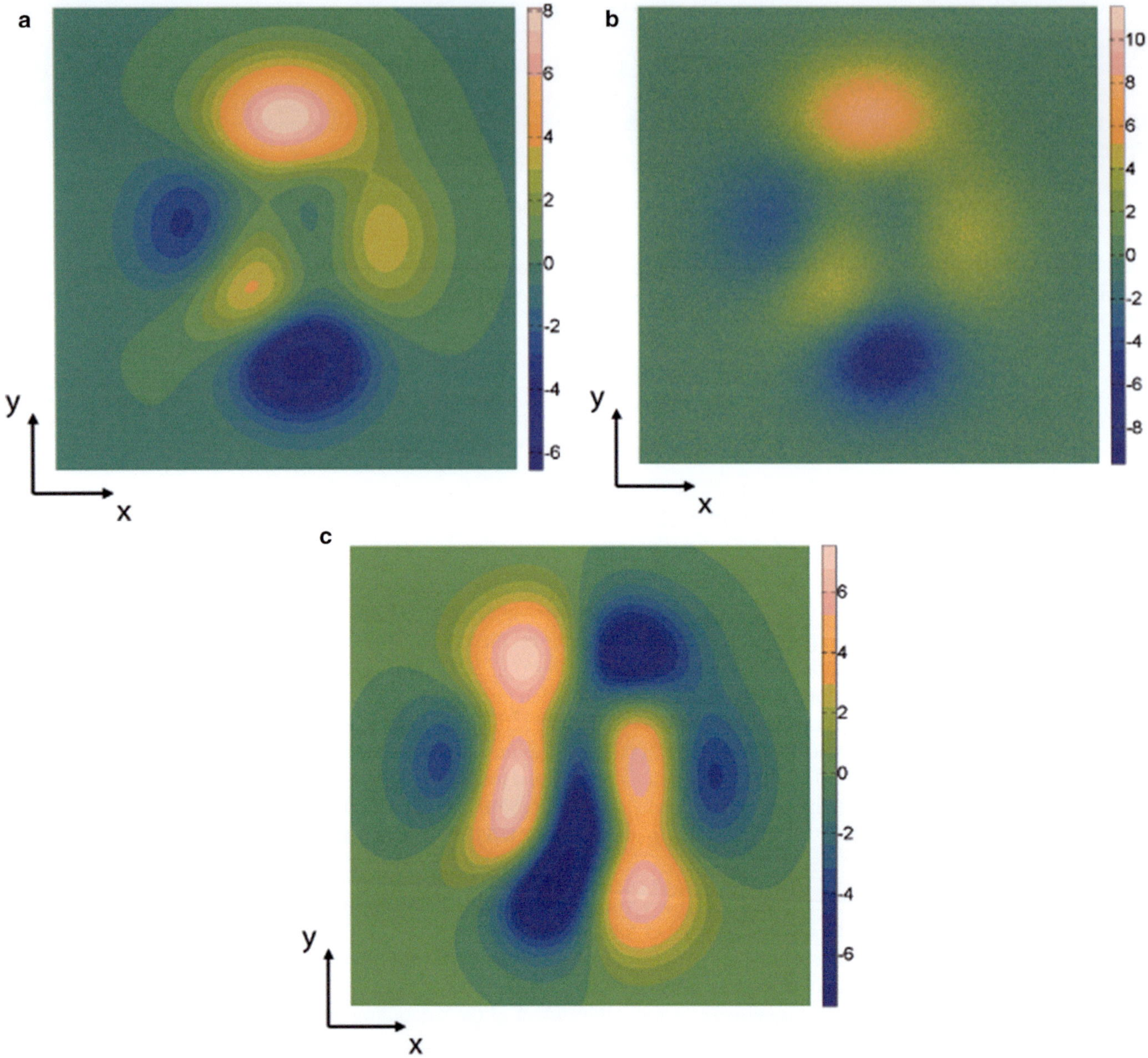

Fig. 5.1 The 1000 × 1000 pixels analytical "peaks" map. (**a**) The function (**b**) the noised function (**c**) its derivative with respect to x

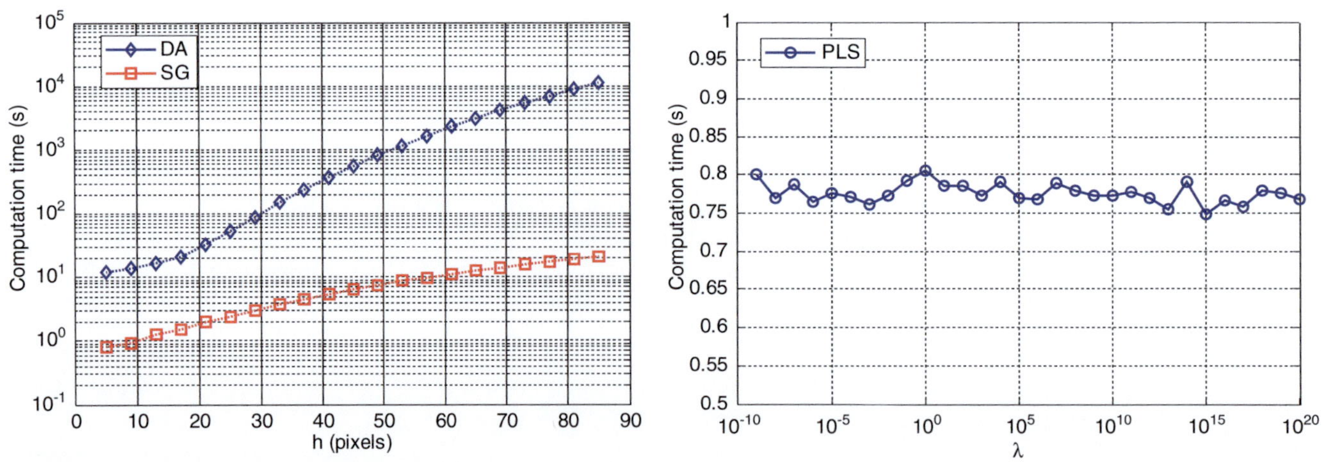

Fig. 5.2 Computation times against the smoothing parameters for DA, SG and PLS

Table 5.1 Performance of the different algorithms with optimized smoothing para. for the noised analytical map

Algorithm	DA	SG	PLS	PLS on raw derivatives
Smoothing para.	$h = 105$ pixels	$h = 87$ pixels	$\lambda = 1.26 \times 10^5$	$\lambda(\partial P/\partial x) = 1.58 \times 10^5$
				$\lambda(\partial P/\partial y) = 1.26 \times 10^5$
$d_{\partial P}$	0.22	0.23	0.12	0.20

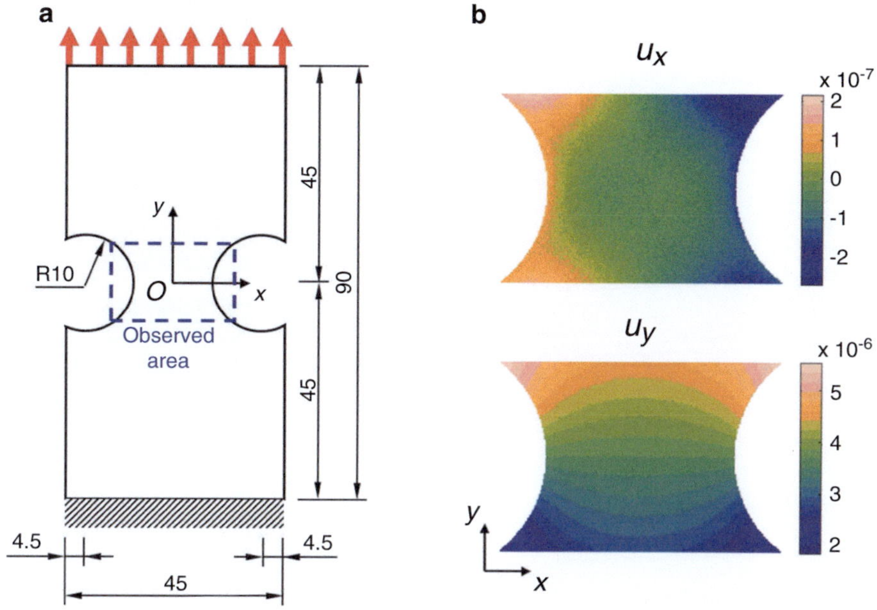

Fig. 5.3 Noised kinematic maps generated by a FE simulation. (a) Geometry of the mechanical test (b) u_x and u_y noised maps

Table 5.2 Performance of the different algorithms with optimized smoothing para. for the noised FE maps

Algorithm	DA	SG	PLS	PLS on raw derivatives
Smoothing para.	$h = 37$ pixels	$h = 31$ pixels	$\lambda(u_x) = 355$	$\lambda(\varepsilon_{xx}) = 1000$
			$\lambda(u_x) = 90$	$\lambda(\varepsilon_{yy}) = 1000$
				$\lambda(\varepsilon_{xy}) = 500$
d_ε	1.47×10^{-6}	2.03×10^{-6}	1.64×10^{-5}	2.13×10^{-6}

5.3.2 *Smoothing of Kinematic Maps from FE Simulation*

The algorithm has been tested a second time on simulated data. For that, a finite element model has been built. The mechanical test was a tensile test on a double notched specimen such as the one used in [11] (Fig. 5.3a).

The finite element model consisted of a plane-stress simulation using standard linear 3-node triangular elements with an isotropic material. The sample was meshed such as the finite element displacement fields could be reasonably considered as the exact ones. To create the synthetic displacement maps, the fields were evaluated on a regular grid of 240×160 pixels by interpolating linearly the nodal displacements so that the data grid spacing is equal to 0.1 mm (which is typically the thinner grid spacing of results that can be obtained from white-light optical measurements). A Gaussian noise with mean zero and standard deviation 10^{-8} (in m = 10 nm) was added to the displacement maps (leading to typical noised displacement maps in terms of signal to noise ratio). Views of these maps are shown in Fig. 5.3b.

The quality of the reconstruction has been investigated using the same kind of criterion as previously and as in [4]:

$$d_\varepsilon = \left\langle \sqrt{\left(\varepsilon_{xx}^{\text{rec.}} - \varepsilon_{xx}^{\text{exact}}\right)^2 + 2\left(\varepsilon_{xy}^{\text{rec.}} - \varepsilon_{xy}^{\text{exact}}\right)^2 + \left(\varepsilon_{yy}^{\text{rec.}} - \varepsilon_{yy}^{\text{exact}}\right)^2} \right\rangle, \tag{5.8}$$

The results obtained are summarized in Table 5.2 and obtained ε_{xx} maps as well as the corresponding error maps $\left(\varepsilon_{xx}^{\text{rec.}} - \varepsilon_{xx}^{\text{exact}}\right)$ are shown in Fig. 5.4.

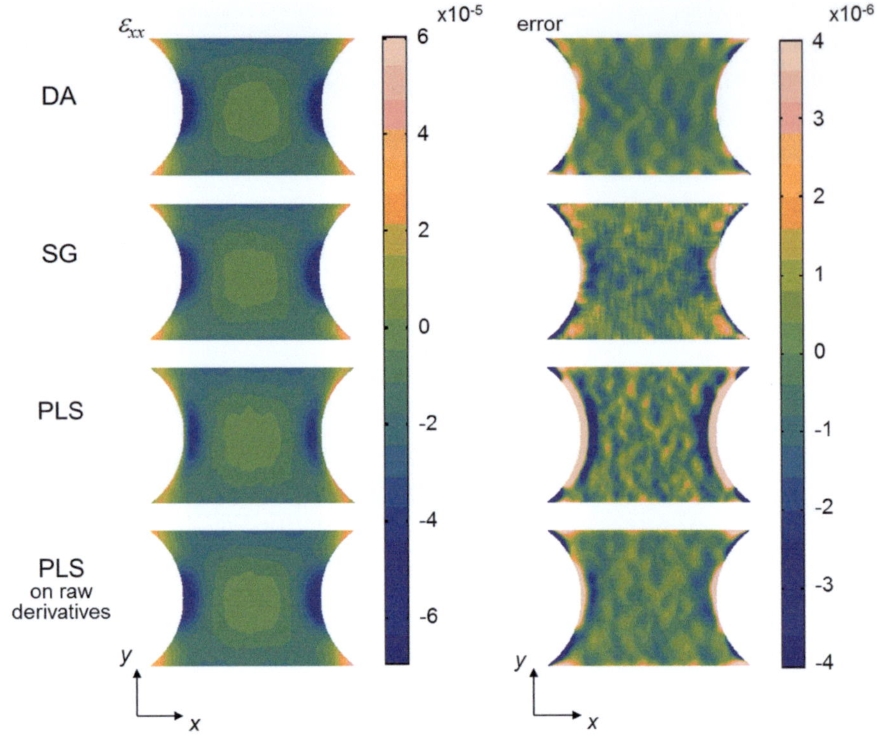

Fig. 5.4 Reconstructed εxx and corresponding error maps for the different smoothing methods

It can be observed that PLS works worse than both DA and SG. Indeed, the obtained average deviation in these conditions is an order of magnitude higher than the ones obtained for the two others algorithms. This counterbalances the conclusions of the previous section concerning the best performances of PLS. It appears especially that PLS does not manage to reconstruct correctly the edges of the circular notches, where the absolute deviation from the exact value of the strain is maximum (Fig. 5.4).

However, when smoothing directly the raw derivatives as previously done for the noised analytical function, the reconstruction by PLS works more efficiently. Indeed, for an optimal set of λ's, the average deviation obtained is of the same range as the one obtained with the two other algorithms.

In conclusion, it can be said that, if PLS is not the best procedure for this problem (possibly because the highest gradients are located on the edges of the maps which was not the case previously, with the analytical function), it shows results comparable with the ones produced by DA or SG when performing the derivation before the smoothing and remains the fastest method.

5.3.3 Smoothing of Experimental Kinematic Maps

Finally, the smoothing procedure by PLS has been tested onto true experimental kinematic maps. The tensile test on a double notched specimen considered in the previous section has been examined. The displacement maps were obtained by the grid method [12]. The grid had a pitch of 100 μm leading to kinematic maps of 159 × 231 pixels^2. The raw displacements fields have been processed by the different algorithms leading to strain maps represented in Fig. 5.5.

In this figure, it can be clearly observed that both DA and PLS directly applied on raw strain maps lead to accurate maps that can be compared to the ones obtained on simulated data even if the experimental results show an asymmetry between the left and right sides due to some misalignment between the axis of the sample and the one of the testing device during the test (Fig. 5.4b). On the contrary, SG and PLS applied before deriving displacement maps do not work. Strain maps resulting from SG show a dithering effect probably due to the important influence of the edges (zones surrounded by dashed rectangles in Fig. 5.5). For the same reason, PLS applied before deriving displacement maps does not provide good results. This is especially the case for ε_{xx} (zones surrounded by dashed ellipses in Fig. 5.5) that is here the most difficult map to reconstruct as the signal to noise ratio is the lowest (the x direction is not highly solicited in this test). So, for such a case, with high

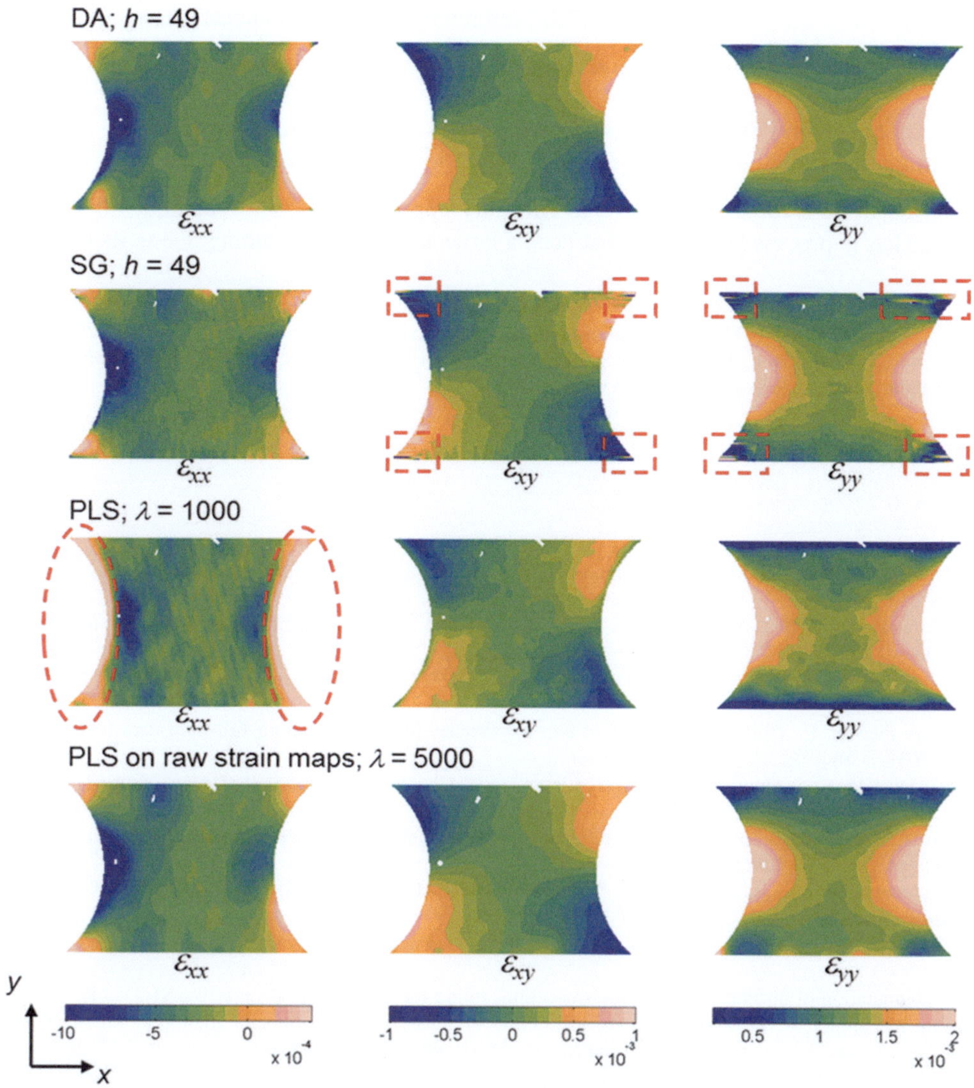

Fig. 5.5 Experimental in-plane strain maps obtained on the double-notched tensile sample by applying different smoothing procedures (size of the region of interest: 16 × 23 mm^2)

gradients on the edges and potentially low signal to noise ratio, it seems preferable to apply a point-to-point differentiation before smoothing by PLS rather than applying the smoothing and then differentiating the smoothed displacement maps: this produces results comparable with the ones of DA.

5.4 Conclusion

In this paper, the possibility to use the penalized least squares smoothing procedure (PLS) as an alternative way to process kinematic full-field measurements has been addressed. At the end of this study, it can be asserted that PLS manages to produce valid strain maps from displacements measurements, with an accuracy comparable to the techniques currently set up such as diffuse approximation (DA) or Savitzky-Golay smoothing filter (SG). Nevertheless, a precaution has to be taken: for kinematic measurements presenting high gradients on the edges of the studied region and showing low signal to noise ratio, it is highly preferable to apply PLS on raw strain fields, after a spatial derivation of displacement fields thanks to a simple point-to-point differences procedure.

Compared to other procedure such as DA or SG, PLS presents two main advantages:

- the computation time of the algorithm is independent of the value of the smoothing parameter and highly lower than both DA and SG;
- the h span parameter of DA and SG is an odd integer that gives only a discrete control of the amount of smoothing to apply; on the contrary, the smoothing parameter λ of PLS is a real continuous value.

Finally, PLS appears to be a robust alternative to currently applied smoothing techniques that can be preferentially selected whenever one has to process large maps that need a large amount of smoothing due to its high computation speed.

Acknowledgments The authors would like to thank Pr. Paul Eilers for some useful exchanges on the principles and developments of the penalized least squares algorithm. They also would like to thank Dr. Marco Rossi for providing the 2D Savitzky-Golay algorithm used in this study. Finally, they would like to thank their colleagues at Arts et Métiers ParisTech (especially, Mr. Alain Prévot, Ms. Virginie Jamar and Mr. Jeremy Blanks) for their help at different stages of the work.

References

1. P. Feissel, From displacement to strain, in *Full-Field Measurements and Identification in Solid Mechanics*, ed. by M. Grédiac, F. Hild (Wiley, Hoboken, 2013), pp. 191–221
2. R.R. Cordero, J. Molimard, F. Labbé, A. Martínez, Strain maps obtained by phase-shifting interferometry: an uncertainty analysis. Opt. Commun. **281**(8), 2195–2206 (2008)
3. R. Moulart, S. Avril, F. Pierron, Identification of the through-thickness rigidities of a thick laminated composite tube. Compos. A: Appl. Sci. Manuf. **37**(2), 326–336 (2006)
4. S. Avril, P. Feissel, F. Pierron, P. Villon, Comparison of two approaches for differentiating full-field data in solid mechanics. Meas. Sci. Technol. **21**(1), 015703 (2010)
5. P.A. Gorry, General least-squares smoothing and differentiation by the convolution (Savitzky-Golay) method. Anal. Chem. **62**, 570–573 (1990)
6. Bohlmann, G.: Ein Ausgleichungsproblem. In: Nachrichten von der Gesellschaft der Wissenschaften zu Göttingen, Mathematisch-Physikalische Klasse, pp. 260–271. German, 1899
7. E.T. Whittaker, On a new method of graduation. Proc. Edinb. Math. Soc. **41**, 63–75 (1923)
8. P.H.C. Eilers, A perfect smoother. Anal. Chem. **75**, 3631–3636 (2003)
9. D. Garcia, Robust smoothing of gridded data in one and higher dimensions with missing values. Comput. Stat. Data Anal. **54**(4), 1167–1178 (2010)
10. M.J. Buckley, Fast computation of discretized thin-plate smoothing spline for image data. Biometrika **81**(2), 247–258 (1994)
11. S. Avril, F. Pierron, Y. Pannier, R. Rotinat, Stress reconstruction and constitutive parameter identification in plane-stress elasto-plastic problems using surface measurements of deformation fields. Exp. Mech. **48**(4), 403–419 (2008)
12. Y. Surrel, Moiré and grid methods: a signal-processing approach, in *Interferometry '94: Photomechanics*, ed. by R.J. Pryputniewicz, J. Stupnicki, vol. 2342 (Proceedings of SPIE, Bellingham, 1994), pp. 213–220

Chapter 6
Speckle Projection Profilometry Using a Diffractive Telecentric Arrangement

Wade Gubbels and Gary S. Schajer

Abstract A modified approach to 3-D shape measurement using the Speckle Projection technique is proposed and demonstrated. Images of a random speckle pattern are obliquely projected onto a reference plane and then onto the specimen surface. The apparent local in-plane image displacements determined by two-dimensional Digital Image Correlation (2-D DIC) combined with triangulation to provide the surface height or shape of the specimen. Here, the projected pattern is created using a custom telecentric projector. This enables the physical projection to follow closely the theoretical assumptions typically used in triangulation and to avoid the artifacts introduced when using the expanding projection created by a typical projector arrangement. In addition, the apparent size of the telecentric projected image remains constant within a significant range of out-of-plane displacements or non-zero surface heights. This feature aligns the apparent in-plane image displacement due to specimen shape to a single direction. Therefore, measurements of any displacements in the second in-plane direction provide an independent measure of changes in camera perspective, effectively reducing the need for pre-measurement calibration. Finally, the use of a diffraction grating creates a parallelogram projection, equalizing the optical path lengths of the entire projected pattern. This maximizes the range of telecentricity or nearly parallel focus of the projected speckle, allowing objects with large surface height variation to be measured accurately. An example optical system is presented and the practical features are explained. Experimental results for a cylinder specimen and also a complex shaped specimen show that the method is effective and accurate. This approach allows for shape measurements to be made efficiently, with minimal calibration and reduced mathematical complexity.

Keywords Digital image correlation • Profilometry • Speckle projection • 3D shape measurement

6.1 Introduction

Shape detection and profilometry measurements are important for many different applications. They can be used to validate the dimensions of low tolerance manufactured parts, quantify surface roughness, and are even used to detect residual stresses using the contour method [1]. There are many approaches to measuring surface profiles, including both contact and non-contact methods. The precision and developed standards associated with contact measurements have made the Coordinate Measurement Machine (CMM) the standard approach in many industries. However, this contact approach is generally expensive and time consuming. In response, different non-contact optical techniques have been developed, including laser scanning and tracking, photogrammetry and several structured light methods. An extensive overview of available optical shape measurement methods can be found in [2].

The present work presents a structured light profilometry method using speckle projection. The technique involves comparing the projected images of a random speckle pattern. The pattern is obliquely projected onto a reference plane and then onto a specimen surface. Two-dimensional Digital Image Correlation (2-D DIC) is used to determine the in-plane displacements of this pattern that are introduced within the measured image by the surface height profile of a specimen. After acquisition of the reference plane image, the method can provide accurate full-field profiles from a single image of the specimen, thus making the method well suited for high-speed applications. This is an advantage when compared to the popular fringe projection method, which typically requires several different patterns be projected onto the surface [3]. Also, while the analysis of fringe images is relatively straightforward, the repeating pattern leads to uncertainties in absolute datum identification, typically requiring further information to provide absolute measurements [4]. Currently,

W. Gubbels
American Stress Technologies, Pittsburgh, PA 15221, USA

G.S. Schajer (✉)
University of British Columbia, Vancouver, BC, Canada
e-mail: schajer@mech.ubc.ca

C. Sciammarella et al. (eds.), *Experimental and Applied Mechanics, Volume 4*,
Conference Proceedings of the Society for Experimental Mechanics Series, DOI 10.1007/978-3-319-22449-7_6

both fringe and speckle projection requires significant initial calibration to determine the geometrical and camera parameters needed to provide accurate shape measurements. However, these calibrations are typically much less involved than those required in other optical shape measurement techniques like stereovision DIC. With the advantages of speckle projection profilometry it appears to be a very useful tool for both research and industry. Thus, the method is worthy of further attention.

The speckle projection method was developed by Gilbert in 1986, in conjunction with the advent of DIC, to help overcome the limitations associated with the classic shadow Moiré method [5]. Early research showed that the method could be used to measure the profiles of relatively simple shapes, as well as out-of-plane displacements, with modest accuracy [6, 7]. One limiting aspect is that the projected pattern is typically modeled as a parallel projection to simplify the triangulation analysis. In actuality the projections used are divergent, thereby leading to significant error when parallel projection is assumed. This effect can be reduced by using very long projector-to-object distances [8], but this makes the equipment very space consuming and sensitive to vibrations.

For accurate measurements, the geometric and viewing parameters of the optical arrangement must be well defined. This typically involves a detailed calibration procedure involving identification of projector-to-object distance, camera-to-object distance and projector-to-camera location. Previous work has been done to create effective calibration techniques [8, 9]. However, they typically require the use of precision targets or translation stages or inclusion of calibration gauges in each image. This generally reduces the practicality of the approach, especially in an industrial setting.

The present work aims to contribute to the development of the speckle projection method by creating an accurate, easy-to-use profilometry system that requires minimal calibration. It uses an inexpensive telecentric projection arrangement. This arrangement has key advantages compared with the common digital projectors typically used in speckle projection. The nearly parallel projection creates a constant angle of illumination. This reduces the required pre-measurement calibration and simplifies the triangulation calculations. The constant size of the projected pattern also restricts the potential sources of subset deformation, making subset matching more efficient. The constant size of the projection also allows for camera perspective error to be independently measured from a single image. In addition, a diffraction grating is used to create the required illumination angle in a parallelogram geometry. This forms equal optical path lengths, a feature that maximizes the range of depths over which accurate measurements can be made. Together, the modifications remove the need for repeated calibrations and increase the flexibility and accuracy of the approach. Measurements obtained using an example system are presented and demonstrate the features of the method.

6.2 Traditional Speckle Projection Profilometry

Figure 6.1 shows a typical speckle projection arrangement. A camera is arranged with its viewing axis normal to a reference plane ***AB***. A projector ("light source") is then aligned to illuminate the reference plane at an oblique angle $\boldsymbol{\alpha}$, relative to the camera viewing axis. When a specimen of general shape, ***A'B'*** moves from the reference plane, it causes the projected pattern to shift within the image captured by the camera. 2-D DIC can then be used to calculate the in-plane shifts within this pattern. If the projection angle is well known, these in-plane shifts can be used to triangulate the out-of-plane surface height of all surface points on the specimen.

The general approach in speckle projection profilometry is as follows. Consider, for example, a point initially registered by the camera at ***b*** in Fig. 6.1. It appears to shift to point ***d*** due to the out-of-plane surface height, ***dz***. From the figure:

$$\mathbf{dx} = \mathbf{dz}\left(\tan\boldsymbol{\alpha}' + \tan\boldsymbol{\beta}'\right) \tag{6.1}$$

where ***dx*** equals the distance between ***b*** and ***d***. Angles $\boldsymbol{\alpha}$ and $\boldsymbol{\beta}$ respectively represent the angles relative to the z axis of the projection and camera viewing rays that meet at point ***f***. From the geometry:

$$\frac{dz}{dx} = \frac{\mathbf{L} + \mathbf{dz}}{\mathbf{D}} \tag{6.2}$$

By placing the camera and projector sufficiently far from the specimen, it is assumed that ***D*** will become much larger than ***dx***. Also, by noticing:

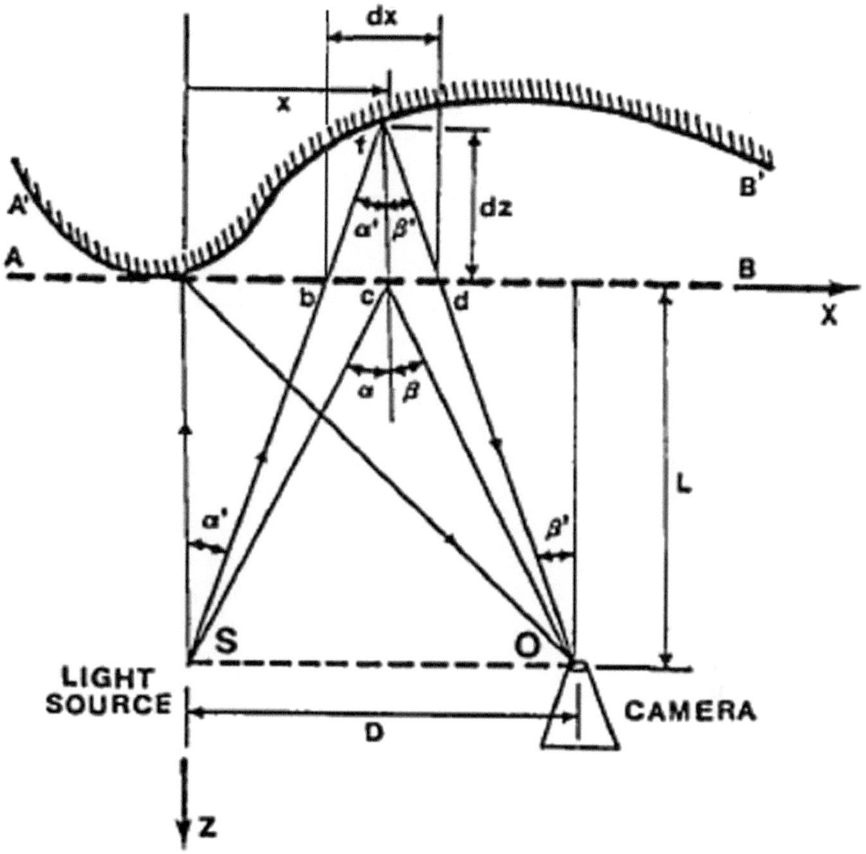

Fig. 6.1 Ray diagram of traditional speckle projection arrangement (adapted from [5])

$$\frac{\mathbf{D}}{\mathbf{L}} = \mathbf{tan}\boldsymbol{\alpha} + \mathbf{tan}\boldsymbol{\beta} \tag{6.3}$$

Equation (6.3) can be simplified to:

$$\mathbf{dz} = \mathbf{dx}(\mathbf{tan}\boldsymbol{\alpha} + \mathbf{tan}\boldsymbol{\beta}) \tag{6.4}$$

Commonly, the camera is arranged to be perpendicular to the reference plane, corresponding to a zero camera viewing angle $\boldsymbol{\beta}$. Therefore, if the projection illumination angle $\boldsymbol{\alpha}$ is known, (6.4) can be used to calculate the out-of-plane displacement ***dz*** from the DIC indicated in-plane shift ***dx*** at each point on the specimen surface. Finally, if the magnification ratio ***M***, of the camera is known, each ***dz*** value can be scaled by this ratio to convert pixel displacements into physical dimensions.

$$\mathbf{dz} = \frac{\mathbf{dx}(\mathbf{tan}\boldsymbol{\alpha})}{\mathbf{M}} \tag{6.5}$$

The concept of speckle projection as explained above is relatively straightforward and this is the typical approach taken by most users today. However, care must be taken in the assumptions and simplifications used when putting the method into practice. Several issues can lead to significant measurement errors and increased analysis complexity. These include:

1. The diverging image configuration (varying $\boldsymbol{\alpha}$) typically used by projector systems significantly changes the in-plane size of the projected speckle pattern. This creates perspective changes that cause local displacements of the projected image in addition to those caused by triangulation, thus introducing geometrical error.
2. A similar perspective error occurs with the camera imaging system. The typically diverging view of a camera (varying $\boldsymbol{\beta}$) causes apparent x and y displacements to occur when the object-to-lens distance varies across a non-planar specimen.

3. The object surface is oriented obliquely to the speckle pattern projector, so differing areas on the object can be at significantly different distances from the projection lens. This can make it difficult to focus the projected speckle pattern accurately over all the illuminated area, leading to decreased correlation effectiveness and lower measurement accuracy.

6.3 Diffractive Telecentric Speckle Projection

A custom diffractive telecentric projector was built to overcome the practical issues discussed in the previous section. The arrangement creates an essentially parallel projection providing an effective solution to Issue 1 above. Figure 6.2 shows a ray diagram for the telecentric projector design. A laser beam is directed onto a diffusion plate that expands the light into a conical beam. This beam is collimated by the first lens, which directs the light though a transparent film on which a random speckle pattern had been applied. This painted transparency forms the object of the double telecentric lens formed by the following two lenses and intermediate aperture. The parallel rays emerging from the final imaging lens project an image that remains constant in size and focus over a significant image distance range known as the telecentric depth. The two plano-convex lenses, lens 2 and 3 in the diagram, making up the telecentric projector have diameters of 4 and 8 cm and focal lengths, f_2 and f_3, of 8 and 16 cm, respectively. Together with the aperture, a telecentric depth (= depth of field) of approximately 8 cm is created. This range enables significant z displacements (varying surface heights) to be measured using triangulation. The telecentric depth depends on the lens focal length and the intermediate aperture.

In the telecentric projector, the laser source was sufficiently coherent to produce substantial laser speckle. This additional speckle greatly impeded the interpretation of the speckle pattern from the painted transparency. The issue was eliminated by rotating the diffusion plate. This smeared out the laser speckle within the image exposure time while leaving the speckle pattern from the transparency intact [10].

For triangulation to be used, the projection system must illuminate the surface at an angle relative to the camera viewing (z) axis. One possibility is to orient the projector as shown in Fig. 6.3a, but this arrangement creates a greatly varying distance between imaging lens and object, Issue 3 above. The variation in the lens-to-object distance could be decreased by reducing the angle theta. However, this would similarly reduce the sensitivity of the measurement.

Variable distance between imaging lens and specimen can be eliminated while maintaining a relatively large projection angle by using a linear diffraction grating, as shown in Fig. 6.3b. Light passing through the diffraction grating diffracts through an angle

$$\boldsymbol{\theta} = \sin^{-1}\left(\frac{\boldsymbol{\lambda}}{\mathbf{d}}\right) \tag{6.6}$$

where $\boldsymbol{\lambda}$ is the wavelength of the light used and $\boldsymbol{d}$ is the line spacing of the diffraction grating [11]. The outgoing diffracted light creates a parallelogram projection. Therefore, aligning the camera and projection arrangement parallel to one another

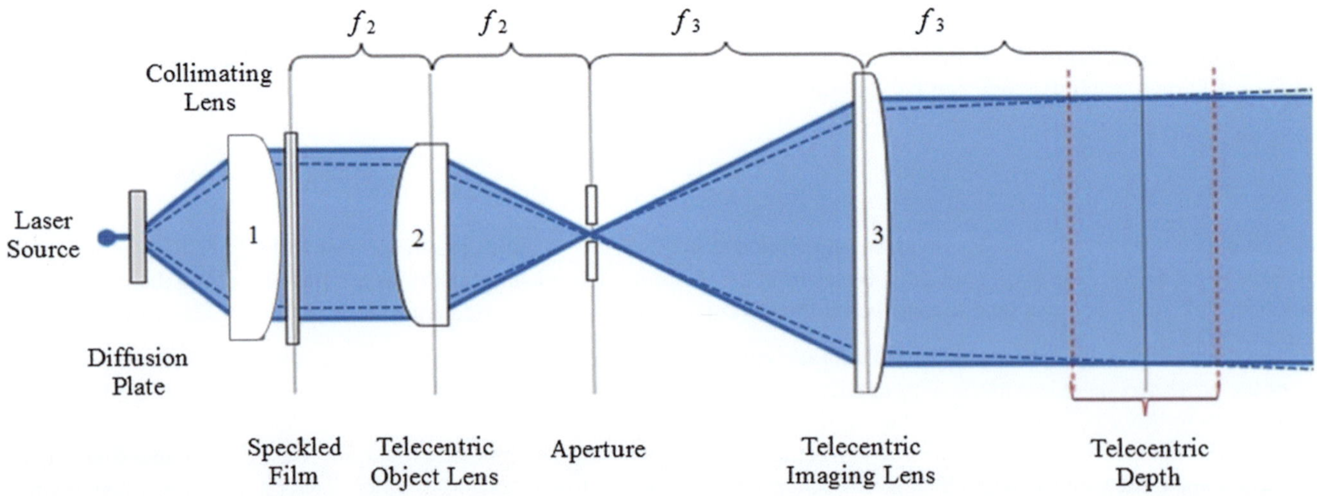

Fig. 6.2 Ray diagram of telecentric projection arrangement

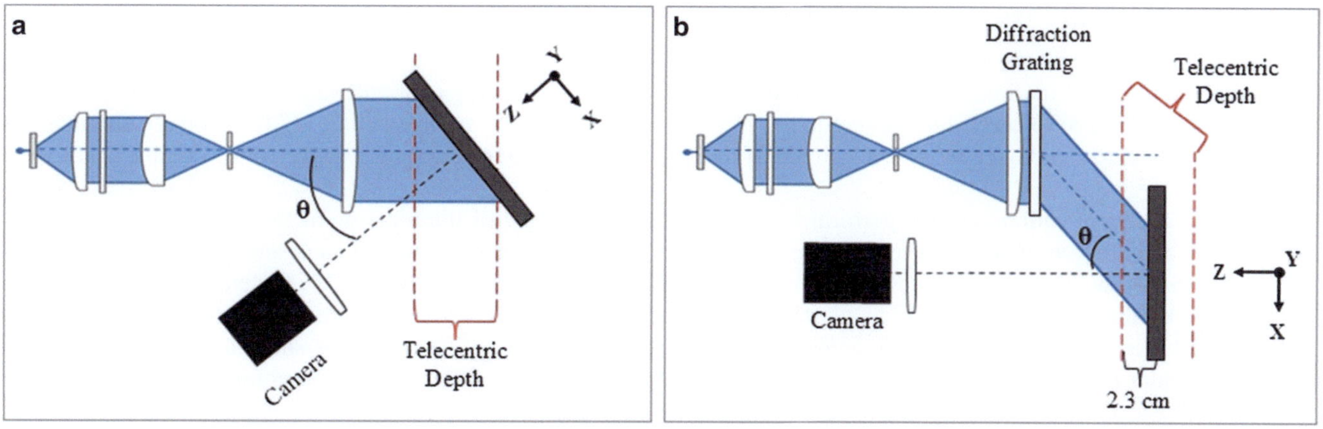

Fig. 6.3 (**a**) Oblique projection without diffraction grating. (**b**) Oblique projection with diffraction grating

ensures that the optical path lengths of the projection remain equal across the specimen surface. This arrangement fixes the diffraction angle $\boldsymbol{\theta}$ to be the illumination angle $\boldsymbol{\alpha}$ defined above.

Here, a $\boldsymbol{\lambda} = 405$ nm blue laser and a diffraction grating with a line spacing $\boldsymbol{d} = 1000$ nm were used in the projection arrangement. Using these values in (6.6), the projection angle is 23.9°. This particular arrangement therefore has the capability of effectively measuring surface height variations within the full range of ±3.5 cm around the center of the telecentric depth.

6.4 Perspective Error Correction

The diffractive telecentric projector enables the projected speckle pattern to maintain a constant size during out-of-plane motion of the measured surface. However, perspective errors will be apparent in the images captured by a camera with a conventional lens, Issue 2 above. Perspective errors appear as a compression/expansion about the center of the image, due to the variation in $\boldsymbol{\beta}$ across the image, and subsequent change in the magnification ratio caused by the change in the camera-to-object distance. This can be handled in two possible ways. One possibility is to use a telecentric lens on the camera as well. This arrangement requires that the external diameter of the telecentric lens be at least as large as the measured object, which would be very costly for specimens larger than about 3–4 cm across. An alternative approach is to use a conventional camera lens and to correct the apparent in-plane displacements using knowledge of the out-of-plane displacements found from triangulation of the projected speckle pattern. This can be done by noting that for an axisymmetric imaging lens the perspective effects are radial and so are equal in both x- and y-directions. Typically, the camera and projector are arranged such that the triangulation is in the x-z plane. Thus, the projected speckle pattern creates only x-displacements. So, any apparent changes in the y-direction in the projected image are solely due to perspective changes. These perspective changes observed in the y-direction can then be subtracted from those in the x-direction to restore the original image scale.

The amount of perspective change measured at each pixel, in both the x- and y-directions, is generally a function of the camera-to-object distance at each pixel as well as the distance from that pixel to the pixel at the optical center on the camera sensor. The pixel locations are explicitly known, therefore the perspective errors encountered due to specimen location relative to the reference plane and specimen geometry can be accounted for. So as not to propagate low level noise, a polynomial surface is fit to the measured y-displacements. This smooth surface is then used to calculate the perspective change at each pixel and subsequently subtracted from the measured x-displacements.

The ability to measure perspective error using self-contained data greatly enhances the flexibility of the system. Specimens may be placed at varying camera-to-object distances and there is no need to obtain the magnification ratio each time or recalibrate the system. Providing that the magnification ratio of the camera is known at the initial reference plane location and the specimen lies somewhere within the telecentric depth of the projector, the varying magnification ratio across the image can be calculated and corrected. Combining this feature with the constant angle of illumination created by the telecentric projector makes the calibration of such a system a one-time procedure. Once the angle of illumination and reference plane magnification ratio are known the system can accurately profile specimens with very little setup time and a single image per specimen.

6.5 Experimental Procedure and Results

Figure 6.4a shows the experimental arrangement used here. For ease of development and validation the arrangement was fixed to an optical table, however this is not required in general. As shown, the arrangement is relatively small, about 60 cm long, and can be placed quite close to the specimen. If needed, the arrangement could be compacted if available space were limited.

A pre-measurement calibration was performed to verify the actual angle of illumination of the projected speckle. This was done by capturing an image of a reference plane (normal to the camera viewing axis) and then translating the planar specimen towards the camera by a known amount using a Newport linear actuator (CMA-25CCL Newport) attached to a precision translation stage. Triangulation of this displacement showed the average angle of illumination was almost uniformly 0.4° less than theoretical value across the entire image. The calibration procedure showed that this was due to a slight rotational misalignment of the diffraction grating and not a variation in the diffraction grating spacing. The slight misalignment was corrected for mathematically as opposed to mechanically.

The planar specimen was placed normal to the camera viewing axis near the center of the telecentric depth and a reference plane image was acquired. The magnification ratio of the camera was acquired at this location and found to be 60 μm per pixel. This reference image was then saved into memory and used for all measurements. The first profile measurement was made on a specimen cut as a segment from a 1.5″ (38.1 mm) radius cylinder. The cylindrical specimen, shown in Fig. 6.4b, was placed within the telecentric depth region and a single image was acquired. This image was then compared to the reference plane image using the process explained above.

A 500 by 500 pixel region of the image was selected, corresponding to a planar area of 30 mm by 30 mm. DIC analysis of the images was done using Ncorr software [12], with circular subsets of radius 28 pixels and first-order shape functions, overlapped so that subset centers were 2 pixels apart, totaling 62,500 subsets. The images were acquired using a machine vision style camera (Prosilica GC2450, AVT, Burnaby, Canada) [13]. The particular camera used had a color sensor of the Bayer type and was chosen to accommodate other needs not part of the present study. In consequence, the projected pattern was only imaged using pixels with blue sensitivity, which comprise only one quarter of the available pixel locations on the camera sensor. This arrangement required that the gaps between these blue pixels be interpolated from neighboring pixels,

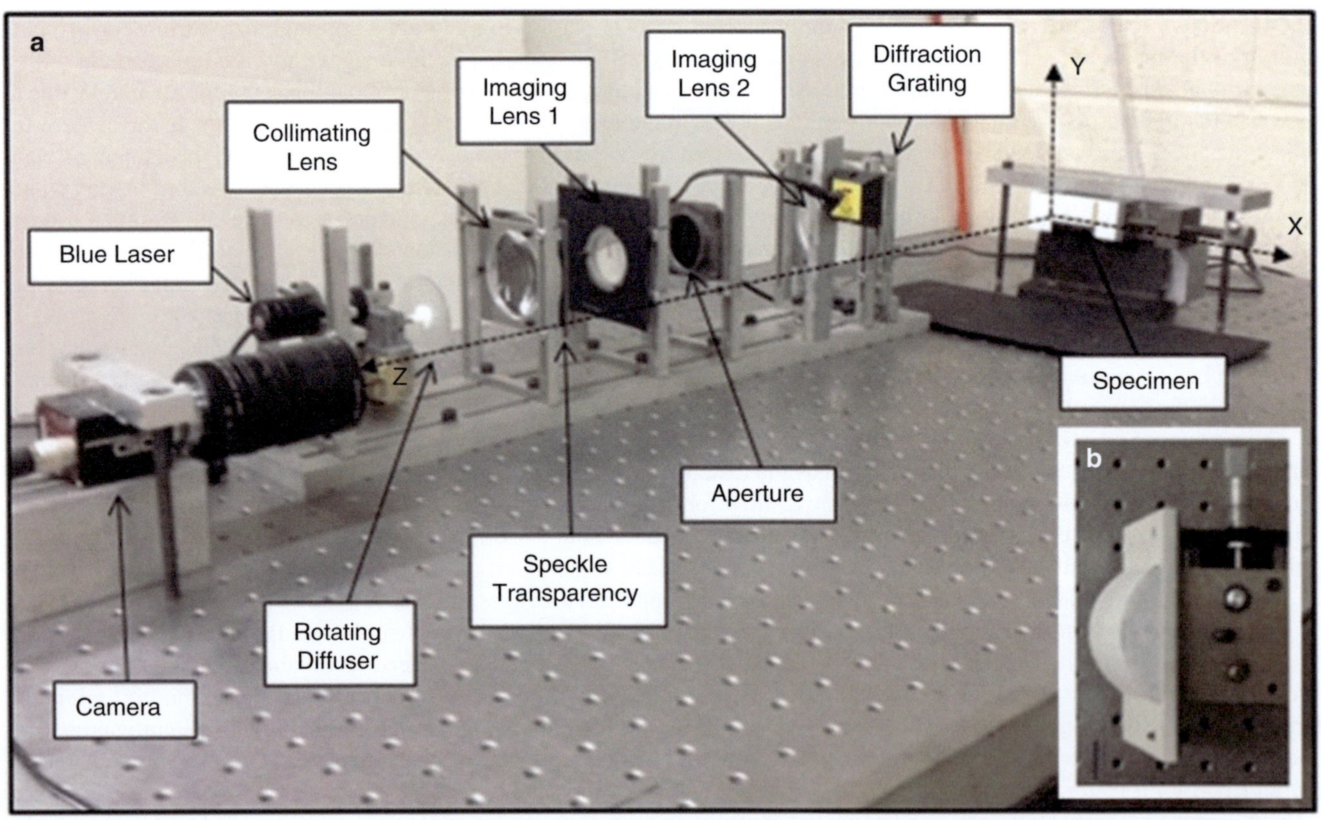

Fig. 6.4 Experimental arrangement of diffractive telecentric speckle projector

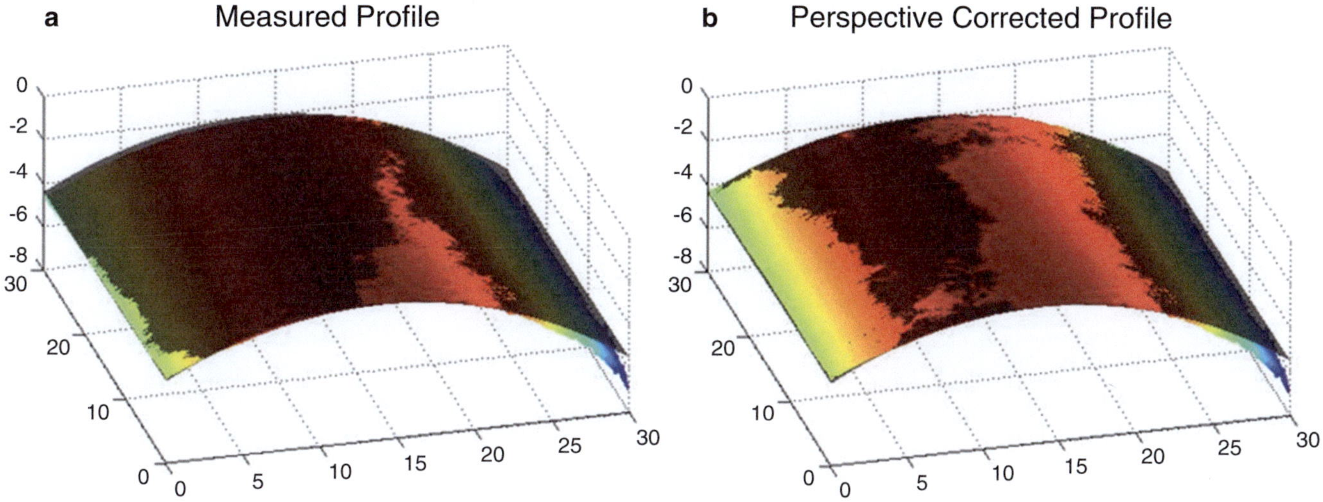

Fig. 6.5 (**a**) Measured surface of cylinder. (**b**) Perspective corrected surface of cylinder

causing some loss of resolution. Thus, the use of a conventional monochrome camera can be expected to produce better DIC precision than reported here. Figure 6.5a shows the initially computed profile of the cylindrical surface, and Fig. 6.5b the profile after perspective error correction.

The superimposed transparent gray surface interlaced with the measured surfaces in Fig. 6.5 is a uniform 38.1 mm cylindrical surface to help visualize the deviations of the measurement from the theoretical values. All surface dimensions are in mm. The z values, or surface heights, are calculated relative to the original location of the reference plane. Therefore, the reported negative values along the z dimension show that the point on the cylinder nearest to the camera was positioned approximately 2 mm behind the original location of the reference plane. This arrangement creates camera perspective errors in which the projected patter appears to compress slightly because of the greater camera-to-object distance and reduced magnification ratios. In this case, the location of the region of interest, relative to the central camera pixel, results in the measured x-displacements being slightly larger in magnitude (more negative) than the true values. However, as shown in Fig. 6.5b, this can be corrected using the information obtained from the displacements measured in the y-direction.

In this measurement the mean absolute error before perspective correction was nearly 120 μm. After perspective correction the mean absolute error was reduced to under 85 μm. In this arrangement 85 μm corresponds to nearly 1.5 pixels. This seems quite high in comparison to the precision of traditional DIC (0.01 pixels). However, it is reasonable to speculate that the extruded cylindrical rod used in this experiment likely did not have a radius that was uniform within 100 μm. There is also a noticeable disagreement between the corrected profile measurement and the theoretical cylinder along the right side of the surface. While there may be a real variation in the actual surface measured, there may be a further source of error. The surface slope on the right side of the cylinder enlarges the relative projection angle and causes large x-strains in the projected speckle pattern. These large strains can create a challenge for the DIC calculations. Further measurements must be made to determine these limitations. Due to lack of availability, the cylindrical surface could not be confirmed with an independent profilometer. However, the comparison of the measured surface to the theoretical surface shows that the deviations are not random, which would be consistent with a smooth variation in surface height of the actual specimen. Regardless, the results show that the system can efficiently measure surface profiles with relatively promising accuracy without any pre-measurement calibration. The results also show that perspective errors can be simultaneously measured and accounted for, increasing the overall accuracy and efficiency of the system.

A second profile measurement was made on a portion of a machined tripod sleeve. Figure 6.6a shows an example of the part and the region of interest indicated by the dashed red box. Figure 6.6b shows the nominal dimensions of the machined part in mm. For this measurement a 400 pixel by 550 pixel region of interest was chosen, corresponding to a planar area of 24 mm by 33 mm. Similar DIC parameters as the first measurement were used, except subsets were spaced one pixel apart. The measured and perspective corrected profiles again compared to the transparent theoretical surface are shown in Fig. 6.7a, b, respectively. All surface dimensions are in mm. In this measurement, when compared to the theoretical surface the mean absolute error before perspective correction was over 120 μm. After perspective correction this was reduced to under 80 μm. In this case, the region of interest straddled the central y-axis of the camera sensor. Therefore, because of perspective errors, the left side of the image measured slightly greater than desired while the right side slightly less. Again, these perspective changes are

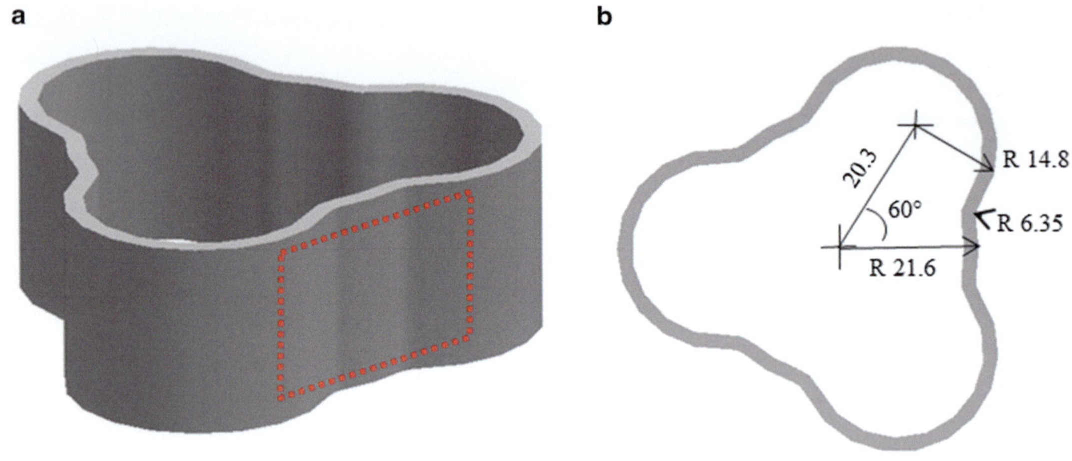

Fig. 6.6 (**a**) Model of measured tripod sleeve. (**b**) Nominal dimensions of tripod sleeve surface

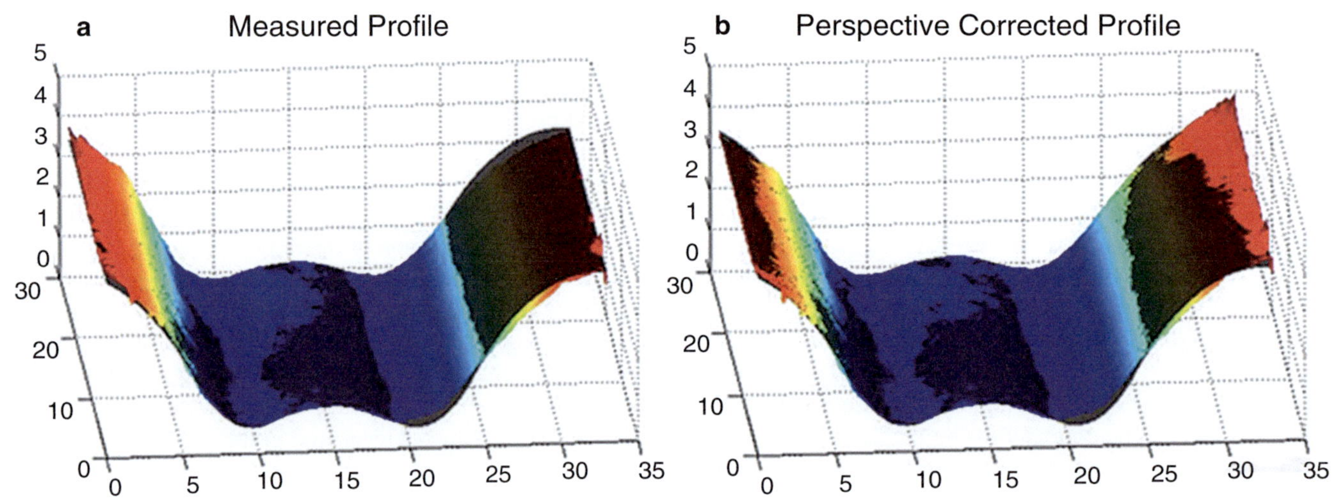

Fig. 6.7 (**a**) Measured surface of tripod sleeve. (**b**) Perspective corrected surface of tripod sleeve

calculated in the y-directional displacements and are used to correct the errors in the profile measurement. For this measurement an average decrease in error of over 40 μm across the entire image is achieved.

While the error, after perspective correction, for both the cylindrical and tripod sleeve specimen is similar, it is most likely a coincidence. The tripod sleeve does not have the same issues with image strain as discussed earlier. It is likely that the thin walled tripod sleeve warped, due to residual stresses, after the large amount of material was removed during machining. However, this is merely a hypothesis as an independent measurement was not accessible.

6.6 Discussion and Conclusion

The results discussed above show that the telecentric speckle projection system provides a promising approach to profilometry measurements that requires minimal calibration. While the telecentric nature of the projection does limit the size of objects which can be measured, it has many advantages when compared to other profilometry techniques. By creating a parallel projected pattern, the physical situation more closely models the theoretical approach to triangulation and reduces the complexity of the needed calculations. By using a diffraction grating with the speckle projector, a parallelogram illumination is created that maximizes the use of the available field-of-view. This maximizes the range of depths at which an object can be measured as well as the range of object sizes that can be captured. Finally it was shown that the approach allows for the direct and simultaneous measurement of camera perspective errors. This is a great advantage that removes the need for repeated pre-measurement calibrations.

Two surface profile measurements were made. Results were obtained for a simple cylinder and a relatively complex machined component. In both cases, perspective correction reduced the average error by nearly 40 μm. Both measurements were made only taking a single image and neither required any further knowledge about the location of the specimen relative to the camera or acquisition of different magnification ratios.

The technique does require that a reference image on a planar surface be acquired before measurements can be made. However, once this image is captured and the magnification ratio at this camera-to-reference plane distance is known this procedure need not be carried out again. This is true so long as the camera-to-projector location remains unchanged. With a traditional projector this is an unreasonable requirement as distances between camera, projector, and specimen can be quite large. However, the telecentric system described here is compact enough that the camera and projector can be fixed into a single unit. This compact arrangement allows the system to be easily portable and to have reduced sensitivity to vibrations during measurements. The resulting diffractive telecentric speckle projection profilometer has the advantage of being easy to use, accurate, portable and requiring only a single image per measured surface.

Acknowledgment The authors sincerely thank the Natural Sciences and Engineering Research Council of Canada (NSERC) and American Stress Technologies, Inc., Pittsburgh, PA, for their financial support of this research.

References

1. G.S. Schajer, *Practical Residual Stress Measurement Methods* (Wiley, Chichester, 2014), pp. 109–135
2. F. Chen, G.M. Brown, M. Song, Overview of three-dimensional shape measurement using optical methods. Opt. Eng. **30**(1), 10–22 (2000)
3. S.S. Gorthi, P. Rastogi, Fringe projection techniques: whither we are? Opt. Lasers Eng. **48**(2), 133–140 (2010)
4. S. Zhang, Recent progresses on real-time 3D shape measurement using digital fringe projection techniques. Opt. Lasers Eng. **48**(2), 149–158 (2010)
5. J.A. Gilbert, D.R. Matthys, M.A. Taher, M.E. Petersen, Shadow speckle metrology. Appl. Optics **25**(2), 199 (1986)
6. M. Sjödahl, P. Synnergren, Measurement of shape by using projected random patterns and temporal digital speckle photography. Appl. Optics **38**(10), 1990–1997 (1999)
7. P. Sriram, S. Hanagud, Projection-speckle digital-correlation method for surface-displacement measurement. Exp. Mech. **28**(4), 340–345 (1988)
8. B. Pan, H. Xie, J. Gao, A. Asundi, Improved speckle projection profilometry for out-of-plane shape measurement. Appl. Optics **47**(29), 5527–5533 (2008)
9. S. McNeill, M.A. Sutton, Z. Miao, J. Ma, Measurement of surface profile using digital image correlation. Exp. Mech. **37**(1), 13–20 (1997)
10. S. Lowenthal, D. Joyeux, Speckle removal by a slowly moving diffuser associated with a motionless diffuser. J. Opt. Soc. Am. **61**(7), 847–851 (1971)
11. M.C. Huntley, *Basic Diffraction Equation, Diffraction Gratings* (Academic, New York, 1982), pp. 22–24
12. J. Blaber, http://www.ncorr.com/index.php/downloads. Accessed 10 Oct 2013
13. AVT, *Prosilica GC Technical Manual, Ver. 2.0.8* (Allied Vision Technologies, Stadtroda, 2013)

Chapter 7
Use of VFM for Heterogeneity Evaluation of Materials Under Uniaxial Tensile Stress

John Considine, F. Pierron, and K.T. Turner

Abstract Identification of spatially varying stiffness is a challenging, but important, research topic especially as materials become more complex and are used in extreme environments. Examination of heterogeneity is often accomplished by unique test fixtures capable of creating the necessary stress configurations for heterogeneous stiffness identification. In this work the Virtual Fields Method (VFM) was used to identify stiffnesses in a simple geometry, namely uniaxial tension, of a simulated material with an inclusion whose stiffness gradually increases from that of surrounding material. Three different VFM analyses were examined and their results compared. Results suggest that cooperative use of these analyses could provide reasonable stiffness identification for heterogeneous materials using uniaxial tensile tests.

Keywords VFM • Stiffness • Heterogeneity • Simulation • Uniaxial tension

7.1 Introduction

Identification of material stiffness is the most frequent use of the Virtual Fields Method (VFM) [1]. A common assumption within VFM is material homogeneity. Often this assumption is quite reasonable, when stiffness spatial variation is small compared to the mean stiffness. Efforts to accurately identify stiffness heterogeneity generally require non-standard test fixtures to activate sufficient full-field strains for numerical analysis. This work examines the common uniaxial tensile test and provides an overview of the use of VFM for heterogeneous stiffness identification.

The objective of this work is to summarize VFM procedures used in heterogeneous stiffness identification, along with the some of their advantages and limitations. VFM procedures can be used cooperatively to provide information on the location and size of heterogeneities. Additionally, we hope to encourage those new to, or just beginning with, VFM to understand its potential usefulness.

Characterization of material heterogeneity is important in process control in order to evaluate and maintain material quality, and in end use environments, to determine stiffness changes due to temperature, moisture, damage, or other external factors. In both occurrences initial assessment of heterogeneity may be physical, e.g., the material looks damaged in a particular location, or another tool, such as ultrasonic evaluation, may be used.

Although challenging in practice, spatially varying stiffness identification can provide an additional tool to examine the extent of material heterogeneity in terms of location(s), frequency, and stiffness gradient. The challenge for the evaluator is coordinated selection of test geometry, stiffness parameterization and appropriate inverse analysis.

The uniaxial tensile test is one of the most used test configurations for mechanical characterization. The test is quick, intuitive, and a part of many quality control processes. With the addition of full-field displacement measurements, such as digital image correlation or the grid optical technique, the uniaxial test and VFM can provide an evaluation of stiffness heterogeneity.

J. Considine (✉)
U.S. Forest Service, Forest Products Laboratory, One Gifford Pinchot Drive, Madison, WI 53726, USA
e-mail: jconsidine@fs.fed.us

F. Pierron
University of Southampton, Southampton SO17 1BJ, UK

K.T. Turner
University of Pennsylvania, 245 Towne Building, 220 S. 33rd St, Philadelphia, PA 19104-6315, USA

C. Sciammarella et al. (eds.), *Experimental and Applied Mechanics, Volume 4*,
Conference Proceedings of the Society for Experimental Mechanics Series, DOI 10.1007/978-3-319-22449-7_7

Here a strip of simulated material, 25 mm × 150 mm, containing a single inclusion with smoothly varying stiffness, under uniaxial tension, was examined with three VFM techniques. Effectiveness of each identification technique was determined by comparison with known stiffness. Given that the uniaxial tensile geometry is not an optimal choice for heterogeneous material evaluation, reasonable identification was achieved.

7.2 Uniaxial Model

A thin rectangular specimen of size $L_1 \times L_2 = 150$ mm × 25 mm and thickness 1 mm was modeled in ANSYS® (ANSYS, Inc, Canonsburg, PA 15317, USA) using 3750 1 mm^2 PLANE42 elements. Origin of the coordinate system was located on the lower left of the specimen as shown in Fig. 7.1. Stiffness was smoothly varying in both the x_1- and x_2-directions and was rotated 60° with respect to x_1-direction. Attributes of the inclusion were selected to simulate actual scenarios and to challenge stiffness identification techniques. Formula for Q_{11} (GPa) is given in Eq. (7.1) and is based upon a 2-D probability curve. Poisson's Ratio was set to 0.3 throughout the model. Although any smoothly varying stiffness distribution could have been selected, those made up of cos(ωx_i) and sin(ωx_i) terms were intentionally excluded because one of the identification techniques is based upon Fourier transforms.

$$Q_{11} = 11.0 + 3.3e^{-(((x_1-75)*\cos(\pi/3)+(x_2-12.5)*\sin(\pi/6))/150*12)^2-(((x_1-75)*\cos(\pi/3)+(x_2-12.5)*\sin(\pi/6))/25*12)^2} \quad (7.1)$$

Figure 7.2 gives the finite element (FE) strains and those from a grid–measurement simulation. Grid simulation is described in [2]. Zero-mean, random gray scale noise was added to the simulated deformed grid image at 1 % of gray scale.

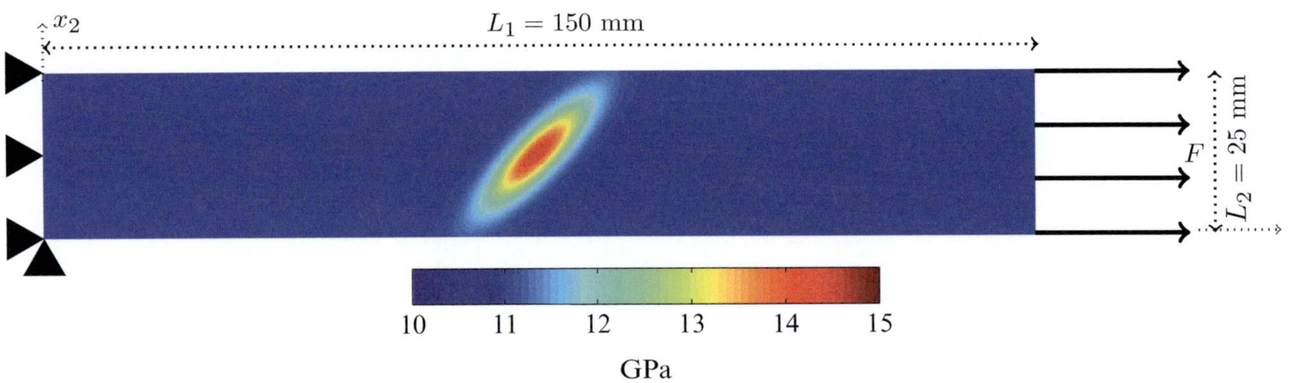

Fig. 7.1 Q_{11} for uniaxial tension model

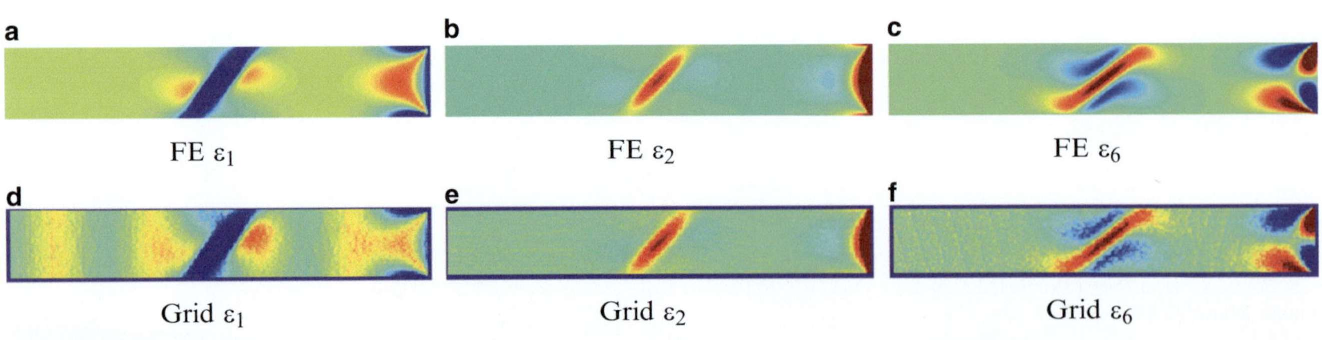

Fig. 7.2 FE strains (**a**)–(**b**) and simulated grid strains (**d**)–(**f**); scale is (%) strain. (**a**) FE ε_1. (**b**) FE ε_2. (**c**) FE ε_6. (**d**) Grid ε_1. (**e**) Grid ε_2. (**f**) Grid ε_6

7.3 Application of VFM to Heterogeneous Materials

Three VFM analyses are presented for the purpose of heterogeneity characterization. The Principle of Virtual Work (PVW) provides the basis for use of VFM to examine heterogeneous materials. For a plane stress problem, PVW can be written as

$$\int_S (\sigma_1 \varepsilon_1^* + \sigma_2 \varepsilon_2^* + \sigma_6 \varepsilon_6^*) dS = \int_{L_f} \overline{T}_i u_i^* dl, \tag{7.2}$$

where S is the area of 2-D domain, σ_i are stresses within S, u_i^* are kinematically admissible virtual displacements, ε_i^* are virtual strains associated with u_i^*, $\overline{T}_i$ are tractions applied on boundary of S, and L_f is the portion of S over which $\overline{T}_i$ are applied.

Assuming a linear elastic isotropic material, the constitutive equation is given by

$$\begin{Bmatrix} \sigma_1 \\ \sigma_2 \\ \sigma_6 \end{Bmatrix} = \begin{bmatrix} Q_{11} & Q_{12} & 0 \\ Q_{12} & Q_{11} & 0 \\ 0 & 0 & \dfrac{Q_{11} - Q_{12}}{2} \end{bmatrix} \begin{Bmatrix} \varepsilon_1 \\ \varepsilon_2 \\ \varepsilon_6 \end{Bmatrix} \tag{7.3}$$

Using Eq. (7.3) in Eq. (7.2) gives an experimentally useful PVW equation as

$$\int_S Q_{11}\left(\varepsilon_1 \varepsilon_1^* + \varepsilon_2 \varepsilon_2^* + \frac{1}{2}\varepsilon_6 \varepsilon_6^*\right) dS + \int_S Q_{12}\left(\varepsilon_1 \varepsilon_2^* + \varepsilon_2 \varepsilon_1^* - \frac{1}{2}\varepsilon_6 \varepsilon_6^*\right) dS = \int_{L_f} \overline{T}_i u_i^* dl \tag{7.4}$$

Equation (7.4) is useful because in many test systems the full-field strains, ε_i, and the external loads, $\overline{T}_i$, are measured. The contribution of VFM is the advantageous selection of the virtual fields, u_i^*, and by strain–displacement relations the virtual strains, ε_i^*, so that the stiffnesses, Q_{ij}, are determined to ensure equilibrium of S.

7.3.1 *Optimized, Piecewise Virtual Fields*

VFM has many advantages over other inverse methods, and many of those advantages are evolved from optimized, piecewise virtual fields that are used to specify the u_i^*. Two of the three VFM techniques described here use them. Virtual meshes are not dense for homogeneous materials; density is specified based on the number of stiffnesses to be identified and expected stiffness gradient. Generally, mesh density is specified such that the number of interior virtual nodes are 2× the number of stiffness values needed, but a dense mesh is used here to reduce the effects of smoothing performed by the virtual fields. The virtual mesh is shown in Fig. 7.3 where each virtual element is 5 mm × 5 mm. A detailed derivation of optimized, piecewise virtual fields is given in Sects. 3.6 and 3.7 of Pierron and Grédiac [1].

7.3.2 *Location of Inclusion Known a Priori*

Assume that the region S in Eq. (7.4) contains an inclusion whose location and boundaries are known. If $S = S_1 \cap S_2$, where S_2 is the inclusion region and S_1 is all of S outside S_2, then Eq. (7.4) can be rewritten as

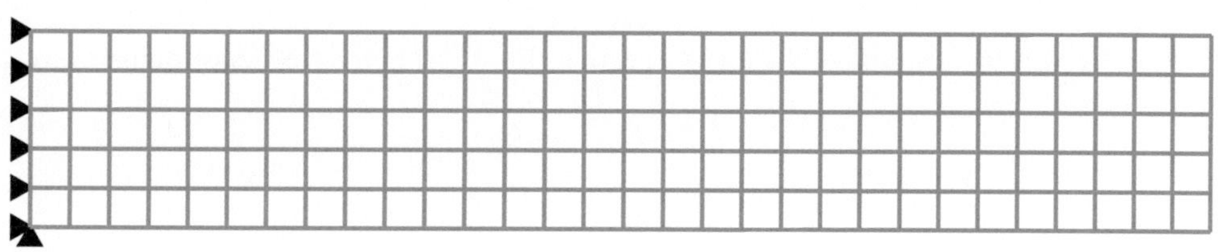

Fig. 7.3 Piecewise virtual field mesh

$$\int_{S_1} Q_{11}^{S_1}\left(\varepsilon_1\varepsilon_1^* + \varepsilon_2\varepsilon_2^* + \frac{1}{2}\varepsilon_6\varepsilon_6^*\right)dS_1 + \int_{S_1} Q_{12}^{S_1}\left(\varepsilon_1\varepsilon_2^* + \varepsilon_2\varepsilon_1^* - \frac{1}{2}\varepsilon_6\varepsilon_6^*\right)dS_1 + \cdots$$
$$+\cdots \int_{S_2} Q_{11}^{S_2}\left(\varepsilon_1\varepsilon_1^* + \varepsilon_2\varepsilon_2^* + \frac{1}{2}\varepsilon_6\varepsilon_6^*\right)dS_2 + \int_{S_2} Q_{12}^{S_2}\left(\varepsilon_1\varepsilon_2^* + \varepsilon_2\varepsilon_1^* - \frac{1}{2}\varepsilon_6\varepsilon_6^*\right)dS_2 = \int_{L_f} \overline{T}_i u_i^* dl \tag{7.5}$$

Accordingly, if k inclusions exist within S, then the left-hand side of Eq. (7.5) would contain $k + 1$ integral terms, where the + 1 refers to the region not contained in the inclusions, i.e., the matrix material. For simplicity, the following development will assume only a single inclusion, but the development for multiple inclusions is straightforward.

If S_1 and S_2 are assumed to be comprised of different linear elastic isotropic materials, then Q_{ij} are constants and can be placed outside the integrals in Eq. (7.5). The new equilibrium equation is

$$Q_{11}^{S_1}\int_{S_1}\left(\varepsilon_1\varepsilon_1^* + \varepsilon_2\varepsilon_2^* + \frac{1}{2}\varepsilon_6\varepsilon_6^*\right)dS_1 + Q_{12}^{S_1}\int_{S_1}\left(\varepsilon_1\varepsilon_2^* + \varepsilon_2\varepsilon_1^* - \frac{1}{2}\varepsilon_6\varepsilon_6^*\right)dS_1 + \cdots$$
$$+\cdots Q_{11}^{S_2}\int_{S_2}\left(\varepsilon_1\varepsilon_1^* + \varepsilon_2\varepsilon_2^* + \frac{1}{2}\varepsilon_6\varepsilon_6^*\right)dS_2 + Q_{12}^{S_2}\int_{S_2}\left(\varepsilon_1\varepsilon_2^* + \varepsilon_2\varepsilon_1^* - \frac{1}{2}\varepsilon_6\varepsilon_6^*\right)dS_2 = \int_{L_f} \overline{T}_i u_i^* dl \tag{7.6}$$

Although no limit on the number of S_k exists, $Q_{ij}^{S_k}$ identification requires sufficient strain data, ε_i, for each region. Practical experience has shown unstable identification for $k > 5$ [3]. Furthermore, inclusion boundaries are rarely "sharp"; an inclusion boundary layer usually exists in which stiffness gradually changes from the matrix to the inclusion. Some judgment by the analyst is needed to define the boundary contour. Conservative use of Eq. (7.6) for multiple inclusions requires that each inclusion be allowed to have unique Q_{ij} initially.

7.3.3 Equilibrium Gap

A further development in the virtual fields method [4] comes from the understanding that PVW can also be used to examine local equilibrium. A similar procedure has been used to examine errors in FE modeling [5]. Discrepancies in local equilibrium are "equilibrium gaps" (EGs) and can be used to locate boundaries of regions that are not in equilibrium with each other when material homogeneity is assumed.

Equation (7.4) can be written as

$$Q_{11}\int_{S_\gamma}\left(\varepsilon_1\varepsilon_1^* + \varepsilon_2\varepsilon_2^* + \frac{1}{2}\varepsilon_6\varepsilon_6^*\right)dS_\gamma + Q_{12}\int_{S_\gamma}\left(\varepsilon_1\varepsilon_2^* + \varepsilon_2\varepsilon_1^* - \frac{1}{2}\varepsilon_6\varepsilon_6^*\right)dS_\gamma - \int_{L_f} \overline{T}_i u_i^* dl = 0 \tag{7.7}$$

Equation (7.7) describes the equilibrium of a region S_γ, where S_γ is a portion of a larger region. If all Q_{ij} are known, usually by identification by homogeneous VFM, and u^* are kinematically admissible, then summation of the integrals should be zero; a non-zero result is termed to be an equilibrium gap. In practice, $\overline{T}_i$ are zero.

7.3.4 Fourier-VFM

A recent contribution to VFM is the solution of the PVW equation (Eq. (7.4) in the frequency domain) by means of Fourier analysis.

The spatially varying Q_{ij} are represented as

$$Q_{11}(x_1,x_2)=\sum_{m=0}^{N}\sum_{n=-N}^{N}a_{m,n}\cos 2\pi\left(\frac{mx_1}{L_1}+\frac{nx_2}{L_2}\right)+\sum_{\substack{m=0\\m+|n|\neq 0}}^{N}\sum_{n=-N}^{N}b_{m,n}\sin 2\pi\left(\frac{mx_1}{L_1}+\frac{nx_2}{L_2}\right) \tag{7.8}$$

$$Q_{12}(x_1,x_2)=\sum_{m=0}^{N}\sum_{n=-N}^{N}f_{m,n}\cos 2\pi\left(\frac{mx_1}{L_1}+\frac{nx_2}{L_2}\right)+\sum_{\substack{m=0\\m+|n|\neq 0}}^{N}\sum_{n=-N}^{N}g_{m,n}\sin 2\pi\left(\frac{mx_1}{L_1}+\frac{nx_2}{L_2}\right) \tag{7.9}$$

Instead of using optimized, piecewise virtual fields described earlier, the virtual fields are unit amplitude sine/cosine waves. A thorough description of this technique is provided by Nguyen et al. [6].

7.4 Results and Discussion

Figure 7.4 gives some results for each VFM analysis technique. When the inclusion is evident, the analyst must define the boundary. For the example shown in Fig. 7.4a, S_2 was defined as the region where $Q_{11} > 11.5$ GPa. In practice the analyst will likely choose many different boundaries and compare the identifications. Accurate identification of the maximum stiffness in this simulation is difficult; as the area of S_2 was reduced to a small region surrounding the peak stiffness, the identification was compromised due to lack of data, i.e., the amount of strain data.

Figure 7.4b shows some results of the EG analysis; each virtual element included 16 (4×4) strain measurements whose physical dimensions were 4.4 mm $\times$ 4.3 mm. An EG analysis provides equilibrium information at the center of the virtual element, so Fig. 7.4b is shorter and narrower by 1/2 the length and height of the EG virtual element. Pragmatic use of EG suggests that results of several sizes will be compared. The size chosen here was used to illustrate a potentially important result, namely the size of the inclusion. Because the EG element scanned within the inclusion, the results indicated the element was near equilibrium. Because Q_{11} and Q_{12} differed only a constant, 0.3 here, Q_{11} was effectively factored out of Eq. (7.7). Therefore, an EG analysis provides information on inclusion boundaries. If an inclusion is not physically apparent, an EG analysis can be used to locate inclusion boundaries which can, in turn, be used to create S_1 and S_2 regions for the VFM technique where location of the inclusion is known a priori.

Figure 7.4c provides example Fourier VFM results. In Eqs. (7.8) and (7.9), $N = 34$, giving total of 2449 coefficients ($a_{m,n}$, $b_{m,n}$) for Q_{11} alone. Results were smoothed with an 88×88 pixel smoothing filter and Q_{11} identification was denoted as NaN at the specimen edges. Because the grid simulated strains were 272×46 individual measurements, additional data were needed in order to determine the large number of coefficients. The grid strains were interpolated to 6000×1000 individual measurements and the limiting factor for the size of N was the amount of data in x_2 direction. Although data interpolation is not a best practice, its use here provided similar information of inclusion size and location, as did the EG analysis and can be similarly used to denote the inclusion boundary for the a priori VFM technique.

Finally, an infinite number of Q_{ij} parametrizations are possible, such as polynomial formulations, e.g., Devivier et al. [4]. Given the relatively small amount of data (272×46 matrix size for each ε_i) and, in this simulation, the small size of inclusion relative to the specimen dimensions, determination of the virtual displacements would lead to an ill-conditioned matrix needed for polynomial Q_{ij} identification.

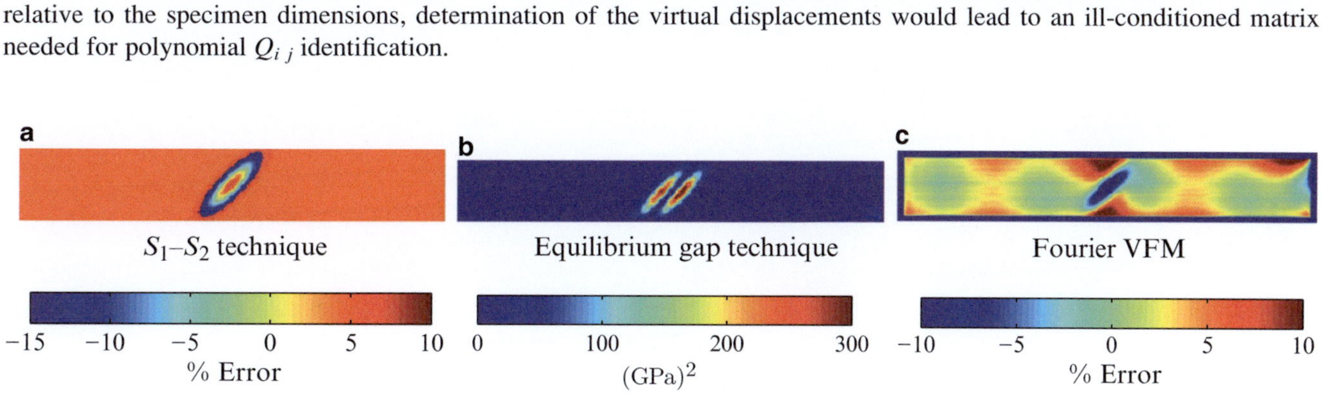

Fig. 7.4 Analysis results for each VFM technique. (**a**) S_1–S_2 technique. (**b**) Equilibrium gap technique. (**c**) Fourier VFM

7.5 Conclusion

Three VFM techniques were used to characterize the simulated, smoothly varying heterogeneous stiffness in a uniaxial tensile specimen. When the location of the inclusion was known, its stiffness could be identified within 10 % error; more accurate identification is possible with increased strain data density. Local equilibrium was examined and used to approximate inclusion size and location. Fourier VFM was used to provide inclusion location and size, but required significant data interpolation. Results of these three techniques suggest that their cooperative use could provide reasonable stiffness identification for heterogeneous materials using a uniaxial tensile test.

References

1. F. Pierron, M. Grédiac, *The Virtual Fields Method: Extracting Constitutive Mechanical Parameters from Full-Field Deformation Measurements* (Springer, New York, 2012)
2. M. Rossi, F. Pierron, On the use of simulated experiments in designing tests for material characterization from full-field measurements. Int. J. Solids Struct. **49**(3–4), 420–435 (2012)
3. J.M. Considine, Characterization of paper heterogeneity. PhD thesis, The University of Wisconsin-Madison (2014)
4. C. Devivier, F. Pierron, M.R. Wisnom, Impact damage detection in composite plates using deflectometry and the Virtual Fields Method. Compos. A Appl. Sci. **48**, 201–218 (2013)
5. P. Ladeveze, D. Leguillon, Error estimate procedure in the finite element method and applications. SIAM J. Numer. Anal. **20**(3), 485–509 (1983)
6. T.T. Nguyen, J.M. Huntley, I.A. Ashcroft, P.D. Ruiz, F. Pierron, A Fourier-series-based virtual fields method for the identification of 2-D stiffness distributions. Int. J. Numer. Methods Eng. **98**, 917–936 (2014)

Chapter 8
Optimized Test Design for Identification of the Variation of Elastic Stiffness Properties of Loblolly Pine (*Pinus taeda*) Pith to Bark

D.E. Kretschmann, J.M. Considine, and F. Pierron

Abstract This article presents the design optimization of an un-notched Iosipescu test specimen whose goal is the characterization of the material elastic stiffnesses of a Loblolly (*Pinus taeda*) or Lodgepole pine (*Pinus contorta*) sample in one single test. A series of finite element (FE) and grid simulations were conducted to determine displacement and strain fields for various ring angles, ring spaces, and proportions of latewood per ring. These displacement and strain fields were utilized to determine the constitutive parameters for the woody material using VFM. Using an error function based on a comparison of the input stiffnesses and those identified by grid simulation information it was possible to narrow the optimum ring angle at which test specimens should be tested. The results of this work suggest that the starting angles for the un-notched Iosipescu test specimens should be approximately 45° to produce the smallest identification error.

Keywords Experimental design • Un-notched Iosipescu test • Virtual fields method (VFM) • Elastic stiffness parameters • Computer simulations

8.1 Introduction

Dramatic differences in properties can be observed in juvenile material, which grows close to the pith of a tree, versus the more mature material that grows near the bark of a tree [1]. Development of techniques to more easily understand the variation of mechanical properties for wood from pith to bark is needed. This information can be used to help explain the large variation in properties and dimensional stability of solid wood products observed when cut from different positions within the tree. Virtual Fields Method (VFM) has proven to be an appropriate technique to determine local, heterogeneous constitutive properties [2]. Xavier and others have demonstrated the possibility of using a VFM to determine the transverse and shear properties of wood in a tree [3, 4]. Work is underway to develop testing procedure that uses VFM as a method of documenting the natural variations in longitudinal, transverse, and shear stiffnesses that occur in Loblolly and Lodgepole pine both within a tree and between trees.

In the portion of the study being reported here, a procedure has been developed to simulate the un-notched Iosipescu test using a finite element (FE) model and a simulated grid pattern to create synthetic displacements and strains. Using this information the VFM was applied to subsequently extract all the earlywood and latewood stiffness parameters for both the FE model and the grid simulation. The uncertainty of the grid measurement, including the systematic and random errors, will be evaluated thoroughly by including several different sources of error. This paper directly reports on one portion of the systematic error. An optimization study was undertaken to look at the impact of ring angle, ring spacing and proportion of earlywood and latewood, and anisotropy ratios (position in tree). This work was conducted to facilitate the selection of ring angle for material property determination. By minimizing the value of an error function at different conditions a range of optimal ring angles for the test configuration was determined.

D.E. Kretschmann (✉) • J.M. Considine
USDA Forest Service, Forest Products Laboratory, One Gifford Pinchot Drive, Madison, WI 53726, USA
e-mail: dkretschmann@fs.fed.us

F. Pierron
University of Southampton, Southampton SO17 1BJ, UK

C. Sciammarella et al. (eds.), *Experimental and Applied Mechanics, Volume 4*,
Conference Proceedings of the Society for Experimental Mechanics Series, DOI 10.1007/978-3-319-22449-7_8

8.2 Methods

8.2.1 Background on Loblolly and Lodgepole Pine Study Material

A study is underway that examines the variation of Loblolly and Lodgepole pine properties from different growth conditions in the South and Western part of the United States [5–8]. The Loblolly pine trees were harvested from two different locations in the state of Arkansas and the Lodgepole pine trees are from four different sites in the U.S. Intermountain West. The western trees included site variations affecting growth, such as elevation, precipitation, and growing seasons. Bolts were extracted from each tree and processed to extract a single, radial-sawn board from each bolt (Fig. 8.1a).

As many 80 × 20 × 5 mm specimens oriented with the growth ring grain pattern, at an angle yet to be determined, will be cut from each board. Variation in the width from pith to bark will result in different number of samples being taken from the North (N) or South (S) side of the board (Fig. 8.1b). For each specimen tree number, sample number, position (N or S) and the ring number from the center of the specimen will be recorded. This information will be used to assess the impacts of position within and between trees.

8.2.2 Experimental Test Methods

Testing will be conducted on an Instron universal test frame configured to run an Iosipescu test with the specimens having the V-notches removed from the classical geometry (Fig. 8.2, un-notched test). It has been shown that a complex and heterogeneous strain field is generated across the central part (L × W Fig. 8.1c) of the Iosipescu test specimen [2]. For these tests the load will be applied at a controlled displacement rate of 1 mm per minute and measured with a 100 kN load cell. The tests will be stopped at approximately 400 N so that the specimen remains in the linear elastic regime. These tests will be carried out at room temperature 23 ± 1 °C and at a relative humidity of 50 ± 2 %. For each sample oven dry density will be determined.

8.2.3 Parameter Identification from Virtual Fields Method

In order to apply the VFM technique full-field displacements measurements must be obtained and the equations for virtual work used to determine the various constitutive parameters of a material. The full-field displacements will be determined using a grid method and digital photography.

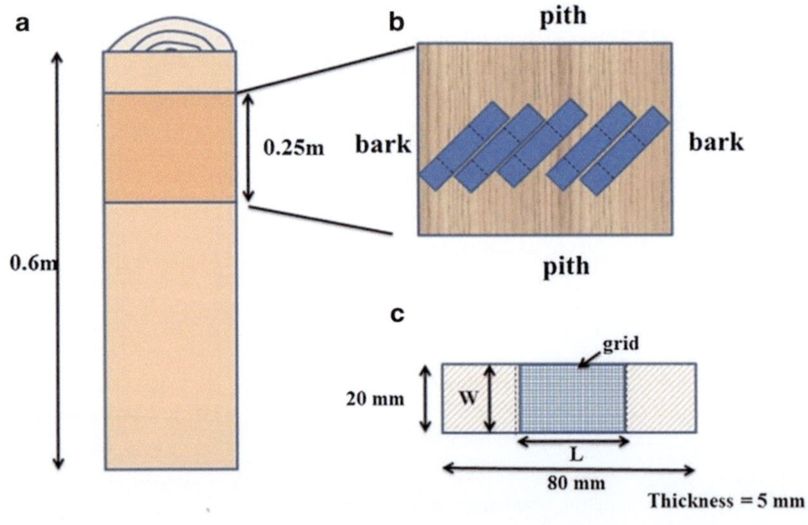

Fig. 8.1 Material sampling: (**a**) log; (**b**) pith to bark test sample cut out pattern; (**c**) dimensions of test specimen

Fig. 8.2 Example of an Un-notched Iosipescu test fixture [9]

8.2.4 Full-Field Displacements Measurement Technique

The grid method measures in-plane full-field displacements by analyzing the deformation of a grid pattern of orthogonal light and dark lines which have been captured by a digital camera. To determine a continuous function that predicts the full-field displacement a series of before and after digital images of the grid pattern are analyzed to determine a polynomial function that fits the deformation pattern observed. Using a spatial phase-shifting method for a given image of the grid a phase function for each image can be evaluated [10]. A phase map can be developed for the image. The difference in phase maps before and after loading is proportional to the displacement field. By using before and after recorded images and the corresponding phase fields the full field displacements can be determined. An additional step is required to calculate the corresponding strain field. Using a least-squares approximations a fit of a two dimensional polynomial function to the raw displacement data is conducted. For previous VFM work with wood a seventh degree polynomial function was found suitable for the least-squares regression scheme [3]. Since the aim of this technique is to identify all the material properties in one single test, the experiment should be designed to induce balanced heterogeneous strain maps so all the components can be identified. A series of simulations have been conducted to select the optimal orientation of the Iosipescu specimen to achieve balanced strain maps.

8.3 Finite Element and Grid Simulations

Three steps are required to determine what ring angle might produce an acceptable balanced heterogeneous strain map: (1) Construction of a FE model that mimics the loading configuration of the un-notched Iosipescu test that provides a numerically-modeled displacement field; (2) Using the FE displacement field as input to a grid simulator [11] grid displacements and strains, based on camera parameters, are created; and (3) Use the constitutive properties identified by VFM for the FE model (no noise numerical data) and the grid simulation (experimentally-simulated data) to determine an optimal ring angle based on evaluation of an error function which minimizes the systematic error between constitutive components.

8.3.1 Finite Element Model

The FE analysis to simulate the un-notched Iosipescu specimen was performed using ANSYS® (ANSYS, Inc., Canonsburg, PA 15317, USA). The dimensions of the model are shown in Fig. 8.3. The model used 608,000 square elements with 400 elements per mm^2 or 20 nodes per millimeter. There were 240,000 square elements in the active area of the model (20 by

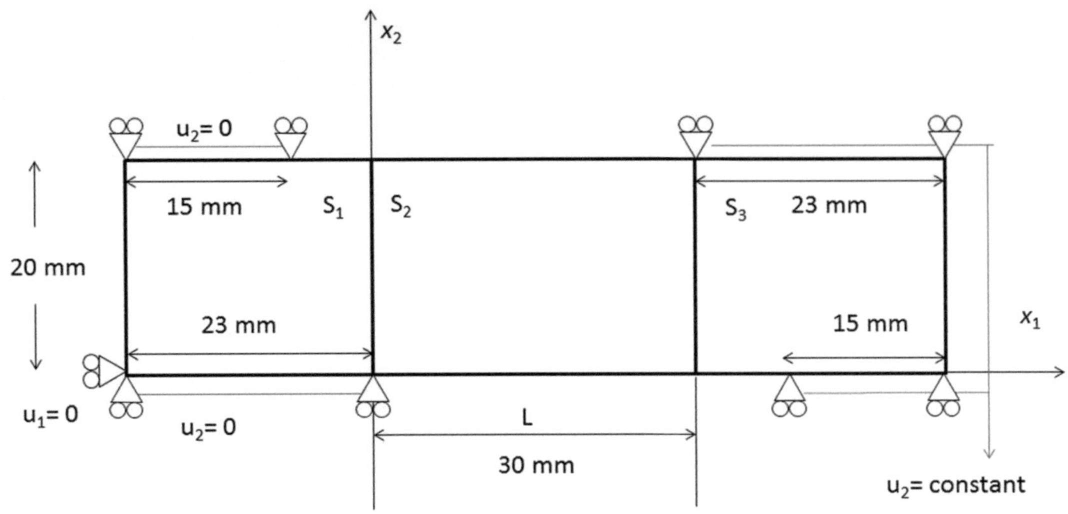

Fig. 8.3 Dimensions of the Iosipescu FE specimen

30 mm sample portion with grid). A 400 N force was applied to the model which was selected based on earlier experimental measurements of the material so that the behavior would remain elastic. The material properties for Loblolly pine were used for these simulations.

8.3.2 *Material Property Input for FE Modeling*

Our selection of property inputs has been based information on obtained through testing of Loblolly pine (*Pinus taeda*) that occurred during 2004 through 2010 [5–7]. The earlywood and latewood Loblolly pine data was gathered on two trips to Loblolly pine plantations in Arkansas. Recommended E_L, E_T, and G values were arrived at after looking at all of this information. It was decided that an appropriate approximation of E_L is 0.60×10^6 (4.14 GPa) and 1.65×10^6 psi (11.38 GPa) for earlywood and latewood respectively. A ratio of E_L to E_T could be taken as 27 for earlywood and of 13 for latewood. Finally, a value of G of 0.11×10^6 psi (0.76 GPa) and 0.25×10^6 psi (1.72 GPa) for earlywood and latewood respectively should be appropriate. It has been shown that latewood closer to the pith is similar in properties to earlywood [5, 6]. As a first approximation of the properties of latewood close to the pith, the latewood properties shown above were cut in half when the simulations were run to mimic material near the pith. Poison's Ratio for all simulations was 0.3 for both earlywood and latewood.

The FE model was constructed so that the angle θ of the rings could be changed (Fig. 8.4). The angles for the model were changes in 10° increments from 0 through 170°. Three different ring spacing were investigated (2, 4, or 6 rings per millimeter); three different proportions of latewood within each ring were investigated (1/2, 1/4, and 1/6 of the ring); and two locations (near the pith and near the bark) were investigated. The simulation test matrix investigated is given in Table 8.1. To this point we have run 210 FE simulations for Iosipescu specimens using the VFM to determine the displacement and strain fields.

8.3.3 *Grid Simulations*

A MATLAB® (MathWorks, Inc., Natick, MA USA) program has been developed to take FE displacements to create a synthetic displaced grid image. The grid simulation used the following assumptions for the grid and CCD camera: a grid pitch = 5 pixels with an array of 1360 pixels horizontally and 1024 vertically; a minimum grey level of 15 and a maximum

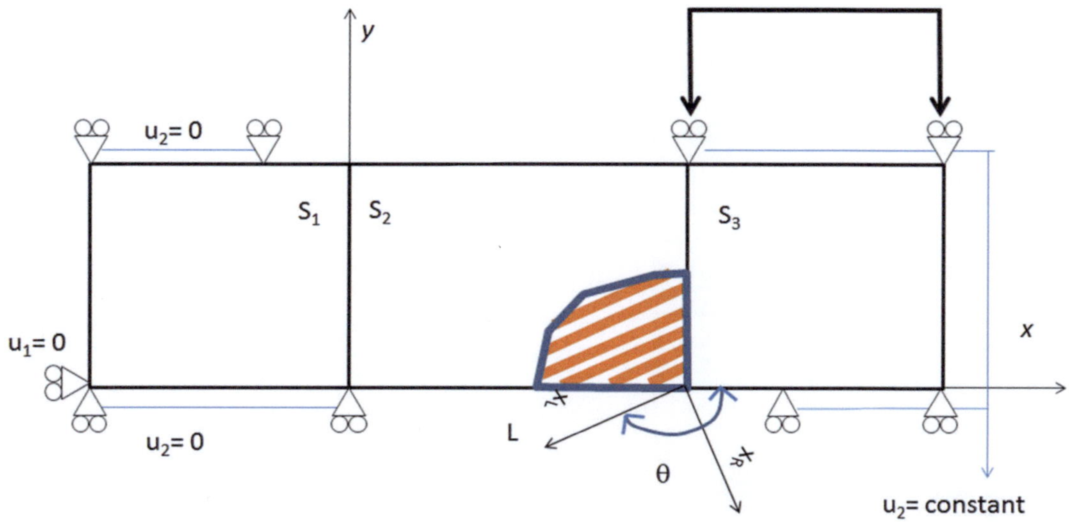

Fig. 8.4 Coordinates system for changing ring angle θ with respect to specimen coordinate system

Table 8.1 Simulation test matrix

		1/2 Ring width		1/4 Ring width		1/6 Ring width	
Ring space	2/1000	0–170	–	0–170	–	0–170	–
	4/1000	0–170	0–170	0–170	–	0–170	0–90
	6/1000	0–170	–	0–170	–	0–170	–

grey level of 4040, a pixel resampling of 5, and a minimum pitch for the grid of 0.1 mm. Figure 8.4 shows the comparisons between the FE strain maps and the simulated grid strain maps for a ring spacing of 2 rings per mm and 50 and 17 % latewood ring width for ring orientations of 0°, 40°, 60° and 90°. The first line of the figures are FE estimated ε_{xx} ε_{yy} and ε_{xy} values and the second line of the figure are grid simulated ε_{xx} ε_{yy} and ε_{xy} values. The strain maps for 40° and 60° orientations are quite similar for all ε values (Fig. 8.5).

8.4 Virtual Fields Method for Constitutive Parameter Identification

VFM for constitutive parameter identification is based on the principle of virtual work which can be written, in the case of plane stress and in the absence of body forces, and assuming a macroscopically homogeneous material with an orthotropic linear elastic behavior, as:

$$Q_{ij} \int_{S} \varepsilon_{j} \varepsilon_{i}^{*} dS = \frac{1}{t} \int_{S_f} T_{\beta} u_{\beta}^{*} dS \tag{8.1}$$

where Q_{ij} are stiffness parameters, ε_j are the actual strain components over S and, ε_i* is the virtual strain field associated with u_β*, S the surface of the interest of the plate, t the thickness of the plate, T_β the distribution of external tractions applied over S_f and u_β* the virtual displacement field.

with,

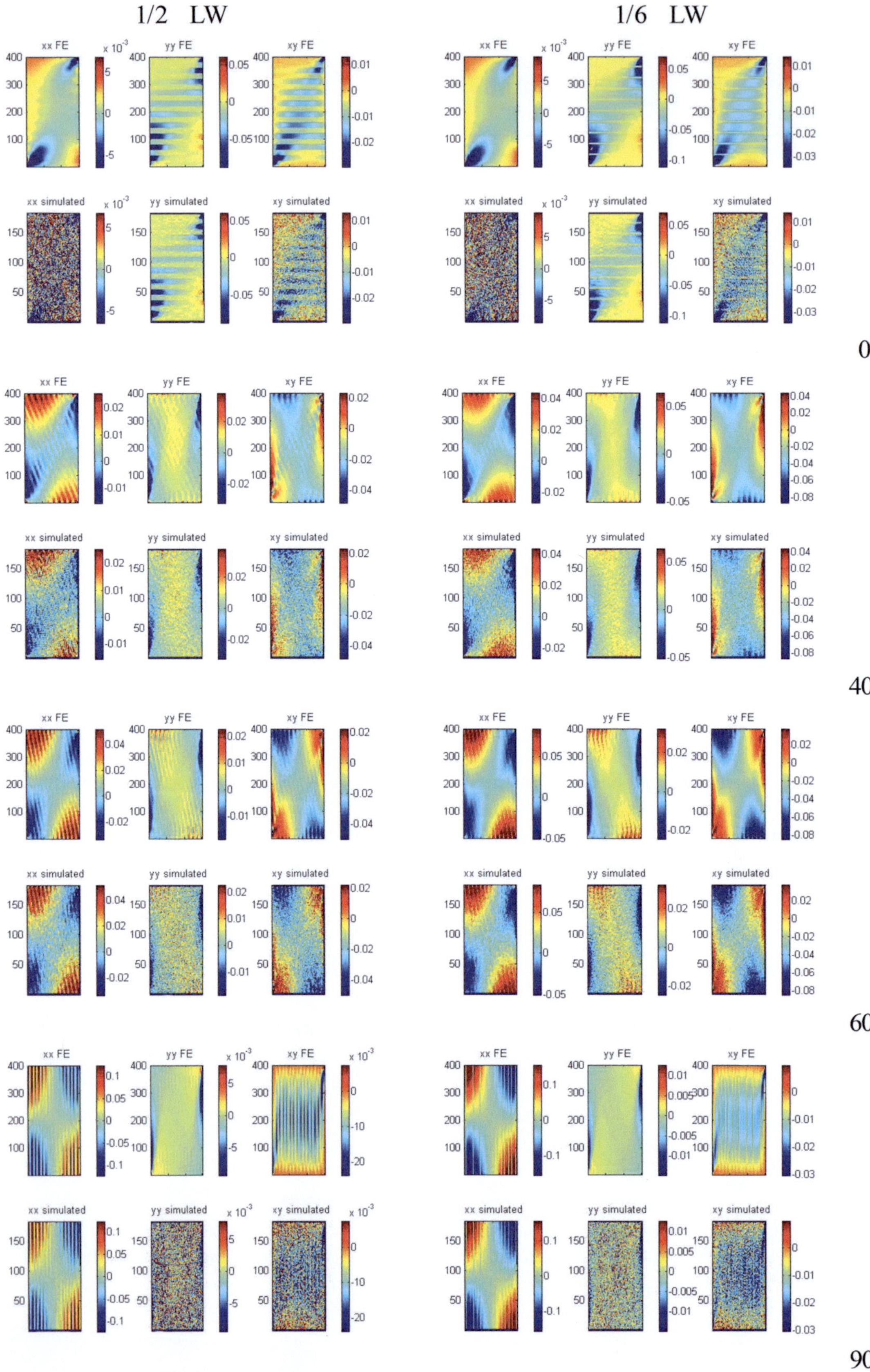

Fig. 8.5 Strains for FE and grid simulations for a ring spacing of 2 rings per mm and 50 and 17 % latewood ring widths at 0°, 40°, 60°, and 90°

$$Q_{ij} = \begin{bmatrix} Q_{11} & Q_{12} & 0 \\ Q_{12} & Q_{22} & 0 \\ 0 & 0 & Q_{66} \end{bmatrix} = \begin{bmatrix} \frac{E_{11}}{1 - \nu_{12}\nu_{21}} & \frac{-\nu_{21}E_{11}}{1 - \nu_{12}\nu_{21}} & 0 \\ \frac{-\nu_{12}E_{22}}{1 - \nu_{12}\nu_{21}} & \frac{E_{22}}{1 - \nu_{12}\nu_{21}} & 0 \\ 0 & 0 & G_{12} \end{bmatrix} \tag{8.2}$$

In (8.2) the symbols E, ν, and G represent, respectively, the modulus of elasticity, Poisson's ratio and the shear modulus and subscripts 1 and 2 denote the orthotropy axes. We have divided the measurement region, S, into earlywood and latewood (8.3).

$$Q_{ij}^{EW} \int_{S_{EW}} \varepsilon_j \varepsilon_i^* dS_{EW} + Q_{ij}^{LW} \int_{S_{LW}} \varepsilon_j \varepsilon_i^* dS_{LW} = \frac{1}{t} \int_{L_{S_2 S_3}} T_\beta u_\beta^* dS \tag{8.3}$$

FE and grid simulation displacements and strains were used to determine the heterogeneous constitutive parameters using a VFM subroutine which had previously been programmed using Python programing language. The various Q values for earlywood and latewood based on FE and grid simulations were tabulated for comparison.

8.4.1 Optimization of Angle Using Simulated Results

The goal for conducting these simulations is determine the optimal ring orientation angle for testing specimens in the Iosipescu test fixture. Using the simulated experimental data, a parametric study investigated the effect of ring orientation as it varied between 0° and 180°. For the analysis reported below, we concentrated on the change between 0° and 90° because of expected symmetry. No random error is considered in this section, noise was not added to the synthetic grid image. This work will be addressed in future sensitivity studies. Error estimates were established for each stiffness estimate by comparing the stiffness parameters Q determined by FE models and the parameters Q determined by grid simulations using (8.4), where Q_{ijFEM} are the Qij results from the FE model and Q_{ijGrid} are the Qij results from the grid simulations.

$$E_{rr} = \frac{Q_{ijGrid} - Q_{ijFEM}}{Q_{ijFEM}} \tag{8.4}$$

An error function was introduced to evaluate the overall systematic error of the VFM identifications (8.5). This error function averages the amount individual error for the four most significant constitutive components 11, 22, 12, and 66. The components 16 and 26 were considered to be extremely small and not included in the function.

$$Error\_Function = \frac{1}{4}(|E_{11}| + |E_{22}| + |E_{12}| + |E_{66}|) \tag{8.5}$$

Figures 8.6 and 8.7 show the results for the range of ring spacing, proportion of ring width, and positions investigated to this point. Results for latewood are shown in red; results for earlywood are shown in black. The general shape of the error function for all variations tested appears to be similar. The error function is largest when away from angles between 30° and 60°. The largest values are for angles greater than 70°. The error function value has more variation as the proportion of latewood per ring increases. Also the most consistent minimum for the earlywood and latewood error function is between 40° and 50°. For the simulations that were conducted to look at the impact of changes in anisotropy (Fig. 8.7, the stiffness of LW was cut in half), there appears to be little impact on the shape of the error function. A summary of the angles with the

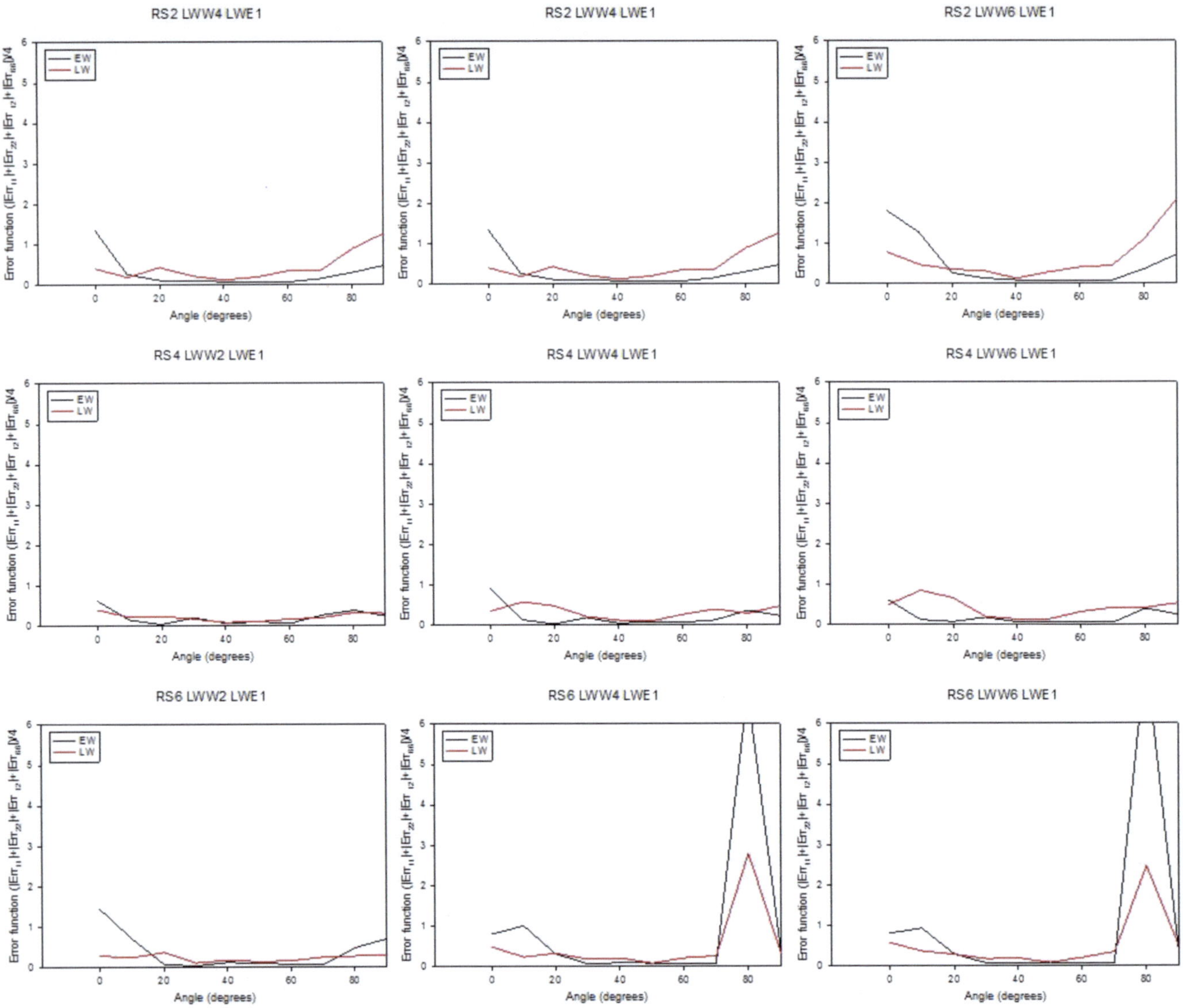

Fig. 8.6 Earlywood and latewood error functions for nine combinations of ring spacing and latewood width [RS is ring spacing (2, 4, or 6 rings per mm), LWW is latewood width (1/2, 1/4, or 1/6), LWE indicates if latewood stiffness values are (near bark 1 or near pith 0.5)]

lowest error functions for the different configurations explored is given in Table 8.2. Results so far would suggest that a starting angle of 45° would be an effective ring orientation for the un-notched Iosipescu test specimen.

8.5 Summary

A series of finite element (FE) and grid simulations were conducted to determine displacement and strain fields for various ring angles, ring spaces, and proportion of latewood per ring. These displacement and strain fields were utilized to determine the constitutive information for the woody material using the VFM. Using an error function based on a comparison of the FE and grid simulation constitutive parameters, it was possible to select the ring angle at which test specimens should be tested for best Q_{ij} identification. The results of this work suggest that 45° ring orientation gives lowest errors for the un-notched Iosipescu test geometry.

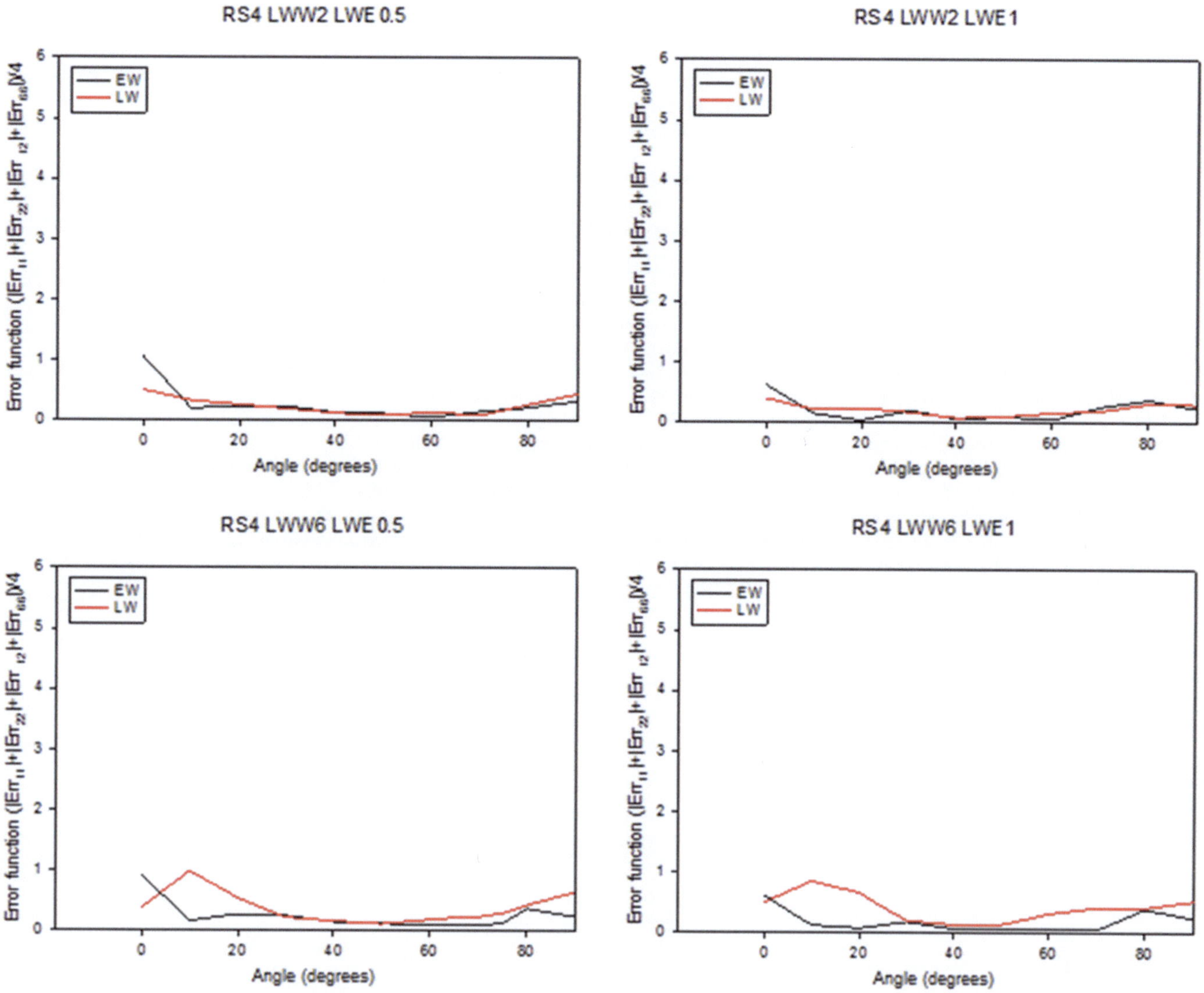

Fig. 8.7 Error function results for pith and bark anisotropy ratios [RS is ring spacing (4 rings per mm), LWW is Latewood width (1/2, or 1/6 of ring), LWE indicates if latewood stiffness values are (near bark 1 or near pith 0.5)]

Table 8.2 Ring angles with the lowest cost functions

		1/2 Ring width		1/4 Ring width		1/6 Ring width	
		Near bark	Near pith	Near bark	Near pith	Near bark	Near pith
Ring space	2/1000	40	–	40	–	40	–
	4/1000	40–50	40–60	40–50	–	40–50	40–50
	6/1000	30–50	–	50	–	50	–

8.6 Future Plans

There will be a continuation of exploring the random and systematic errors of this measurement technique. A series of experimental tests using un-notched Iosipescu test fixture (Fig. 8.2) will begin in the spring of 2015. The initial testing will be conducted on Loblolly pine and then progress on to Lodgepole pine.

References

1. P.R. Larson, D.E. Kretschmann, A. Clark III, J.G. Isebrands, Juvenile wood formation and properties in southern pine. *General Technical Report, FPL GTR 129*, U.S. Department of Agriculture, Forest Service, Forest Products Laboratory, Madison, 2001, p. 42
2. M. Grediac, F. Pierron, S. Avril, E. Toussaint, The virtual fields method for extracting constitutive parameters from full-field measurements: a review. Strain **42**, 233–253 (2006)
3. J. Xavier, S. Avril, F. Pierron, J. Morais, Novel experimental approach for longitudinal-radial stiffness characterization of clear wood by a single test. Holzforschung **61**(5), 573–581 (2007)
4. J. Xavier, S. Avril, F. Pierron, J. Morais, Variation of transverse and shear stiffness properties of wood in a tree. Compos. Part A **40**(12), 1953–1960 (2009)
5. S.M. Cramer, D.E. Kretschmann, R. Lakes, T. Schmidt, Earlywood and latewood elastic properties in loblolly pine. Holzforschung **59**(5), 531–538 (2005)
6. D.E. Kretschmann, S.M. Cramer, The role of earlywood and latewood properties on dimensional stability of wood. Chapter. *The Compromised Wood Workshop 2007*, University of Canterbury, Christchurch, 29–31 January 2007, pp. 215–236
7. D.E. Kretschmann, The influence of juvenile wood content on shear parallel compression and tension perpendicular to grain strength and mode I fracture toughness of loblolly pine at various ring orientations. For. Prod. J. **58**(7/8), 89–96 (2008)
8. T.M. Gorman, D.E. Kretschmann, Characterization of juvenile wood in Lodgepole pine in the intermountain west. In: *Proceedings of the 55th International Convention of Society of Wood Science and Technology*, Beijing, 27–31 August 2012
9. F. Pierron, The virtual fields method, extracting constitutive mechanical parameters from full-field deformation measurements short course. In: *16th International Conference on Experimental Mechanics, ICEM 16*, Cambridge, 6 July 2014
10. Y. Surrel, Fringe Analysis, in *Photomechanics*, ed. by P. Rastogi. Topics in Applied Physics (Springer, Heidelberg, 1999), pp. 57–104
11. M. Rossi, F. Pierron, On the use of simulated experiments in designing tests for material characterization from full-field measurements. Int. J. Solids Struct. **49**(3–4), 420–435 (2012)

Chapter 9
Investigation of the Hydric Transfer Phenomenon in Wood at the Ring Scale with the Grid Method

Djily Dang, Evelyne Toussaint, Michel Grédiac, and Rostand Moutou Pitti

Abstract The water diffusion phenomenon and its effect on the deformation of wood samples are studied here. A suitable experimental device is designed to impose a direct contact on one side of the specimen with water. Strain maps are measured over time on the other side. The grid method is used for this purpose. White fir of Massif Central (*Abies alba* Mill) is studied here. During the experiments, strain gradients clearly appear between early and late woods, highlighting the different diffusion properties at the scale of the rings. The results also show changes in the moisture diffusivity of the material during the test. A cycle of adsorption/desorption is also studied. Results highlight the existence of residual strains at the wood ring scale.

Keywords Adsorption • Desorption • Early wood • Gradient • Grid method • Late wood • Strain

9.1 Introduction

Wood is a biological material produced by trees at the level of roots, trunk and branches. It is a highly heterogeneous material both in composition and architecture. The degree of heterogeneity of wood appears at the microscopic, mesoscopic and macroscopic scales. At the macroscopic scale, a cross section of the trunk consists in alternating dark and light zones (usually very distinct in most species) which correspond to annual growth rings. The chemical composition of wood reveals that it is mainly composed of cellulose, hemicellulose and lignin. These chemical components and their proportions are at the origin of the viscoelastic behavior of wood at the macroscopic scale [1]. It is important to mention that each of the chemical components of wood is highly sensitive to moisture content changes. Indeed, under the fiber saturation point (25–30 %), any variation of wood moisture content directly influences its mechanical and physical characteristics. In the present work, the aim is to study wood material under the water diffusion phenomenon using a full-field measurement technique: the grid method [2, 3]. Such a measurement technique reveals, within certain limit, heterogeneities in strain fields that are caused here by the heterogeneous nature of the material. The materials and methods are described in the first section of the paper. The results are then presented and discussed. A conclusion and some perspectives close the paper.

9.2 Materials and Methods

The goal is to study the phenomenon of water diffusion and its effect on the deformation of wood samples. A suitable experimental device, presented in Fig. 9.1, was designed for this purpose. The wood samples used here are made of white fir of Massif Central. The dimensions are $50 \times 49 \times 8$ mm^3 (see Fig. 9.2a). One side of the specimen was directly in contact with water, and the opposite face was equipped with a cross grid featuring a pitch equal to 200 μm (see Fig. 9.2b). Grid images were recorded at different times (every 30 min) by a CCD camera to monitor the diffusion phenomenon of the material under adsorption. Then, water was removed and the desorption phenomenon was studied using the same process. Strain maps corresponding to several steps of water diffusion were deduced from the grid images using an in-house dedicated software.

D. Dang • E. Toussaint (✉) • M. Grédiac • R.M. Pitti
Clermont Université, Université Blaise Pascal, Institut Pascal, BP 10448, Clermont-Ferrand 63000, France

CNRS, UMR 6602, Institut Pascal, Aubière 63171, France
e-mail: evelyne.toussaint@univ-bpclermont.fr

C. Sciammarella et al. (eds.), *Experimental and Applied Mechanics, Volume 4*,
Conference Proceedings of the Society for Experimental Mechanics Series, DOI 10.1007/978-3-319-22449-7_9

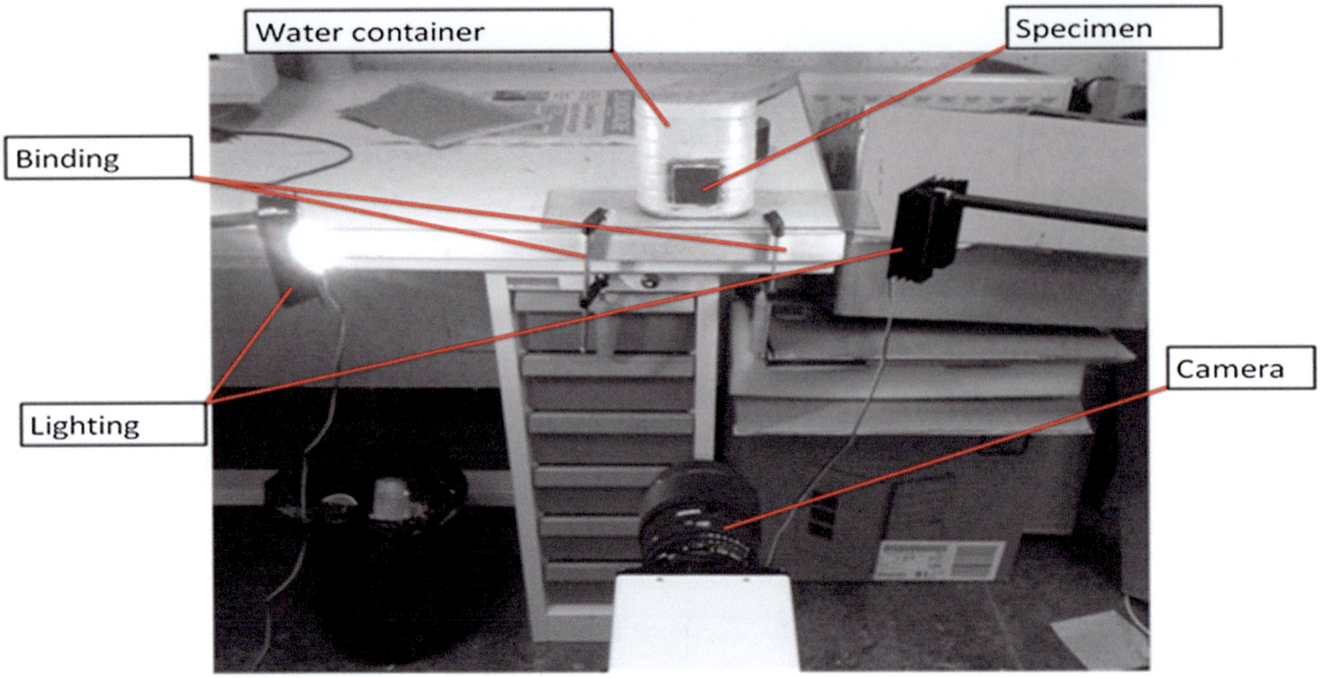

Fig. 9.1 Experimental set-up for water diffusion test

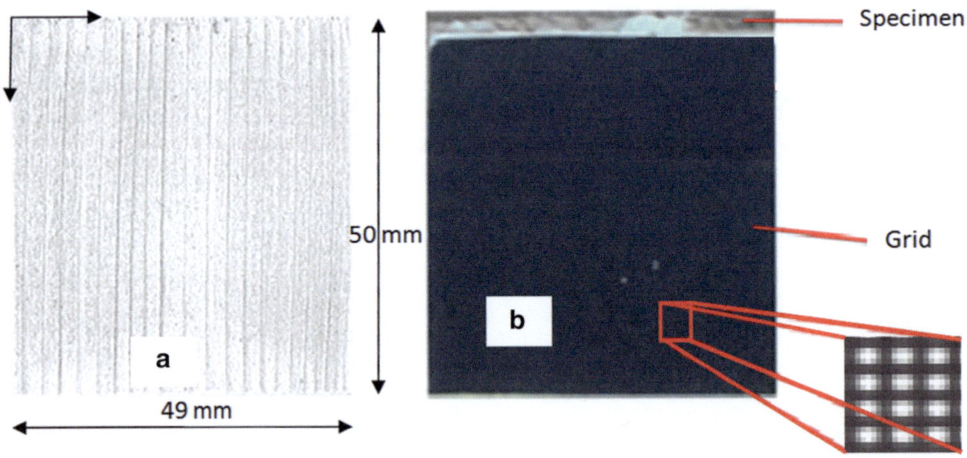

Fig. 9.2 Grid transfer procedure: (**a**) view of the face on which the grid is transferred, (**b**) transferred grid

9.3 Experimental Results and Discussion

Figures 9.3 and 9.4 show typical examples of radial strain maps obtained at different levels of diffusion during adsorption and desorption. One can clearly distinguish early wood (EW) and late wood (LW) in these maps.

Mean radial strains are calculated for EW and LW by using a mask (built from the scan of the surface of the specimen on which the grid was transferred). This enables us to extract the strain values in both EW and LW from the whole strain maps. Mean radial strain for the whole sample is also deduced.

During adsorption, strain maps for images taken respectively after $t_1 = 30$ h (map N° 40) and $t_2 = 96$ h (map N° 172) of exposure allow to investigate the kinetic of diffusivity. Indeed, between times t_1 and t_2, the mean radial strain of EW goes from 1.2E−02 to 1.49E−02; thus increasing by 24.2 % during 66 h (strain increase rate: 4.4E−03 %/h). Between times t_1 and t_2, the mean radial strain of LW drops from 1.84E−02 to 2.29E−02; thus increasing by 24.6 % during 66 h (strain increase rate:

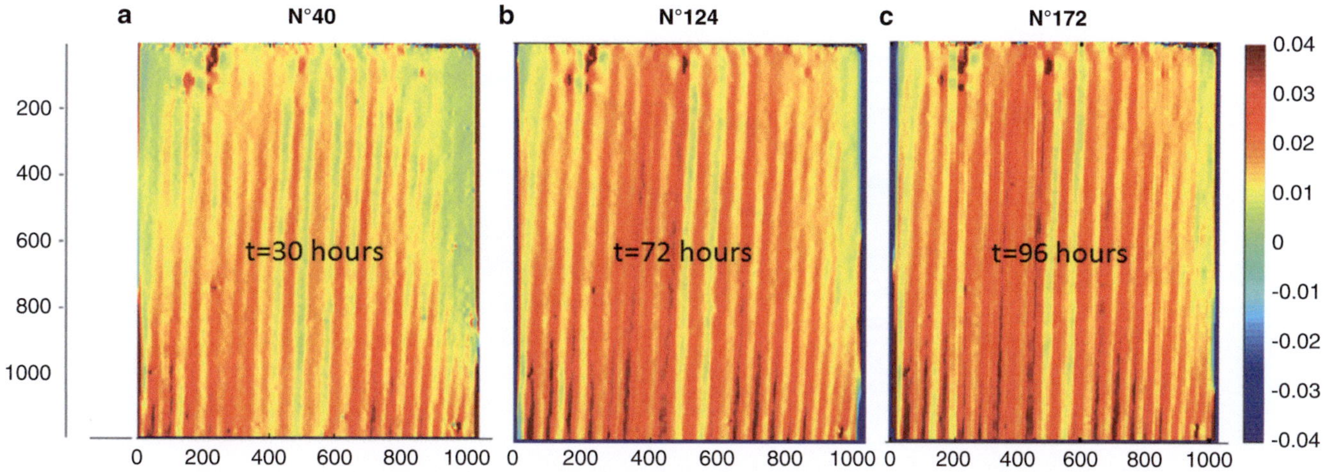

Fig. 9.3 Radial strain maps during adsorption: (**a**) after 30 h exposure. (**b**) After 72 h exposure. (**c**) After 96 h exposure

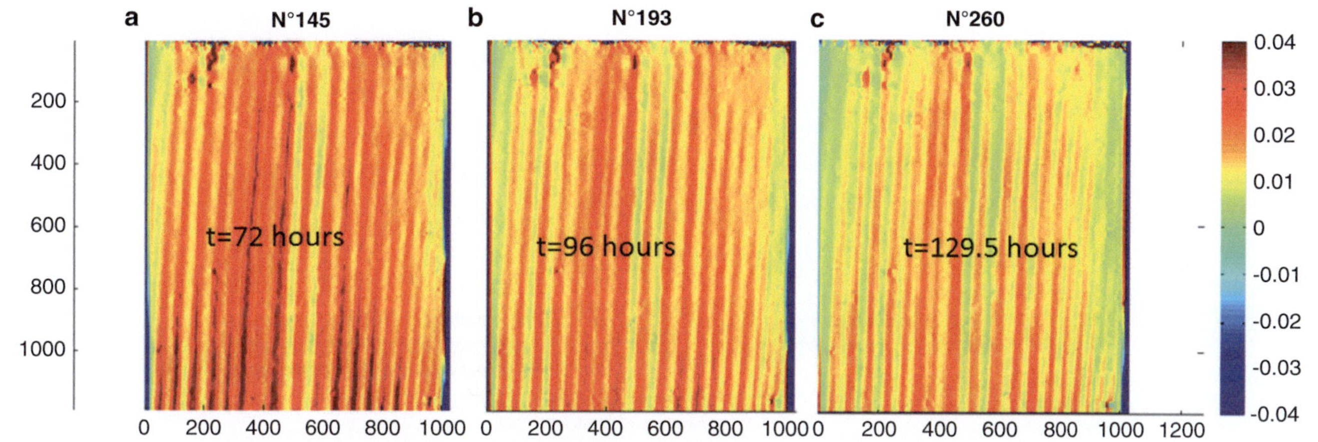

Fig. 9.4 Radial strain maps during desorption: (**a**) after 72 h exposure. (**b**) After 96 h exposure. (**c**) After 129.5 h exposure

6.8E−03 %/h). If we consider the mean radial strain of the whole sample between t_1 and t_2: this latter varies from 1.51E−02 to 1.87E−02; thus increasing by 23.9 % during 66 h (strain increase rate: 5.45E−03 %/h), see Figs. 9.3 and 9.5.

Strain maps during desorption are shown in Fig. 9.4. They were obtained at $t_1 = 72$ h (map N° 145) and $t_2 = 129.5$ h (map N° 260). Such maps enable us to study the kinetic of diffusivity during this part of the test. Between times t_1 and t_2, the mean radial strain of EW goes from 1.55E−02 to 0.98E−02; thus decreasing by 36.7 % during 57.5 h (strain decrease rate: 1E−02 %/h). Between t_1 and t_2, the mean radial strain of LW drops from 2.33E−02 to 1.56E−02; thus decreasing by 35 % during 57.5 h (strain decrease rate: of 1.34E−02 %/h). Let us now consider the mean radial strain of the whole sample between t_1 and t_2: the strain drops from 1.93E−02 to 1.26E−02; thus decreases by 34.7 % during 57.5 h (strain decrease rate: 1.16E−02 %/h).

The analysis of results from adsorption and desorption confirms that the diffusivity of EW is greater than that of LW. Despite this remark, it is obvious to note that the mean radial strain increase and decrease rates of LW are higher than that of EW. This can be justified by the fact that expansion coefficient of LW is higher than that of EW. The analysis also highlights that during adsorption and desorption, the mean radial strain rate lies between LW's one and EW's one. In addition, one can note that the kinetic of diffusivity during desorption is higher than the kinetic of diffusivity during adsorption.

Figure 9.5 presents the mean radial strain for early and late woods. As expected, the mean radial strain in LW is higher than the one reached in EW. Obviously the mean strain of the whole specimen lies between these two values.

Figure 9.6 shows the evolution of mean radial strain rates for the whole sample, EW and LW. In accordance with Fig. 9.5, the three mean strain rate curves decrease along the adsorption phenomenon. One can also conclude that after 30 h of

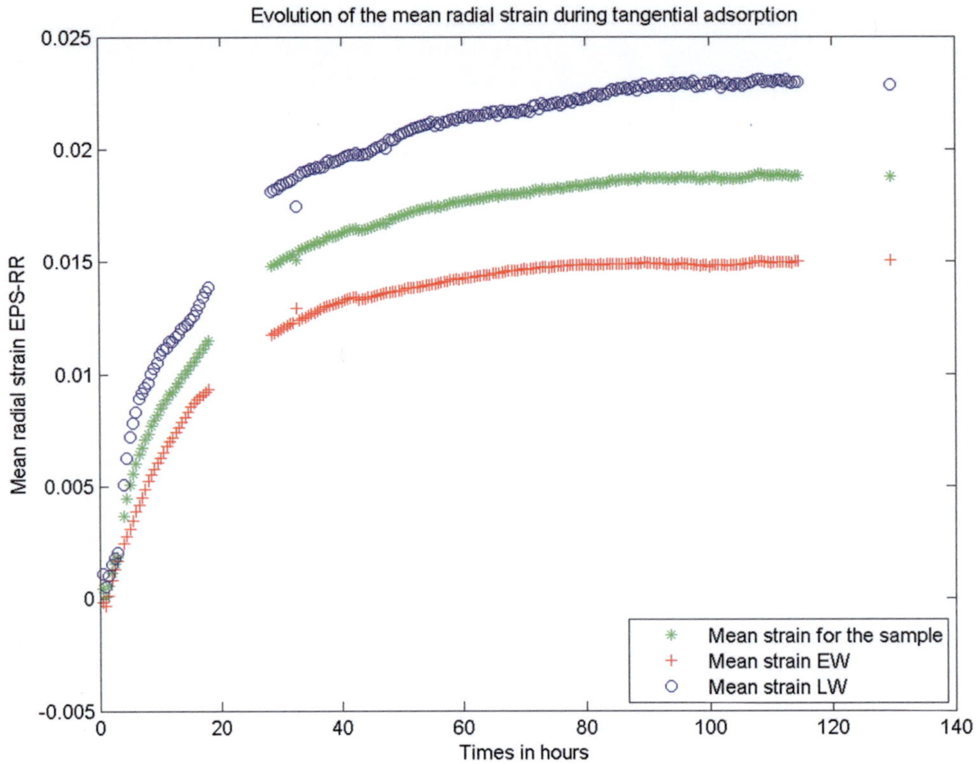

Fig. 9.5 Evolution of mean radial strain at the scale growth ring during adsorption

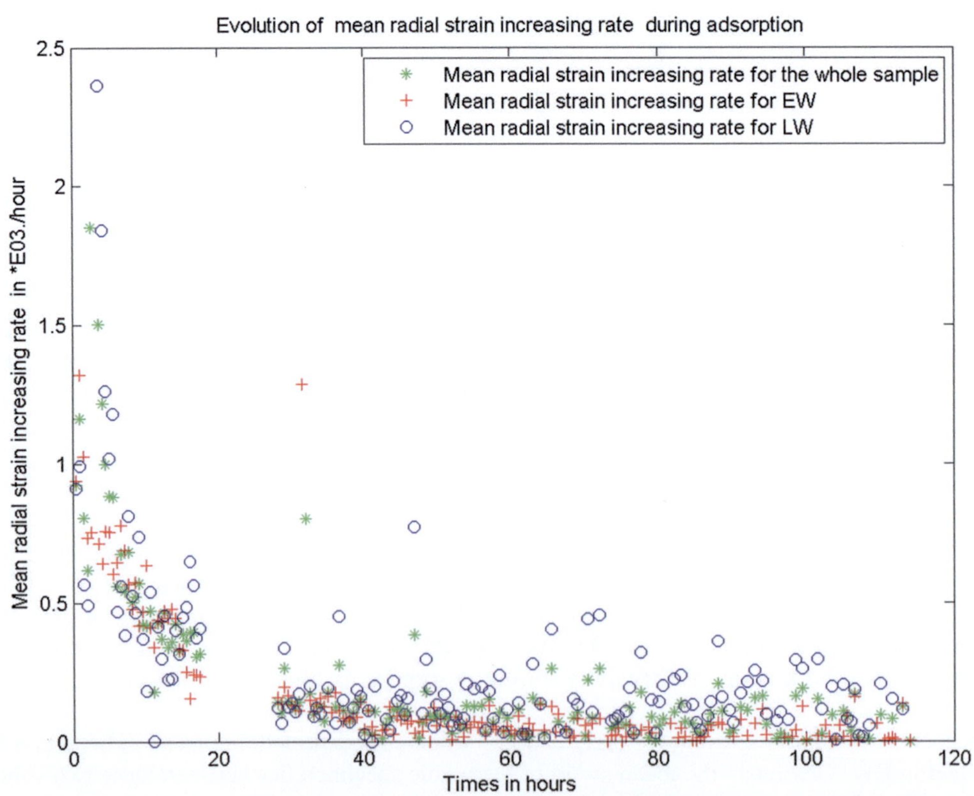

Fig. 9.6 Evolution of mean radial strain rates during adsorption for the whole sample

adsorption, the mean strain rate for LW is higher than that of the whole sample, the latter being also higher than that of EW. This is in accordance with results presented above.

9.4 Conclusion

During this study, we developed an experimental device that allowed us to determine radial strain maps during a cycle of water absorption and desorption. The grid method was used for this purpose. Results showed the difference in terms of strain levels and strain rates between early and late woods during water diffusion phenomenon.

Work is underway to process these experimental results to identify both the diffusion and the expansion coefficients at the ring scale.

References

1. G. Almeida, C. Assor, P. Perré, The dynamic of shrinkage/moisture content behavior determined during drying of microsamples for different kinds of wood. Drying Technol. **26**(9), 1118–1124 (2008)
2. D. Dang, R. Moutou Pitti, E. Toussaint, M. Grédiac, Experimental evidence of water diffusion gradient in wood using the grid method. In *Final Cost Action FP0904 Conference*, 19–21 May 2014
3. C. Badulescu, M. Grédiac, J.-D. Mathias, Investigation of the grid method for accurate in-plane strain measurement. Meas. Sci. Technol. **20**(9), 1–17 (2009)

Chapter 10
Extension of the Non-linear Virtual Fields Method to Inertial Heterogeneous High Strain Rate Tests

S. Dreuilhe and F. Pierron

Abstract This paper presents the development of a novel inertial test for the identification of non-linear parameters for elasto-plastic constitutive models at high strain rates. After briefly presenting the principle of the approach (stress reconstruction from acceleration fields measured using an ultra-high speed camera), the methodology is detailed including the development of the identification routine and the experimental implementation. Results will be discussed and future work presented.

Keywords High-strain rate • Ultra-high speed camera • VFM • Non-linear model identification • Inertial test

10.1 Introduction

The identification of the mechanical behaviour of materials at high strain rates is still limited by the current testing techniques. The Split-Hopkinson-Pressure-Bar (SPBH) is a technique based on one-dimensional wave propagation theory that reduces the field of experimental exploration to small specimens. Inertia effects at strain rates higher than $10^4\ s^{-1}$ cause difficulty in reaching equilibrium conditions, which is a key assumption of the technique. As an alternative to this method, the recent use of cameras to record fields of displacement and strains over the surface of specimens under loading has been made possible through white light imaging techniques such as Digital Image Correlation (DIC) or the Grid Method, including at high rates by using high speed cameras. With the Virtual Field Method it has recently been shown [1] that identification of elastic parameters of rectangular composite specimens heterogeneously loaded could be realised based on the fields of strain and displacement obtained with recent ultra-high speed cameras. This very promising method does not require any load measurement, presenting an advantage over current methods based on Hopkinson bars which are affected by inertia effects at high rates. This approach uses instead the field of acceleration as a load cell, derived from the time-resolved displacement field. This has the advantage of relieving the constraint on specimen size as force equilibrium is not required anymore, even unwanted. The objective of this paper is to present the extension of the work in [1] to elasto-plastic laws.

10.2 Principle

As already detailed in [1], the dynamic equilibrium of a thin specimen loaded on both sides such as in Fig. 10.1a, writes according to (10.1), at each time:

$$F_1 + F_2 = \int_V \rho a_1 dV \tag{10.1}$$

where a_1 is the acceleration in the loading direction. Acceleration and stress fields are assumed to be homogeneous through the small thickness t_h.

The acceleration field is obtained by double differentiation of the time-resolved displacement field, obtained with DIC or the Grid Method applied on the images from an ultra-high speed camera. Given the large number of points the integral can

S. Dreuilhe (✉) • F. Pierron
Engineering Materials Research Group, Faculty of Engineering and the Environment, University of Southampton, Southampton SO17 1BJ, UK
e-mail: smd2e13@soton.ac.uk; f.pierron@soton.ac.uk

C. Sciammarella et al. (eds.), *Experimental and Applied Mechanics, Volume 4*,
Conference Proceedings of the Society for Experimental Mechanics Series, DOI 10.1007/978-3-319-22449-7_10

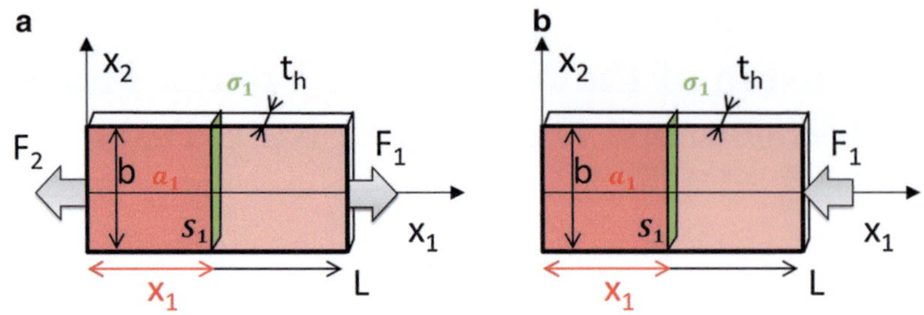

Fig. 10.1 (**a**) 'Anvil' and (**b**) 'Inertial' test configurations considered for thin heterogeneously loaded specimens

be approximated by a discrete sum over the measurement points at the surface. Assuming constant density, (10.1) can be expressed as:

$$F_1 + F_2 = \rho t_h \times Lb \frac{1}{n}\sum_{i=1}^{n} a_1{}^{(i)} = \rho V \overline{a_1}^{S_{tot}} \tag{10.2}$$

In particular, if the one end is free as in Fig. 10.1b ($F_2 = 0$), (10.3) is obtained, where the impact force F_1 is directly obtained through the acceleration:

$$F_1 = \int_{-\frac{t_h}{2}}^{\frac{t_h}{2}}\int_{-\frac{b}{2}}^{\frac{b}{2}} \sigma_1 dx_1 dx_3 = t_h b \overline{\sigma_1}^{x_2} \Rightarrow \overline{\sigma_1(x_1 = L, t)}^{x_2} = \rho L \overline{a_1}^{S_{tot}} \tag{10.3}$$

It is valid at any section of coordinate x_2 along the specimen length, hence $\overline{\sigma_1(x_1,t)}^{x_2} = \rho x_1 \overline{a_1(x_1,t)}^{S_1}$. Average stress profiles along the width can therefore be reconstructed and since the strain field is also available at any time, stress-strain curves can be constructed in all transverse sections of the specimen [1]. For non-linear behaviour, the identification is realised through minimization of a cost function over the material parameters such as in [2].

10.3 Methodology

10.3.1 Validation on Finite Element Simulated Measurements

In order to validate the method in the case of non-linear behaviour, a simple elasto-plastic material model with linear isotropic hardening has been implemented by finite element. The strains and accelerations of a rectangular model of dimensions 40 mm × 30 mm × 3.6 mm impacted on one edge by a purely elastic rectangular impactor at a speed of 20 $m \cdot s^{-1}$ have been extracted. The properties are provided in Table 10.1.

In the case of non-linear behaviour, the Virtual Fields Method consists in minimizing a cost function built up from the squared differences between the average stress in all sections calculated from the average acceleration over surface S_1 on the one hand, and from the strain field which depends on the material parameters. The stress reconstruction from strains uses the return-mapping algorithm as described in [3]. It is used for the reconstruction of the three stress components in the case of strains measured at a surface (only three components of strain available), simplified version of full 3D finite element reconstruction algorithms. Although iterative, this method is one order of magnitude less computationally expensive than finite element model updating, for instance, as there is no need for a full FE computation to evaluate the cost function. The procedure is summarized in the flowchart in Fig. 10.2.

Table 10.1 Finite element model parameters

Specimen	
Dimensions	$40 \times 30 \times 3.6$ mm^3
Density	8000 kg·m^{-3}
Elastic properties	
Young's modulus	E = 200 GPa
Poisson's ratio	$\nu = 0.3$
Plastic properties (bilinear isotropic hardening)	
Yield stress	$\sigma_0 = 180$ MPa
Hardening modulus	H = 3 GPa
FE model	
Mesh density	0.5 mm (60×80 elements)
Element type	Abaqus CPS4R, reduced integration plane stress elements with thickness (linear elements)
Impactor	
Dimensions	$300 \times 40 \times 5$ mm^3
Density	8000 kg·m^{-3}
Elastic properties	
Young's modulus	E = 210 GPa
Poisson's ratio	$\nu = 0.3$
Impact conditions	
Speed	20 m·s^{-1}
Contact type	Hard contact (normal and tangential)

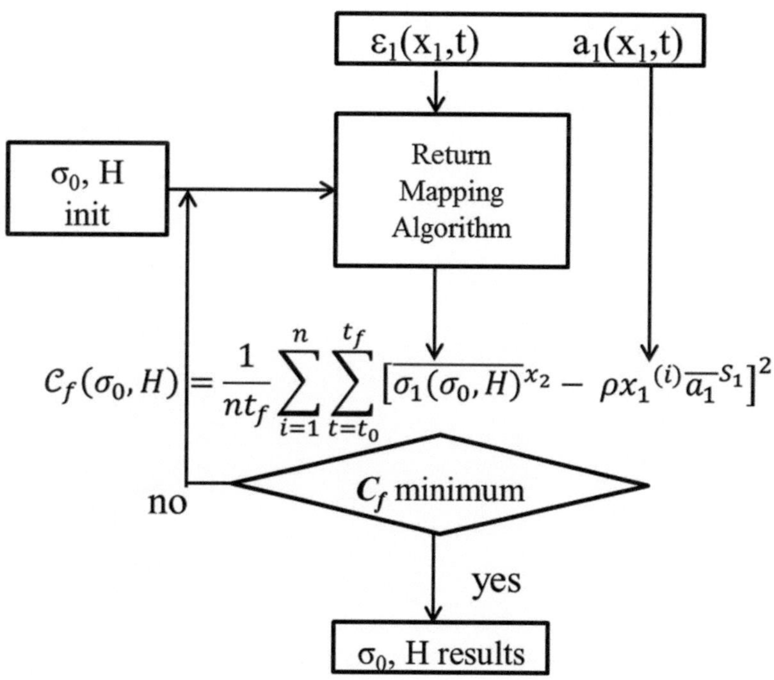

Fig. 10.2 Identification procedure based on the Virtual Fields Methods, further described in [4]

10.3.2 *Test Design*

Investigation of pulse shape: idealized surface loads on one edge of the specimen have been simulated to investigate the effect of the duration and amplitude of the impact pulse on the heterogeneity of the stress and strain states in the specimen.

Investigation of specimen geometry: different specimen shapes and sizes have been investigated to find a configuration that 'activates' the two plastic parameters over as large a portion of the specimen as possible.

10.3.3 Experimental Implementation

From the test design study, a specimen of dimensions 80 mm × 30 mm × 3.6 mm was retained. Al-6061-T6 test specimens are tested. The camera used for the tests is the Specialized Imaging Kirana ultra-high speed cameras. The grid method is used to provide time resolved displacement maps from which strain and acceleration are derived as in [1].

10.4 Results

Current numerical results show that the method is feasible for simple elasto-plastic constitutive models. Figure 10.3 shows the average stress reconstructed from both acceleration fields and strain fields with the return mapping algorithm. Average stress values at different sections and different times generally match with no more than 1.5 % error, despite some higher errors over 5 % at the impacted edge where spatial convergence was not attained. From these data, the cost function C_f expressed in Fig. 10.2 can be minimized following the scheme presented. The results are reported in Fig. 10.4 in the form of values of the cost function for a range of parameters around the reference values. Coarser meshes lead to even better identification results, with improved convergence of the model.

The results of the test design section lead to choose longer specimens in order to increase the proportion of specimen length deformed plastically. The presence of notches or a tapered shape could increase the increase the state of heterogeneity and therefore, improve identification though this has not been tested yet.

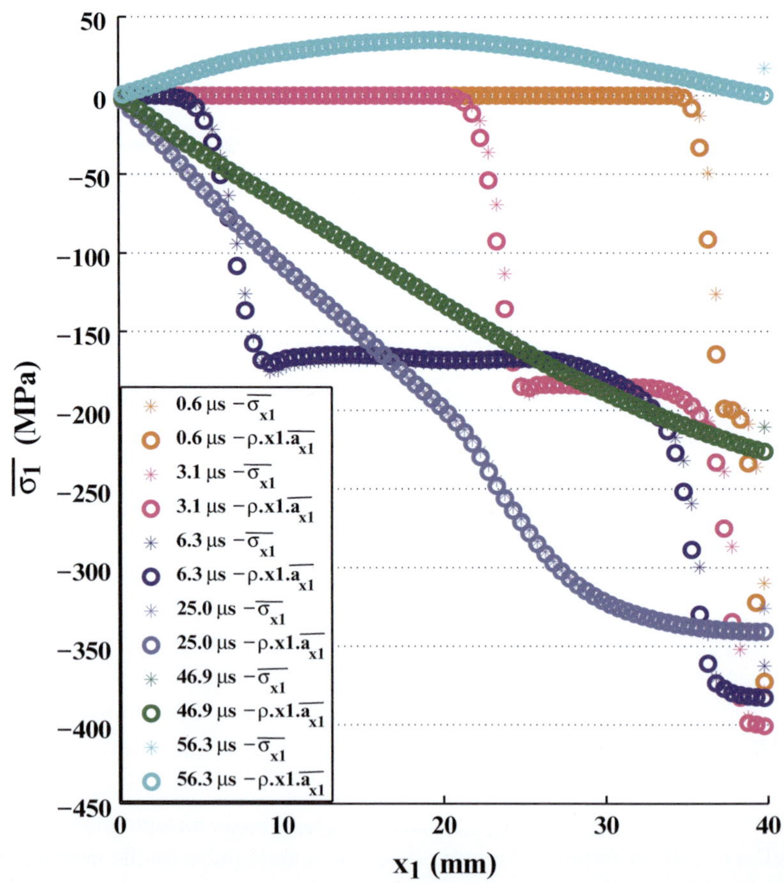

Fig. 10.3 Stress reconstructed along specimen length at different times from both acceleration and the return mapping algorithm

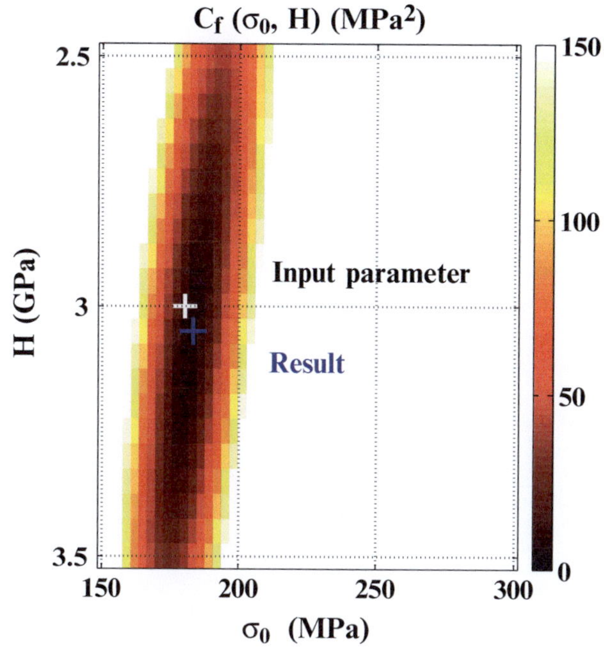

Fig. 10.4 Mapped cost function values at the surrounding of input values of the finite element model

10.5 Conclusion

The procedure developed for the identification of two non-linear elasto-plastic parameters from only strain and acceleration maps processed from ultra-high speed images shows to be feasible on simulated data. The experimental implementation will seek to confirm the validity of the method on the identification of the behaviour of Al-6061. Future work will focus on uncertainty quantification in the measurement via modelling of the imaging technique for strain measurements, as in [5]. This will enable to simulate the test more realistically and provide a good basis to optimize the settings (specimen geometry, loading, frame rate, spatial resolution etc.)

Acknowledgements This material is based on research sponsored by the Air Force Research Laboratory, under agreement number FA8655-13-1-3041. The U.S. Government is authorized to reproduce and distribute reprints for Governmental purposes notwithstanding any copyright notation thereon.

The views and conclusions contained herein are those of the authors and should not be interpreted as necessarily representing the official policies or endorsements, either expressed or implied, of the Air Force Research Laboratory or the U.S. Government.

The authors would like to thank Dr. Jason Foley (AFOSR) and Dr Matt Snyder (EOARD) for supporting this work.

References

1. F. Pierron, H. Zhu, C. Siviour, Beyond Hopkinson's bar. Philo. Trans. R. Soc. A Math. Phys. Eng. Sci. **372**(2023), 1471–2962 (2014)
2. S. Avril et al., Stress reconstruction and constitutive parameter identification in plane-stress elasto-plastic problems using surface measurements of deformation fields. Exp. Mech. **48**(4), 403–419 (2008)
3. M.A. Sutton et al., Determination of elastic-plastic stresses and strains from measured surface strain data. Exp. Mech. **36**(2), 99–112 (1996)
4. M. Grédiac, F. Pierron, Applying the virtual fields method to the identification of elasto-plastic constitutive parameters. Int. J. Plast. **22**(4), 602–627 (2006)
5. M. Rossi, F. Pierron, On the use of simulated experiments in designing tests for material characterization from full-field measurements. Int. J. Solids Struct. **49**(3), 420–435 (2012)

Chapter 11
Experimental Mechanics for Graduate Students

Christian Franck

Abstract This talk provides a description of a current graduate course in experimental mechanics taught at Brown University. This course is designed for graduate students with a strong engineering background who want to learn more about the physical and mathematical principles behind experimentation. It includes the fundamentals of optics, optical interferometry, and imaging systems including Digital Volume Correlation and Particle Image Velocimetry. It will also include signal processing, and spatial and temporal filtering techniques. The goal of the course is to give students a quantitative appreciation of commonly used experimental methods in mechanics, their applications, limitations and theories. Students will be able to apply the theories learned in lectures to carefully designed laboratory experiments. As such the course is intended for all students with interests in continuum mechanics. The purpose of this talk is to give the audience an overview of how experimental mechanics can be integrated into the pedagogy of the classroom.

Keywords Interferometry • Graduate education • Digital image correlation • Diffraction theory • Particle image velocimetry

11.1 Introduction

Integration of research and education, a hallmark mission of the National Science Foundation as well as higher level academic institutions across the country, is imperative for the successful training and mentoring of the next generation workforce of scientists and engineers in a rapidly evolving technological world. Although research and education are often portrayed to be mutually exclusive, they are in many ways more interconnected than it might seem. This is especially true for graduate courses, which offer both the faculty member as well as the students the opportunity to explore active research questions within the context of a classroom. By actively integrating current state-of-the-art experimental and theoretical approaches both parties benefit from the mutually synergistic learning experience. The students become familiar with state-of-the art approaches for addressing mechanics problems, whereas the faculty member has the opportunity to challenge students to tackle actual research problems, and prepare them for their Ph.D. research. This proceeding outlines the approach to familiarize engineering graduate students with the experimental methods commonly employed in the field of continuum mechanics. The described course, ENGN2320, is part of Brown University's course offering for graduate studies in the mechanical engineering graduate curriculum. The goal of the course, through the integration of intuitive-based teaching with concurrent research topics and active research, is to instill a rigorous, physics and math-based understanding of the application of experimental mechanics methodologies to active research questions. The key elements of the course consist of regular lectures providing the theory behind experimentation, as well as case studies of ongoing research questions, and four distinct laboratory modules to allow the students to design and build state-of-the art experiments.

11.2 Course Description

ENGN2320 provides a brief introduction and overview of the fundamental principles of optics, optical interferometry, and imaging systems including Digital Volume Correlation and Particle Image Velocimetry. It also includes mathematical approaches to signal processing, and spatial and temporal filtering techniques. The main components of the course are illustrated graphically in Fig. 11.1. The goal of the course is to give students a quantitative appreciation of commonly used

C. Franck (✉)
Brown University, Providence, RI 02912, USA
e-mail: franck@brown.edu

C. Sciammarella et al. (eds.), *Experimental and Applied Mechanics, Volume 4*,
Conference Proceedings of the Society for Experimental Mechanics Series, DOI 10.1007/978-3-319-22449-7_11

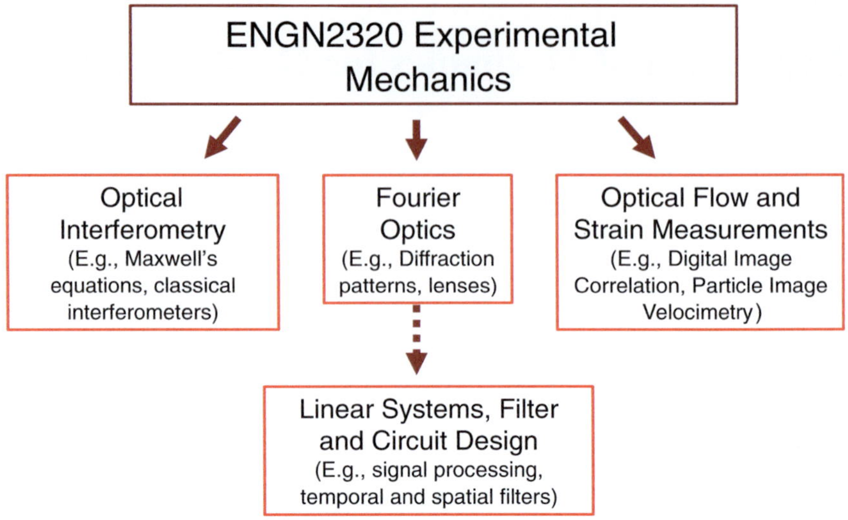

Fig. 11.1 ENGN2320 experimental mechanics course content overview

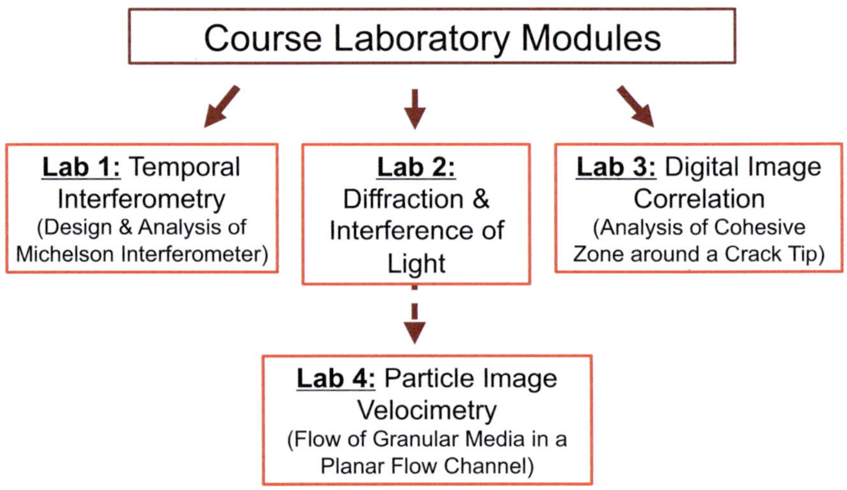

Fig. 11.2 Featured course laboratory modules

experimental methods in mechanics, their applications, limitations and theories. Students will be able to apply the theories learned in lectures to carefully designed laboratory experiments (Fig. 11.2).

11.2.1 Optical Interferometry

The course begins with a brief introduction of Faraday's law of induction, Ampere's circuital law, and Gauss' laws for electromagnetic fields cumulating in Maxwell's equations to finally derive the governing equations for electromagnetic waves [1]. Analogies of Maxwell's equation are drawn to the conservation of mass and momentum in continuum mechanics to provide students that lack a background in optics with simple analogies in mechanics. Several geometric solutions to the wave equation are provided with the central focus of the course on harmonic plane waves. Prior to the construction of the central interference equation of two mono-chromatic interfering beams, students are familiarized with the notion of refraction, optical path length, accumulated phase, wave propagation vector and different in-plane polarization states of the traveling harmonic electromagnetic field waves. Finally, the key result of the first section of the course, namely the two-beam interference detector intensity formulation is derived, as summarized by (11.1):

$$I = I_1 + I_2 + 2\sqrt{I_1 I_2}\hat{m}\,\hat{l}\,\cos(\Phi) \tag{11.1}$$

The first two terms present the amplitudes of the two interfering beams whereas the last expression in (11.1) describes the interference term, where $\hat{m}$ and $\hat{l}$ denote the respective in-plane polarization states, whereas $\Phi = \left(\overline{k_1 \overline{r}} - \overline{k_2 \overline{r}}\right) + \phi_1 - \phi_2$, describes the physical alignment between the two intersecting wave fronts given by the propagation vectors ($\mathbf{k}_1, \mathbf{k}_2$) and their change in accumulated phase, $\Delta\phi$. Via use of (11.1) students learn the construction process of several amplitude splitting interferometers including Michelson, Mach-Zehnder, and Moire interferometers. The section concludes with lab module one (Fig. 11.2) entailing the design and execution of a Michelson interferometer.

11.2.2 Introduction to Fourier Optics

This section of the course, again starting from Maxwell's equations, derives the Helmholtz-Kirchhoff diffraction integral, and eventually arrives at the Fresnel-Kirchhoff diffraction integral [2]. Using this integral relation, students learn to understand that the diffraction of electromagnetic waves can be understood as a series of Fourier Transforms operating on the incident optical disturbance. Simple diffraction patterns through rectangular, circular and periodic apertures are analytically derived, and the physical operations of lenses are introduced along with the notion that a detector image is a simple mathematical and physical convolution of the original object through the imaging optics. The section concludes with lab module two (Fig. 11.2), which teaches students on how to construct and quantify various diffraction patterns in the far-field to their analytical counterparts. In addition, students gain an appreciable understanding of the notion of the optical point spread function (PSF) of an optical setup.

11.2.3 Digital Image Correlation and Particle Image Velocimetry

This section introduces the mathematical framework behind image motion and template matching algorithms that are commonly employed in digital image correlation (DIC) and particle image velocimetry (PIV) code packages [3]. Construction of image pair matching cost functions in the real and Fourier domains are derived for various levels of sophistication [3]. These include use of higher order shape functions (e.g., the deformation gradient, subset curvature effects) as well as various normalization procedures for lighting changes during the image acquisition stage. Construction of sub-pixel algorithms and general DIC sensitivity analyses are provided. Finally, application of DIC in the use of inverse problem solutions and the virtual field method are presented and discussed. This section of the course concludes with the completion of the two laboratory modules three and four (Fig. 11.2), where students will program their own DIC and PIV code packages in Matlab (Mathworks, Natick, MA).

11.2.4 Linear Systems, Filter and Circuit Design

The last section of the course introduces students to the construction of signal *kernels* as a class of linear operators for the integration, differentiation and filtering of various spatial and temporal signals. The section begins with a brief reminder that the generation of diffraction pattern can be represented by an optical transfer function, or kernel, operating on an input signal, or incident optical disturbance [2]. Students gain appreciation in the proper treatment of signal differentiation in the presence of noise and the use of convolution kernels, which allow for quick execution via FFTs in the Fourier domain. In addition to digital filtering and filter design, the course provides an overview of the construction of simple analog LRC and operational amplifier circuits that function as commonly employed signal filters including the Butterworth filter.

11.3 Course Laboratory Modules

The course is divided into four distinct laboratory modules (lab 1–4) in which students gain significant and valuable experience in the design, execution and analysis of experimentation (Fig. 11.2). Particular attention is paid to the quantification of noise in each experiment, the analytical validation of the experimental design, and the instrument's sensitivity analysis. Students work in small groups (2–3 students per group) and submit individual pre-lab reports, as well as one joint, 5-page lab report. Professional formatting following peer-reviewed journal templates is required, and the use of LaTex as the main text editor is strongly encouraged. Students submit all of their work electronically to the course website in forms of pdf documents.

11.4 Student Assessment

Students are assessed via the following course grading scheme (out of 100): Class participation (10), Lab modules (60), and the Final Exam (30). The Final Exam is cumulative encompassing all of the material covered throughout the course.

11.5 Course Case Study Example

The following adapted excerpt (showing the introduction, theory and discussion sections in italicized fonts) provides an example of the typical student-drafted laboratory report as part of the course. The specific lab report addresses lab module 2.

11.6 Experimental Mechanics ENGN2320 Lab-2 Diffraction Lab

Xiaopeng Lai, Max Monn, and Insun Yoon

11.6.1 Introduction

Chemical or crystallographic analysis such as infrared spectroscopic analysis, visible spectroscopic analysis and X-ray crystallography are examples that the diffraction phenomenon is applied to. Likewise, diffraction is widely used in current research.

Fraunhofer diffraction is a diffraction phenomenon when the light source and image plane are in infinite distance. Practically when the observation distance (the distance between aperture and image plane) is much larger than diffraction pattern and wavelength of light, we call the diffraction Fraunhofer diffraction. Fresnel diffraction is different from Fraunhofer diffraction at the point that the observation distance is finite. In this lab, Fraunhofer diffraction patterns from different aperture shapes were obtained from an experimental setup. The intensity of a laser beam was modeled as a Gaussian distribution and under Fraunhofer condition, point spread function was constructed and diffraction patterns are computationally obtained. As a result, actual and computational diffraction patterns are compared. Effect of intensity distribution, effect of aperture shape and attempt to measure observation distance was further discussed.

11.6.2 Theory

11.6.2.1 Gaussian Distribution

Here, the method for fitting Gaussian distribution from given data is suggested. The Gaussian distribution is expressed as:

$$I(x) = ae^{-\frac{(x-b)^2}{2c^2}} \tag{11.2}$$

where a corresponds to the amplitude and b corresponds to the center of distribution. By taking the natural log of both sides, (11.2) *can be converted to:*

$$ln(I(x)) = ln(a) - \frac{(x-b)^2}{2c^2} \tag{11.3}$$

which is in the form of a parabola. From data ln(I(x)) and using the least square method via a polynomial fitting function, the unknown values of a, b, c are determined in terms of the parameters in a quadratic (11.3).

11.6.2.2 Fraunhofer Diffraction

Features of Fraunhofer Diffraction are listed below. The general form of Fresnel Diffraction is expressed by:

$$E_0(\underline{x}) = \frac{k}{2\pi i}\int_{\xi} E_0^I\big|_{\xi}\left\{\frac{\cos\left(-\underline{r_1}\cdot\underline{e_1}\right) - \cos\left(\underline{r_1}\cdot\underline{e_1}\right)}{2}\right\}\frac{e^{i\underline{k}\cdot\underline{r}}}{r}d\underline{\xi} \tag{11.4}$$

where r is defined as $\sqrt{Z^2 + (x_1-\xi_1)^2 + (x_2+\xi_2)^2}$. *Here* x_1, x_2 *represent the coordinates of image plane,* ξ_1, ξ_2 *represent the coordinates of data plane, and Z represents the observation distance (the distance between the two planes). In case of Fraunhofer diffraction, the observation distance Z is taken to be very large. From this,* $\frac{cos\left(-\underline{r_1}\cdot\underline{e_1}\right) - cos\left(\underline{r_1}\cdot\underline{e_1}\right)}{2}$ *can be approximates to* 1, *and the Fresnel diffraction integral is simplified to:*

$$E_0(\underline{x}) = \frac{k}{2\pi i}\int_{\xi} E_0^I\big|_{\xi}\frac{e^{i\underline{k}\cdot\underline{r}}}{r}d\underline{\xi} \tag{11.5}$$

$E_0\big|_{\xi}$ *in* (11.5) *is product of the incident wave* $\hat{E}_0\left(\underline{\xi}\right)$ *and transmittance function* $t\left(\underline{\xi}\right)$. *For a monochromatic wave,* $\hat{E}_0\left(\underline{\xi}\right)$ *is determined by the intensity distribution from the source. In this lab, the intensity distribution was fit to a Gaussian distribution. The transmittance function* $t\left(\underline{\xi}\right)$ *is defined as the shape of the aperture in the data plane. That is,* $E_0\left(\underline{\xi}\right)$ *is obtained by multiplying the Gaussian distribution and aperture shape function point by point. The final intensity in the image plane can be calculated as:*

$$I_0(\underline{x}) = E_0(\underline{x})\cdot E_0(\underline{x})^* = \mathscr{F}\left\{E_0\left(\underline{\xi}\right)\right\}\cdot\mathscr{F}\left\{E_0\left(\underline{\xi}\right)\right\}^* = \mathscr{F}\left\{\hat{E}_0\left(\underline{\xi}\right)\cdot t\left(\underline{\xi}\right)\right\}\cdot\mathscr{F}\left\{\hat{E}_0\left(\underline{\xi}\right)\cdot t\left(\underline{\xi}\right)\right\}^* \tag{11.6}$$

Finally the point spread function (PSF) is defined as the normalized intensity.

$$PSF \hat{=} \frac{I_0}{I_{max}} \tag{11.7}$$

11.6.3 Discussion

11.6.3.1 Gaussian vs. Delta Source Intensity Profile

During the calculation of the theoretical PSF and diffraction pattern, the beam intensity profile was computed as a Gaussian distribution. Relative to the size of the slits analyzed, the effective beam width was large. Intuitively, this implies that a

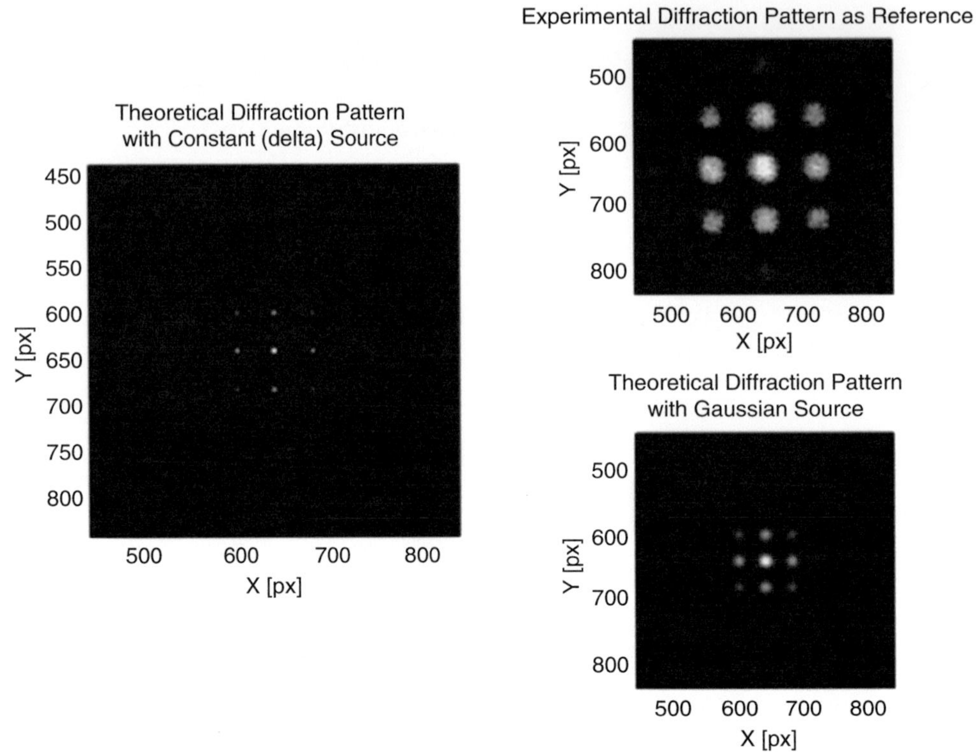

Fig. 11.3 (***a***) *Diffraction pattern of a constant source incident on a 5 × 5 lattice of circular holes.* (***b***) *Experimentally observed diffraction of light through the lattice-type aperture plate.* (***c***) *Diffraction pattern of a source with Gaussian intensity distribution incident on a 5 × 5 lattice of circular holes*

similar result could be achieved by replacing the Gaussian intensity profile with a constant intensity over the effective width of the beam. To test the validity of this assertion, we calculated the diffraction pattern for the lattice-type slit model using point-by-point multiplication of the aperture function by a constant intensity value rather than the Gaussian distribution. The result of this computation, shown in Fig. 11.3, contradicts our original assertion.

The diffraction pattern of this configuration (Fig. 11.3a) shows sharpening of the image features when compared to the use of a Gaussian beam profile (Fig. 11.3c). This observation can be explained by the way that sizes and shapes map from real space to fourier space. The circular truncation of the entire lattice pattern manifests as small Gaussian-like features (blurred dots) in the image plane (Fig. 11.3c).

11.6.3.2 Source Centering

In the experiment we assumed that the beam was centered with respect to the aperture used. For a source of invariant spatial intensity (constant over beam width), the resulting diffraction pattern would not change with small deviations from assumption. However, the Gaussian distribution of beam intensity results in increased sensitivity to beam location (Fig. 11.4). In order to assess this sensitivity, the center of the Gaussian source function was shifted by $\frac{w}{2}$ *where w is the width of the aperture being analyzed. The resulting PSF and diffraction pattern for the double slit configuration (Fig. 11.4b and c respectively), shows both attenuation of higher order terms and blurring of image features.*

By spatially shifting the Gaussian source function, the widths of the aperture features are modified, increasing the effective width of one slit while truncating the other. Once again, the resultant blurring of the diffraction pattern can be explained intuitively by considering the way that features map from the spatial domain to the frequency domain. By decreasing the effective width of one slit, larger scale features are created in the frequency domain. Furthermore, the symmetry of the function on the data plane is lost, which results in the loss of well-defined directionality and sharpness in the image plane.

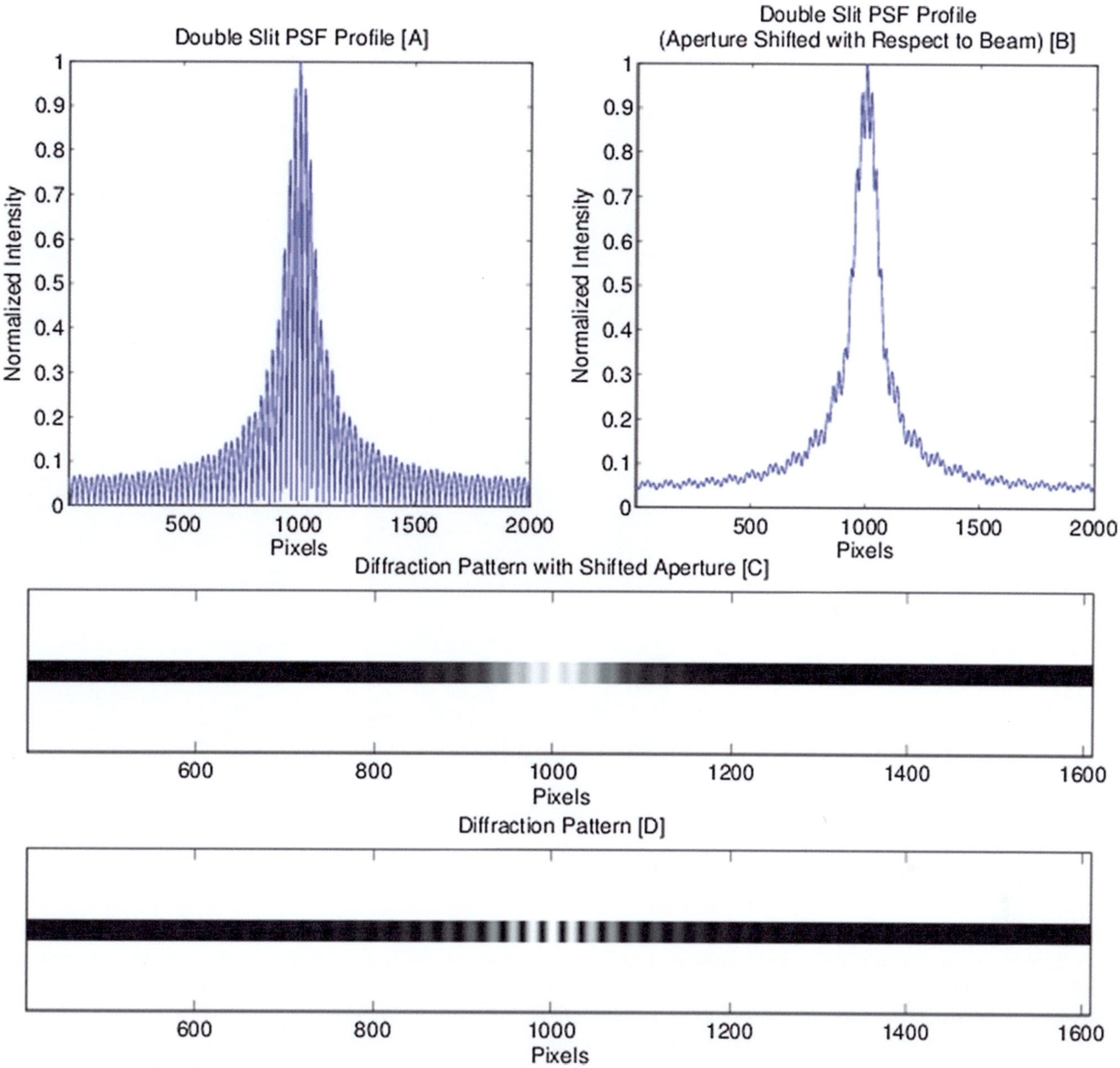

Fig. 11.4 *(**a**) Intensity profile of PSF using double slit aperture with beam centered over the aperture (**b**) intensity profile of PSF using double slit aperture with beam center offset by* $\frac{w}{2}$ *(**c**) diffraction pattern as a result of shifted beam center (**d**) diffraction pattern corresponding to a centered beam*

11.6.3.3 Effect of Aperture Shape on Diffraction Pattern Symmetry

We can use diffraction through a lattice as a means to explore the mapping of features in the spatial domain to features in the frequency domain by modifying the shape of the elements in the lattice and observing the resultant diffraction pattern. Although present, the circular symmetry of diffraction through a lattice of circles (Fig. 11.5) is not immediately obvious. To better demonstrate the large scale symmetries on the image plane due to the shape of small features on the data plane, we calculated the theoretical diffraction patterns through lattices composed of cross-like and annular features (Figs. 11.5 and 11.6 respectively).

In Fig. 11.5 one can observe a cross-like symmetry in the diffraction pattern aligned with the vertical and horizontal axis. Similarly, in Fig. 11.6 there is annular symmetry of the diffraction pattern. In both instances, details in the shape of the aperture were mapped to whole-image symmetries in the calculated diffraction patterns.

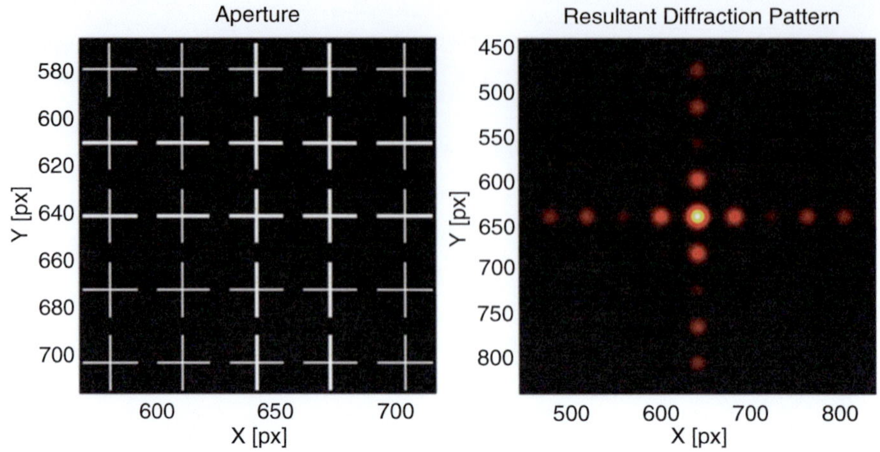

Fig. 11.5 *Aperture of a 5 × 5 cross-type lattice and the resulting theoretical diffraction pattern showing large scale cross-like symmetry*

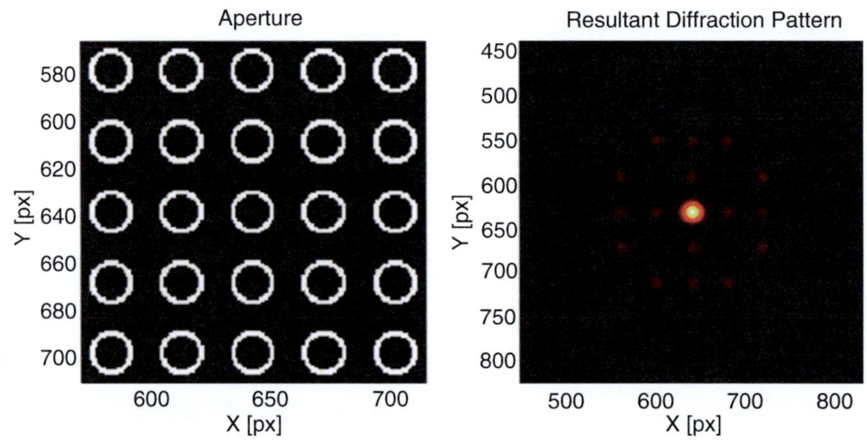

Fig. 11.6 *Aperture of a 5 × 5 lattice populated by annular shapes and the resulting theoretical diffraction pattern showing a bullseye symmetry*

11.6.3.4 Observation Distance

Actual distance from the aperture to CCD camera was 432 *mm, and calculated distance was* 531 *mm (Sect. 5.5). This result shows* 22.9 *% of error. Fraunhofer approximation can be presumable for the reason of the error. From the assumption of infinite observation distance, was established. In practice, the observation distance is not infinite. In this lab, the ratio between the distance of the first minimum from single slit diffraction pattern was* $\frac{0.97}{100}$*. i.e. the distance in image plane was about 1 % of the observation distance. With this ratio, it is not assumable for the observation distance to be infinite compared to the distance in the image plane. Miscellaneous measurement errors also can be arisen for the explanation for the error. Dimension of single slit and the distance of the first minimum may contain error and effect on the observation distance calculation.*

*This result implies that with proper analysis of error, accurate measurement and known observation distance, this system can be used for measuring wavelength of a certain monochromatic light by means of (*11.8*):*

$$\lambda = Lsin\left(\frac{|x|}{Z}\right) \tag{11.8}$$

11.7 Conclusions

This proceeding provides a brief summary of the course description, and the experimental methodologies that are introduced to graduate students taking ENGN2320 Experimental Mechanics at Brown University. The goal of this proceeding and the accompanying talk is to provide an example of the integration of the teaching of fundamental experimental mechanics-principles with state-of-the art research methodologies. Going forward it is important to note that a course such as the one presented here is always evolving, and strives to be the target of continuous improvement. As such, the contemporary flavor of the course per given time are of lesser importance compared to a rigorous theoretical and physical foundation (e.g., Maxwell's equations) that motivate the experimental problem at hand. In this light, the integration of teaching and research follows naturally as one employs the same methods of analytical validation, design and analysis in the lab and classroom. The syllabus to the course can be found online (http://franck.engin.brown.edu/~christianfranck/ENGN2320.pdf).

Acknowledgements I would like to thank Professor Michael Mello at CalTech for graciously providing most of the content of the interferometry teaching section. I would also like to thank him for his technical feedback regarding the course development and content in general. I would like to acknowledge Professor Paul Dimotakis at CalTech, on whose notes my section on signal processing and filtering techniques are based on. I would like to acknowledge my colleagues at Brown, Professors Guduru, Kim, Breuer, and Zia who I have worked closely with to develop the course. I would like to acknowledge Xiaopeng Lai, Max Monn, Insun Yoon as the graduate authors of the case study example presented in this proceeding.

References

1. E. Hecht, *Optics*, 4th edn. (Addison-Wesley, Boston, 2001)
2. J. Goodman, *Introduction to Fourier Optics*, 3rd edn. (Roberts and Company, Englewood, 2004)
3. M.A. Sutton, J.J. Orteu, H. Schreier, *Image Correlation for Shape, Motion and Deformation Measurements: Basic Concepts, Theory and Applications*, 1st edn. (Springer, Heidelberg, 2009)

Chapter 12
Bridge Foundation Depth Estimation Using Sonic Echo Test

Saman Rashidyan, Tang-tat Ng, and Arup Maji

Abstract Nondestructive Tests are capable of determining unknown characteristics of the bridge foundations especially the depth of the piles and drilled shafts. Sonic Echo—Impulse Response (SE/IR) is one of the methods used in determining the depth of wooden and concrete piles and drilled shafts. When the top of the pile is inaccessible, sonic waves in the pile would be generated by either horizontally striking the side of the pile or striking vertically on a block attached to the pile surface. In the present research, in order to have reasonable criteria to determine the depth of piles, the data acquired by the sensors should be interpreted properly. In addition, while a sensor is mounted somewhere far from the top of the pile, the sensor receives different echoes at different times and reflection of the waves from the top of the piles and superstructures makes the case more complicated. In order to develop a methodology for testing and Interpretation, tests were conducted and analyzed in the laboratory and on railway bridges in New Mexico.

Keywords Sonic Echo • Impulse response • Pile • NDT • Sensor

12.1 Introduction

Different Non-destructive Testing (NDT) methods have been used to control the quality of deep foundations during construction process [1–3]. Various NDT techniques have been developed over the past two decades towards condition assessment of civil infrastructure [4, 5]. A more compelling need has arisen for the methods to determine the conditions of unknown foundations. The Sonic Echo (SE)/Impulse Response (IR) method is an inexpensive and simple method to determine the conditions of unknown foundations. They are normally used to test existing piles [2].

A typical velocity-time response plot for the SE test is shown in Fig. 12.1. The echo is the reflection from the pile toe. The time difference between the impulse and the echo is used to determine the length of the pile.

The stress wave is transmitted through the pile at a velocity (v). The time lapse, t, is obtained between the hammer impulse and the arrival of the echo to determine the length L of the pile:

$$L = \S t^{*} v \tag{12.1}$$

Frequency-domain analysis can also be performed in the IR test. The length of the pile is obtained from the frequency response:

$$L = \S v/f \tag{12.2}$$

Δf is the difference in the frequency between the resonant responses. Although the test procedure and the theory seems straightforward, precise interpretation is required in the presence of different factors that might affect the results. In order to develop a methodology for testing and Interpretation, tests were conducted on a typical timber column as well as timber foundations of a railway bridge in New Mexico. The results were analyzed and the effect of different parameters such as the hammer tip and strike direction was investigated.

S. Rashidyan (✉) • T.-t. Ng • A. Maji
Civil Engineering Department, University of New Mexico, Albuquerque, NM, USA
e-mail: srashidyan@unm.edu; tang@unm.edu; amaji@unm.edu

C. Sciammarella et al. (eds.), *Experimental and Applied Mechanics, Volume 4*,
Conference Proceedings of the Society for Experimental Mechanics Series, DOI 10.1007/978-3-319-22449-7_12

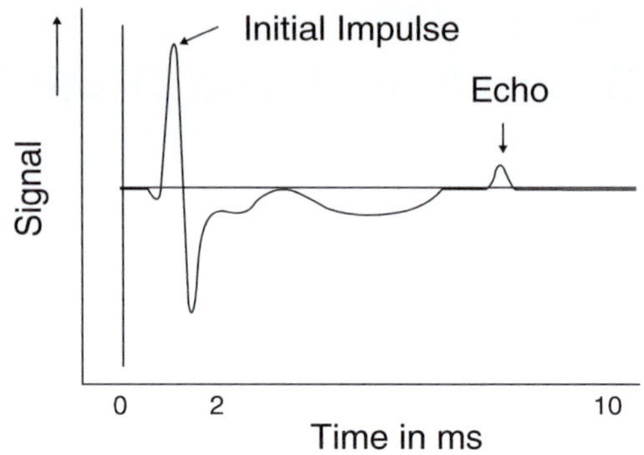

Fig. 12.1 Typical examples of Sonic Echo method graph

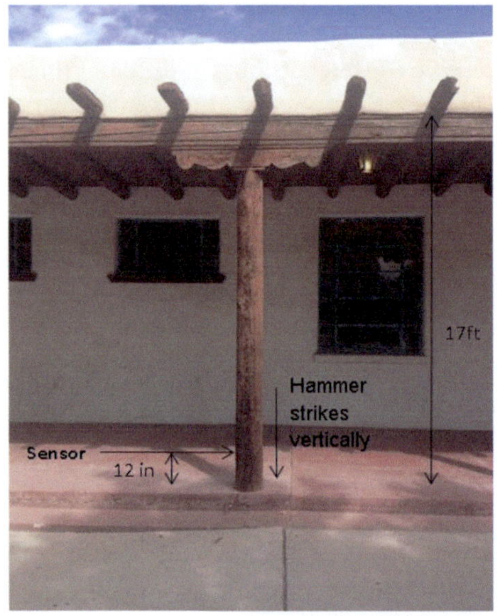

Fig. 12.2 The source and receiver locations on the wooden column

12.2 SE Tests on a Wooden Column

Prior to conducting in-field tests, SE tests were carried out on a timber decorative column of the Biology Annex building on the University of New Mexico (UNM) campus. The column and the locations of the strike and sensor (an accelerometer mounted vertically) are shown in Fig. 12.2. SE tests have been carried out with a hard tip hammer. The source is produced by striking the hammer vertically at the concrete pavement next to the column. Although the hammer is not striking on the column, the reflected wave will travel up the column in this case.

Figure 12.3 shows the velocity time-history (integrated from the acceleration time history). The low-frequency fluctuations in accelerometer graph indicate the reflection of the wave that propagates along the column. The Initial wave and its subsequent echo are recognizable on the graph.

The time between the Initial wave and Echo can be identified (Δt) which is approximately equals to 3 ms and could be used to measure the length as follows:

$$L = \frac{v\text{æ}t}{2} = \frac{10,000\text{æ}3\text{æ}10^{3}}{2} = 15\,\text{ft}$$

v=10, 000 ft/s, the compressive wave velocity in wood.

The measured length is close to the actual length of the column.

The Impulse Response (IR) analysis of the data supports the above calculation. A typical frequency-domain plot of the accelerometer is shown in Fig. 12.4.

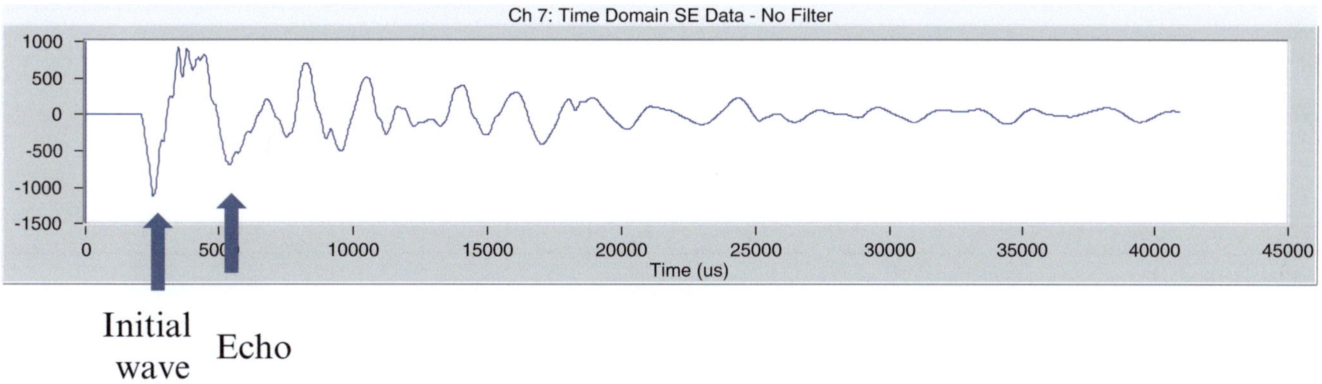

Fig. 12.3 Velocity time-history graph

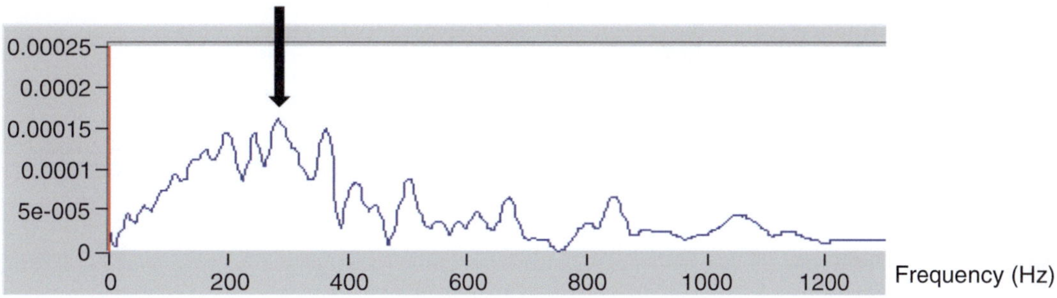

Fig. 12.4 Frequency-domain plot

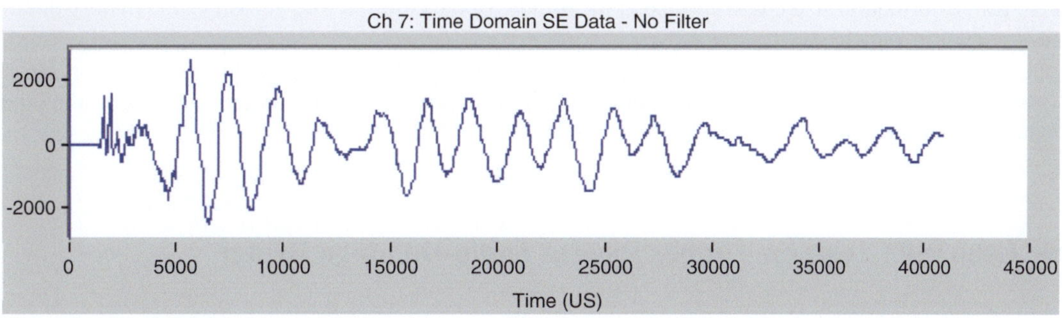

Fig. 12.5 Velocity time-history graph

The interpretation is not such straightforward. This may be due to the fact that the direct wave is not started on the column. The first peak region which is the lowest dominant frequency of vibration is shown by the arrow in the Fig. 12.4. The pertinent frequency is approximately equal to 300 Hz. Using this frequency the measured length will be:

$$L = \frac{v}{2^*f} = \frac{10,000}{2^*300} = 16.7\,\text{ft}$$

The calculated length is similar to that determined by the time domain analysis, and the difference illustrates the extent of accuracy in SE/IR methods.

Two more SE tests were conducted to investigate the effect of a horizontal strike. In the first test a horizontal strike was applied on the column at a point close to the sensor. The velocity time history obtained from the accelerometer and the amplitude time history of the hammer are shown in Figs. 12.5 and 12.6 respectively.

Figure 12.5 shows that the Impact close to the sensor results in spurious reflections with frequencies close to that of the hammer. Likely the sensor receives the local direct vibration due to hammer rather than longitudinal vibrations. Therefore to

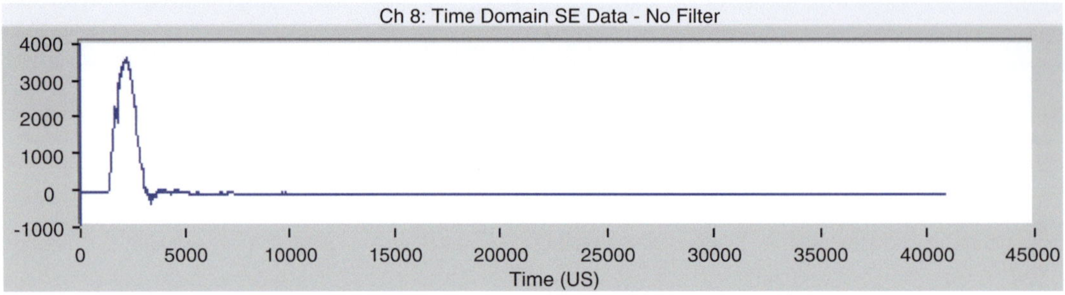

Fig. 12.6 Amplitude time-history graph of the hammer

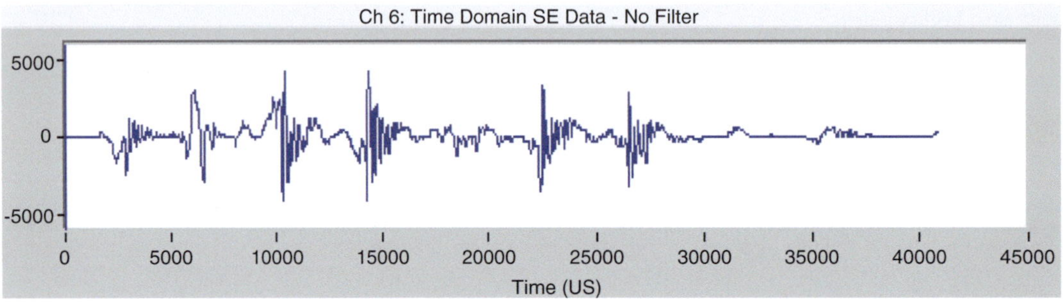

Fig. 12.7 Velocity time-history graph

avoid mentioned problem the sensor should be placed at a sufficient distance with respect to the source. Since the initial wave and the Echo related to the reflections along the column are not recognized, the length of the column could not be measured in this case.

In another SE test, a horizontal strike was applied at the middle of the column. The velocity time history obtained from the accelerometer is shown in Fig. 12.7.

The overall shape of the graph in Fig. 12.7 shows that the lateral impact near the middle of the column creates unwanted resonances. Therefore due to interfering of significant lateral vibrations with the longitudinal vibrations, the interpretation of the results would be quite difficult.

12.3 Sonic Echo (SE) Tests on Timber Piles of Santo-Domingo Bridge

After conducting preliminary tests, a bridge named "Santo Domingo" was provided by NMDOT for the in-field SE test investigations. The bridge was supported by timber piles. The coordinates of the bridge are at 35.488752 N, 106.385058 W. Figure 12.8 shows the street view of the bridge.

SE tests were conducted on the timber pile. Vertical hammer strikes on a metal block attached to the pile were used as the source. The accelerometer was mounted on a metal block vertically. The test setup and test specifications are shown in Fig. 12.9 and Table 12.1 respectively.

The results of the velocity time-history graphs are represented. The velocity time histories of the Test 1~6 and the initial waves and Echoes are shown in Fig. 12.10.

The Test 1 result is used to demonstrate the determination of the embedment depth of the pile (L_B). As shown in Fig. 12.10, the Impulse and Echo points can be identified easily. The embedment depth of the pile can be determined as:

$$t = 79002400 = 5500\,\text{s}$$

$$L_B = \frac{v\text{æ}t}{2}\,2.5 = \frac{10,000\text{æ}5500\text{æ}10^6}{2}\,2.5 = 25\,\text{ft}$$

Fig. 12.8 View of the Santo Domingo bridge

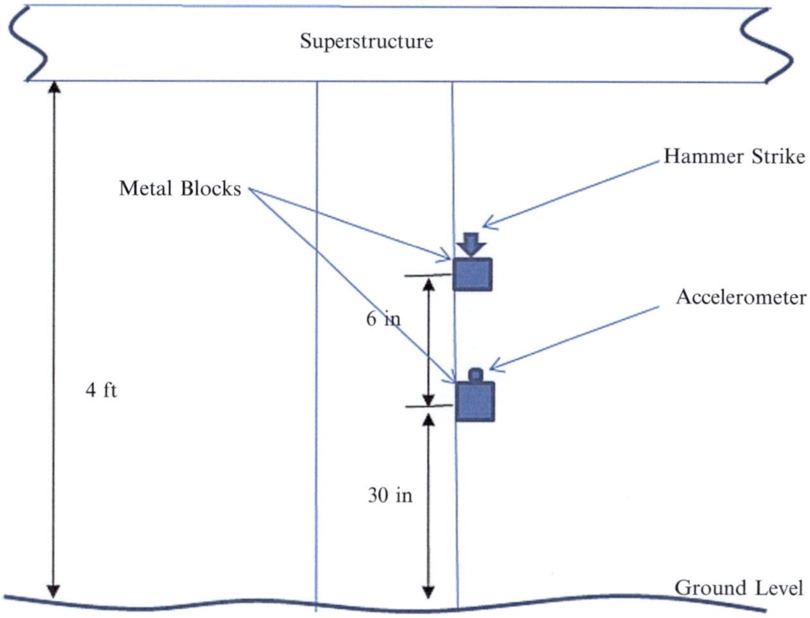

Fig. 12.9 SE tests setup at Santo Domingo bridge (not to scale)

Table. 12.1 SE tests on Santo Domingo bridge specifications

Test no	Hammer tip	Direction of hammer strike
1	Hard	Vertical
2	Hard	Vertical
3	Hard	Vertical
4	Medium-hard	Vertical
5	Medium-soft	Vertical
6	Hard	Horizontal

$$v = 10,000 \text{ ft/s}$$

, the assumed wave velocity

Table 12.2 shows the results of all tests. As shown in Table 12.2, the result is affected by the hammer tip type and strike direction. Good results are found when using a hammer with hard tip (Tests 1, 2, and 3). The results are not good when using a hammer with softer tip. The general embedment depth of the Pile is about 25 ft. The accuracy of determining the embedment depth depends on the clarity of Impulse and Echo in the graphs. More distinct Impulse and Echo points yield more confidence on the estimated length.

The Impulse Response (IR) analysis of the data supports the above calculation. The mobility (frequency-domain) plot of the acceleration for tests 1 and 3 are shown in Figs. 12.11 and 12.12.

Fig. 12.10 The velocity time histories of the Test 1–6

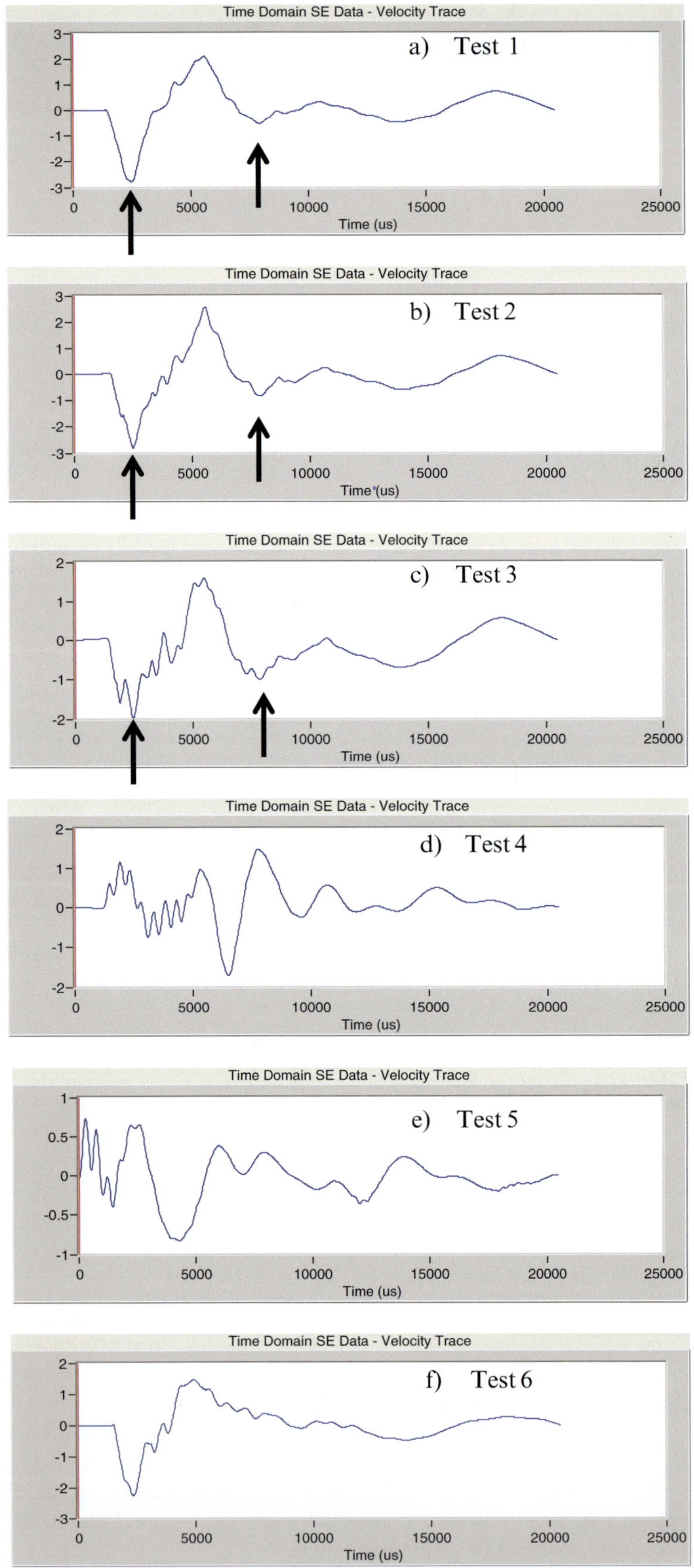

Table. 12.2 Calculated wave periods and pile lengths for tests 1–6

Test no	Clarity of impulse and echo	Δt (μs)	Embedment depth (ft)
1	Good	5500	25
2	Good	5420	24.6
3	Good	5400	24.5
4	Bad	Not clear	Not clear
5	Bad	Not clear	Not clear
6	Bad	Not clear	Not clear

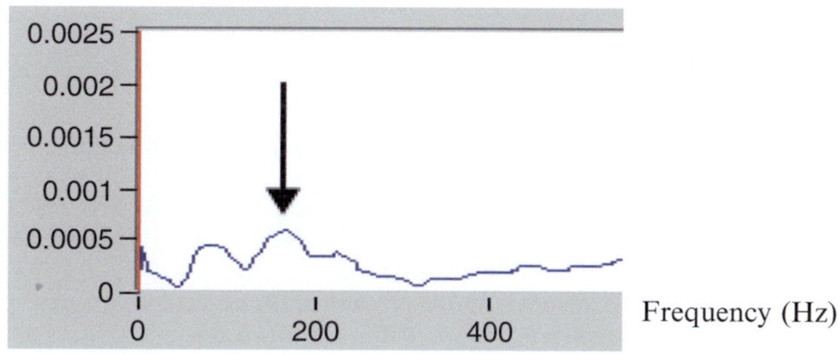

Fig. 12.11 Mobility plot of Test 1

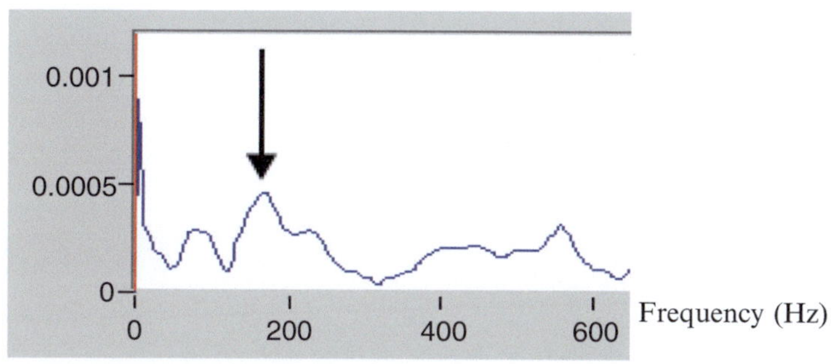

Fig. 12.12 Mobility plot of Test 3

The first peaks which are the lowest dominant frequency of vibration are shown by the arrows in the Figs. 12.11 and 12.12. The lowest dominant frequency is approximately equal to 170 Hz; subsequent peaks are due to the impulse duration of the hammer strike and numerical conversion to frequency domain. Using the 170 Hz frequency the measured length will be:

$$L = \frac{v}{2^*f}2.5 = \frac{10,000}{2^*170}2.5 = 26.9\,\text{ft}$$

The calculated length based on the lowest frequency is close to that of determined from the time domain analysis. Therefore the buried length of the pile could be determined using both time-domain and frequency domain analysis.

12.4 Conclusions

The results of the primarily SE tests on the wooden column indicate that:

- Useful SE tests can be conducted by indirect striking (not striking on the pile).
- The frequency analysis of the data could assist the time domain analysis in determining the length of the column.
- The Impact close to the sensor results in spurious reflections with frequencies close to that of the hammer.
- The Lateral Impact near the middle of the column creates unwanted resonances. In the presence of significant lateral vibrations interpretation of the graphs would be quite difficult

The results of the SE testing at the Santo Domingo bridge provided observations such as:

- The frequency analysis of the data could assist the time domain analysis in determining the length of the column
- Hammer tip type, strike direction, and sensors placement affect the result.
- Softer tips does not work well with timber piles
- Horizontal strike produces results for which interpretation is quite difficult.

Acknowledgements The writers would like to thank the NMDOT project technical panel (Michelle Mann and Dr. Thiet Nguen) for their valuable suggestions during the previous technical meeting. The valuable service and time of Ms. Keli Daniell, who is the project manager, and Mr. Bob Meyers, who is the sponsor of this project are greatly appreciated.

References

1. A.G. Davis, C.S. Dunn, From theory to experience with the nondestructive vibration testing of piles, Paper 7764. Proc. Inst. Civ. Eng. **57**(2), 571–593 (1974)
2. A.G. Davis, Nondestructive evaluation of existing deep foundations. ASCE J. Perform. Constr. Facil. **9**, 57–74 (1995)
3. L.D. Olson, F. Jalinoos, F. Aouad, *Determination of unknown subsurface bridge foundations, Geotechnical guideline no. 16, a summary of the NCHRP 21-5 interim report*, FHWA (1998)
4. G.A. Washer, Developments for the nondestructive evaluation of highways bridges in the United States, in *Proceedings of Non-Destructive Testing in Civil Engineering*, vol. 2, (NDT-CE '97) (Liverpool, 1997) pp. 543–552
5. S.K. Kaushik, A.K. Maji, S.P. Shah (eds.), *Nondestructive Evaluation of Concrete: US-India Workshop* (Indian Concrete Institute, Roorkee, 1997), p. 280

Chapter 13
Detonation of Small Charges Buried in Cohesionless Soil

Eve Roger, Benjamin Loret, and Jean Paul Calvel

Abstract As a preliminary to an experimental program involving larger charges, the detonation of 100 g of the explosive C4 has been analyzed by varying several soil and geometry parameters.

The particle size distribution, mass density and water content of the tested soils have been controlled before and after the shots. Particular emphasis was laid on the influence of the water saturation, and tests were performed on dry and saturated soils. Mass density and water content were measured both by densitometry and by standard laboratory test methods.

The preparation of the soil in the tank, including the protocols of filling the tank with the soil, wetting and compacting steps, has been standardized to improve repeatability. Three depths of burial have been considered. The propagation of the shock wave is visualized via shadowgraphy, which also provides indications on the cone of ejection of the soil particles and ejecta, despite the sand dust.

The speed of propagation of the pressure wave and its time course are recorded through five gauges located on a shaft at the vertical of the charge. The vertical force induced by the explosion is also recorded by a transducer placed at the bottom of the tank. The craters left by the explosion are measured by a 3D scanner.

Keywords Buried explosion • Shadowgraphy • Crater • Shock wave • Water content

13.1 Introduction

When an aerial detonation occurs, a shock wave which is a discontinuity of pressure, density and particle velocity propagates. It drives a sudden pressure increase which occurs within microseconds. The intensity of the pressure peak is all the more important as the location of the measure is close to the explosive.

Soil can be defined as a three phase medium composed of grain particles and voids filled with air and water. A dry soil contains no water while air has been expelled in a saturated soil. Between these two extreme states, the soil is said to be partially saturated, or equivalently, non saturated. Landmines can be found in any type of soil. The state of the soil depending on its properties such as the density and the water content influences the energy transmitted by the mines. The study here focuses on recording the aerial pressures and shock wave propagation due to small quantities of explosives buried at several depths in dry and saturated standardized soils. The setting of experiments is adapted from the works of [1, 2].

A standardized soil, the STANAG soil is used. This is a sandy gravel defined by the NATO document referred to as AEP 55 [3] which imposes three requirements on the granulometry curve: 100 % must pass the 40 mm sieve, between 60 and 40 % must pass the 5 mm sieve, while maximum 10 % must pass the 80 μm sieve. By definition, it must be saturated, with a required density of 2200 ± 100 kg/m^3. As for the water content, the humidity level must be 1.5 % below the Proctor optimum. In our tests, the granulometry of the soil fits the STANAG requirement while various densities and water contents are tested to assess their influence on the air pressures and speed of propagation of the shock wave. Twelve tests were shot in two steps: the five first shots were carried out in April 2013 while the five last ones were shot in January 2014. More details of these experiments performed at DGA TT are reported in [4].

E. Roger (✉) • J.P. Calvel
DGA Techniques terrestres, Rocade Est, Echangeur de Guerry, Bourges Cedex 18000, France
e-mail: eve.roger@intradef.gouv.fr

B. Loret
Laboratoire Sols Solides Structures, Domaine Universitaire, BP53, Grenoble Cedex 9 38041, France

C. Sciammarella et al. (eds.), *Experimental and Applied Mechanics, Volume 4*,
Conference Proceedings of the Society for Experimental Mechanics Series, DOI 10.1007/978-3-319-22449-7_13

13.2 Overview the Experimental Setup

The experimental campaign targets the detonation of small explosive masses (100 g of C4) buried at three different depths (0, 3 and 8 cm) in dry and in saturated STANAG soil. The propagation of the shock wave, the pressures and the resulting craters are recorded. The setup shown in Fig. 13.1 is composed of a soil tank in which the explosive is buried, a shaft where pressure gauges are mounted and a reflective panel.

A soil tank of small dimensions (1 × 1 ×0.5 m) has been chosen in order to prepare the smallest quantity of soil possible to ensure sufficient repeatability while avoiding spurious reflections from the edges. Two moisture contents are tested: the soil (stored outdoors) is either tested with its natural water content, or saturated.

The charge is a 20 mm height cylinder of C4 with a diameter of 64 mm. It is surrounded radially and on the bottom by polystyrene with a density of 31 kg.m^{-3} and a thickness of 25 mm. The polystyrene is required to prevent spreading because C4 does not have enough stiffness. The explosive charge is initiated at the bottom in the center.

Pressures can be measured in two different ways depending on the orientation of the gauges as sketched on Fig. 13.2. The pressure gauge may be parallel to the direction of the shock wave: in this case, the measure is said to be a side-on pressure. If the pressure gauge is embedded in some rigid material and positioned perpendicularly to the direction of the shock wave, the measured pressure is called a reflected pressure. For the experiments reported here, five side-on pressure gauges were placed on a 2 m long crossbar as in [1]. The lowest gauge was positioned at 30 cm above the soil surface while the four other gauges were placed above every 40 cm. To prevent drifting signals due to heat and light from the explosion, some pressure gauges are covered with fat and aluminium powder (Fig. 13.3).

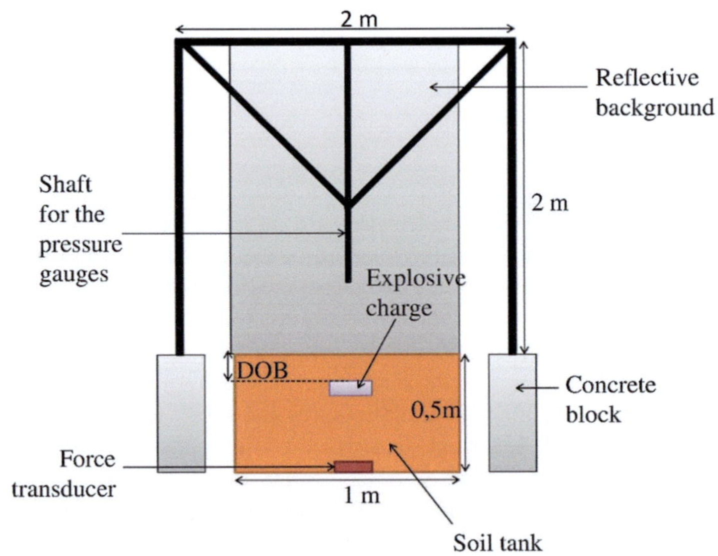

Fig 13.1 Experimental setting for experiments using small charge. The size of the soil tank is chosen as small as possible but still large enough to optimize repeatability and avoid boundary effects. Three depths of burial (DOB) are tested. Pressures gauges are mounted on a shaft at the vertical of the explosive

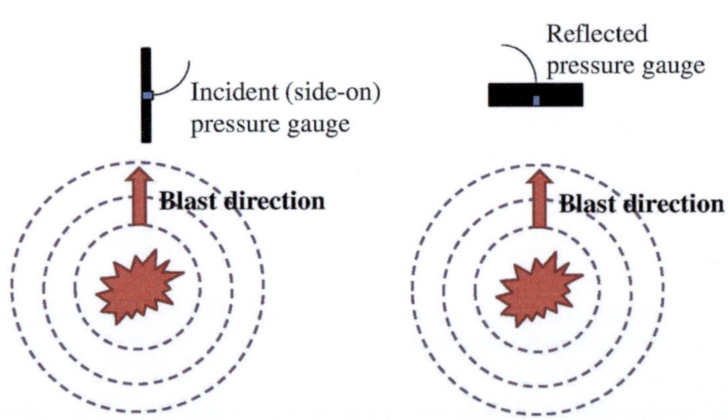

Fig 13.2 Side-on pressures and reflected pressures. A side-on pressure is recorded if the pressure gauge is parallel to the direction of the shock wave whereas a pressure gauge embedded in rigid material perpendicularly to the direction of the shock wave provides a reflected pressure

Fig 13.3 The five side-on pressures mounted on a vertical shaft right above the explosive

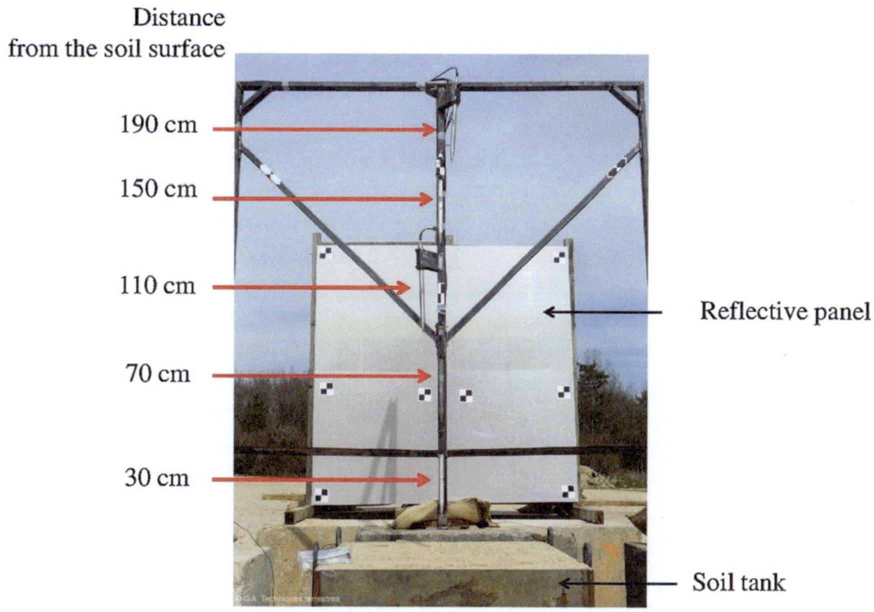

In view of obtaining information on the stresses within the soil tank, a gauge force was placed at the bottom of the tank. The gauge has a diameter equal to 70 mm and it can measure forces up to 120 kN.

The shock wave drives a density jump. Therefore its propagation can be recorded exploiting the principle of shadowgraphy: its projection on a reflective panel placed at the background of the device is tracked by a high speed camera.

To synchronise all measurements, two aluminium sheets were placed on the explosive. At detonation, the two sheets are put in contact which closes an electronic circuit, thus providing the time of detonation.

13.3 Soil Tank Preparation and Control of the Soil

13.3.1 Step by Step Soil Tank Preparation

A protocol for the preparation of the soil is very important to ensure repeatability. A rectangular soil tank with a height of 0.5 m and length and width of 1 m is used. The soil is poured in a first layer of half the height of the tank. When a saturated soil is used, water is added to saturate the first layer. The layer is compacted to densify the soil as required by AEP 55. The tamping is performed with a grooming machine. Next, the second layer is prepared in the same way as the first one: some soil is poured; water is added in the case of saturated soil, and the entire soil tank is compacted. In this second layer, a container of 2.8 L is placed in order to determine later the density, the moisture content and to check the granulometry. Once the soil tank is ready, the container is extracted from the soil and transported to the laboratory for analyses. Soil is added to complete the soil tank. Just before the shot, density and water content measures are controlled at five specific locations with a nuclear gauge.

A hole is then dug in the center of the soil free surface to place the explosive. Its vertical position is reckoned by the depth of burial, which is the distance between the top of the explosive and the free surface of the soil tank. The explosive is centered in relation with the highest pressure gauge. Finally, for a buried explosive, soil is placed on the top of the charge.

After the test, the shape of the crater is recorded with a 3D scan, and soil density and moisture content are measured at control points. The test tank is finally emptied and ready to be prepared for the next shot.

13.3.2 Density and Water Content Controls of the Soil Prior and After the Shots

Density and moisture content are measured with two methods: with traditional laboratory analysis (weight of a specific volume of soil to obtain its density, and oven dried to obtain its water content), and with a nuclear gauge at several locations and depths before and after each shot.

The nuclear density method is non destructive. It allows to determine both density and water content of the soil. The densitometer is composed of four parts: the device itself which is shielded with two distinct emitter probes which are located in the soil to be tested, a detector and a counter on the device.

To determine the density of the soil, a radioactive isotope source (Cesium 137) is sent from the probe in the soil. Photons and more specifically gamma rays are emitted. They are scattered when colliding with soil particles. The higher the density of the soil, the more the gamma rays spread. The difference between the emission and the reception from the scatter is measured by a Geiger counter and related to soil density.

As for the water content, a radioactive source (Americium 241/Beryllium) is released from the probe. The fast neutrons are slowed down when colliding with hydrogen atoms. The number of slow neutrons can be directly linked to the number of hydrogen atoms close to the source, and thus to the water content of the soil since hydrogen in a soil can be mostly found in water.

The main advantages of the nuclear density method are: (a) the speed (a few minutes) at which the results of density and water content are obtained, and (b) he non-destructive aspect of the technique. At variance, for the traditional laboratory analyses, the results are known after the test. Moreover the measures with the nuclear gauge can be made at various locations and various depths whereas the soil sampling for laboratory analyses provides results at one location only.

13.3.3 Comparison of the Two Methods of Soil Measures

For each shot, an average value of density and water content measured by the nuclear gauge is calculated. It is compared in Fig. 13.4 to the value delivered by the laboratory method. Except for the third shot, the two methods provide close results. The outputs for saturated and dry shots are rather close. Besides, since the soil was stored outdoors, the weather in the weeks preceding the tests appears as a significant player in the process. The seven first shots were carried out in April 2013 while the five last ones occurred in January 2014. The week before the shots in April, it was raining all day long, but during the seven shots, the weather was sunny and hot. The measured water contents confirm this forecast: the water content of dry soil tanks progressively decreases from the first to the seventh shot.

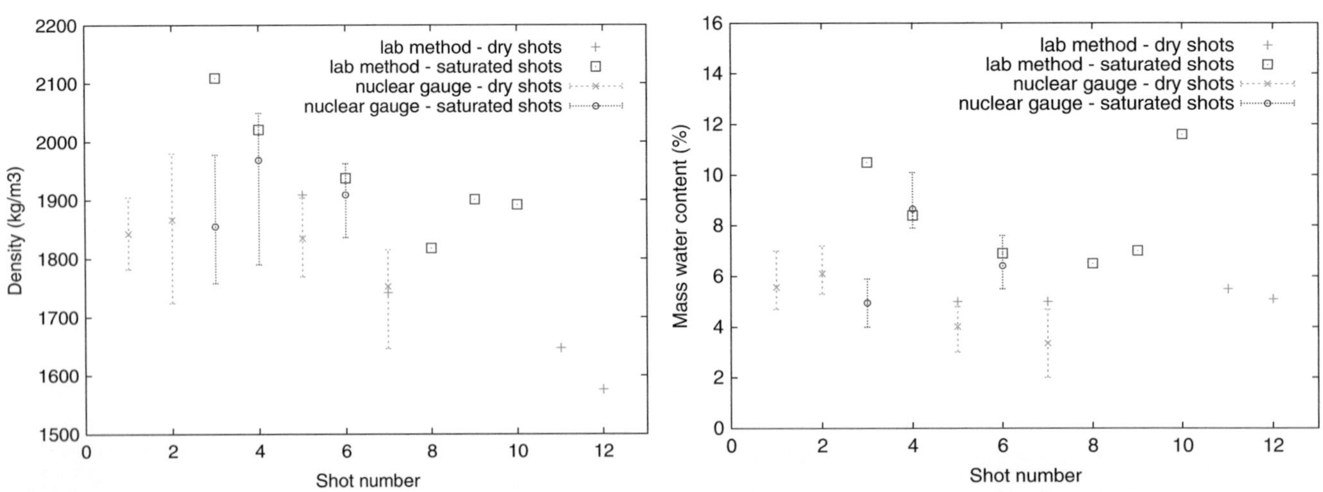

Fig 13.4 (*Left*) and (*right*): Measures of density (on the *left*) and of mass water content (on the *right*) before the shots using the laboratory method and the nuclear gauge

13.4 Experimental Results

13.4.1 Propagation of the Shock Wave

The propagation of the shock wave can be easily followed thanks to the high speed camera with a rate of 20,000 pictures per second and to the reflecting panel. The shock wave being accompanied by a large difference of density, the propagation of the light is affected and changed, and the shock wave is visible.

The Fig. 13.5 shows the propagation of the shock wave emanating from the detonation of the explosive charge up to the formation of the detonation products and the projection of the ejecta for a flush charge and for an explosive buried at 3 cm.

In the first set of pictures, three steps can be distinguished: a fire ball expands first, the shock wave follows and finally detonation products spread in air. The sand projections are not well visible.

At variance, for the buried charge, the fire ball is not visible since the detonation occurred in the soil where less oxygen is available. The soil then rises above the explosive creating a kind of "camouflet". The shock wave is progressively visible and finally ejecta are clearly distinguishable. The velocity of these soil projections has been evaluated up to 900 m/s.

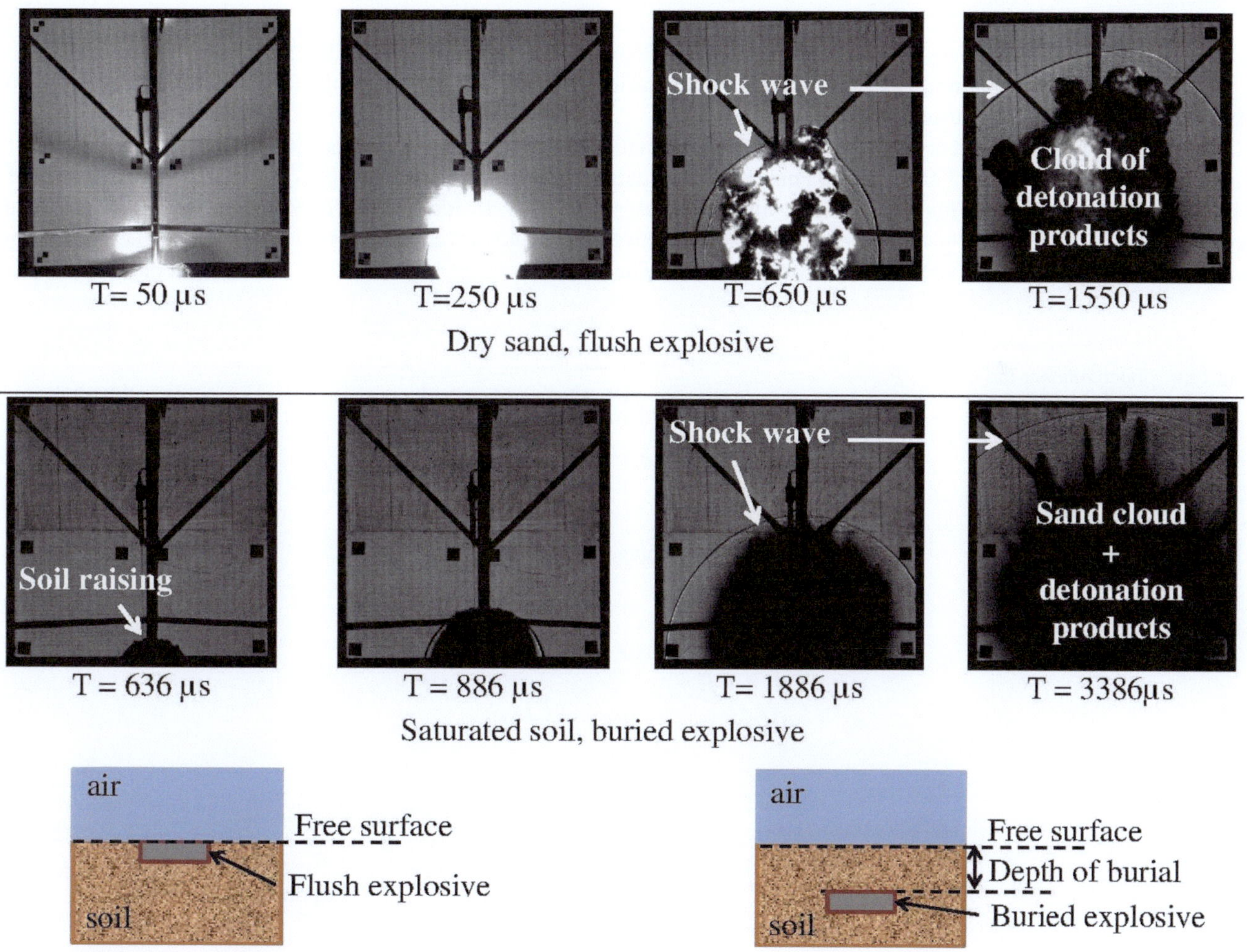

Fig 13.5 Pictures of the detonation of the explosive and the propagation of the shock wave for a flush charge (*top line*) and a charge buried at 3 cm in a saturated soil (*bottom line*)

13.4.2 Measured Craters

In our experiments, after each shot, the crater was scanned to create a mesh of the crater. The knowledge of the elevation of mesh points provides sections of the crater, diameters, heights of the lips and crater volumes. All of these quantities increase with the depth of burial. The crater volumes for all shots are shown in Fig. 13.6. For a depth of burial of 3 cm, the volume is twice as the one of a flush charge and it is three times smaller than for a charge buried at 8 cm.

The influence of the state of the soil (whether it is dry or saturated) increases with the depth of burial: for a flush charge, negligible difference can be noticed whereas, for the depth of burial of 3 and 8 cm, shots in dry soil provide larger craters. The water in saturated soils acts like a binder between soil particles.

13.4.3 Main Pressure Results

The pressures were measured at five different heights along a vertical line every 40 cm, starting at 30 cm above the free surface of the soil up to 190 cm. The maximum pressures P (in kPa) plot linearly as a function of the distance d between the gauge and the explosive charge (in meter), namely

$$P = k/d^n \tag{13.1}$$

The coefficient k and exponent n depend on the soil and depth of burial as shown in Table 13.1.

The experimental maximum pressures are plotted in logarithmic scale on Figs. 13.7 and 13.8 for the three depths of burial together with the approximation of (13.1).

The maximum pressures between shots in dry and saturated soils differ but not by much. Moreover the reproducibility is higher for a charge buried at 8 cm than at 3 cm. The small quantity of explosive (only 100 g) compared to the grain sizes of the soil which can be up to 4 cm may contribute to this lack of reproducibility. In addition, the experiments with a smaller depth of burial are more dependent on the way the soil is placed over the explosive charge: tamping is not feasible and the grain arrangement obtained by pluviation is hardly controllable. The issue is perhaps less important as the depth of burial increases.

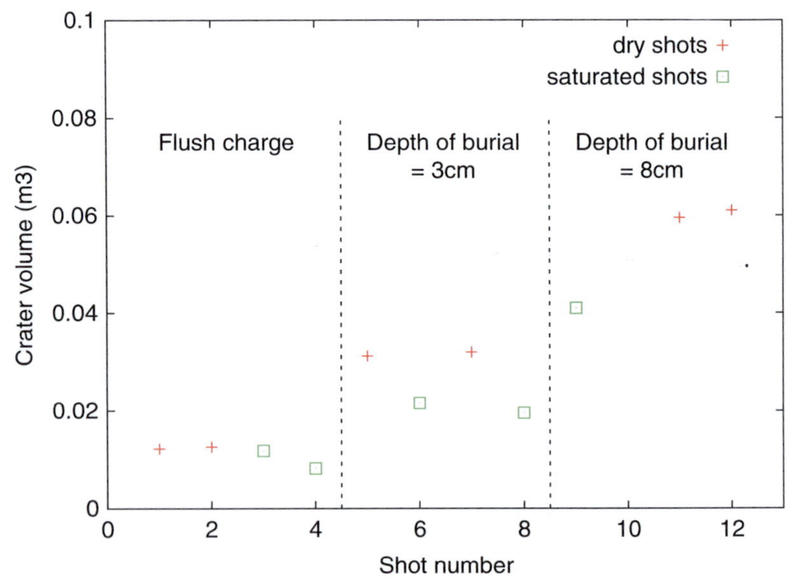

Fig 13.6 Crater volumes for the three depths of burial: the volumes increase significantly with the depth of burial

Table 13.1 Parameters of (13.1) for the three depths of burial (DOB)

DOB (cm)	0	3	8
k	682.14	203.1	79.09
n	3.03	2.04	1.39

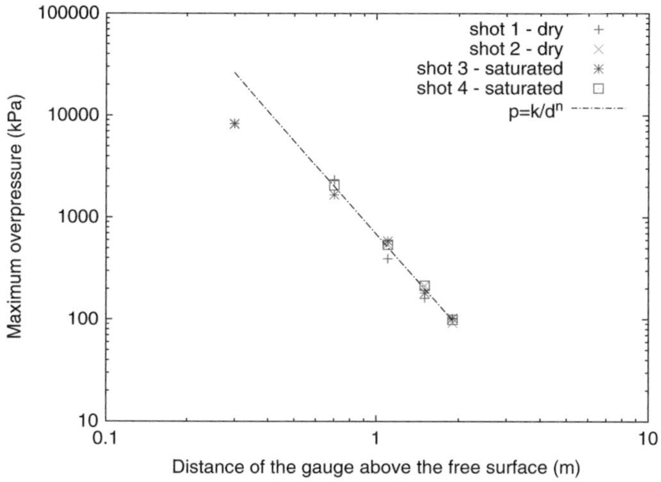

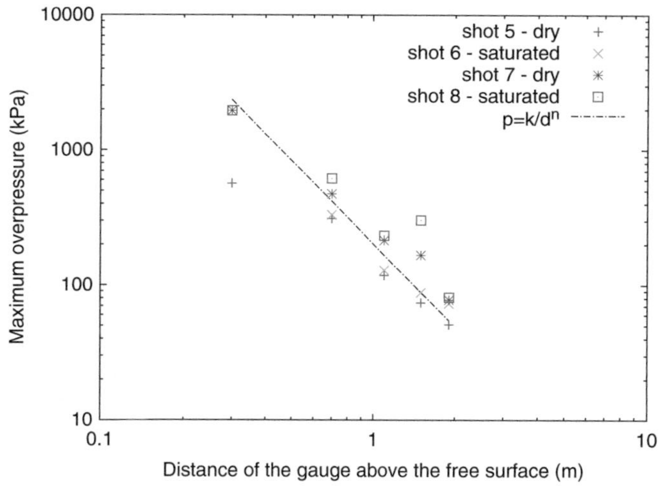

Fig 13.7 (*Left*) and (*right*): Maximum overpressure at the five gauges for the shots with a flush charge (*left*) and with a charge buried at 3 cm (*right*)

Fig 13.8 Maximum overpressure at the five gauges for the shots with a charge buried at 8 cm

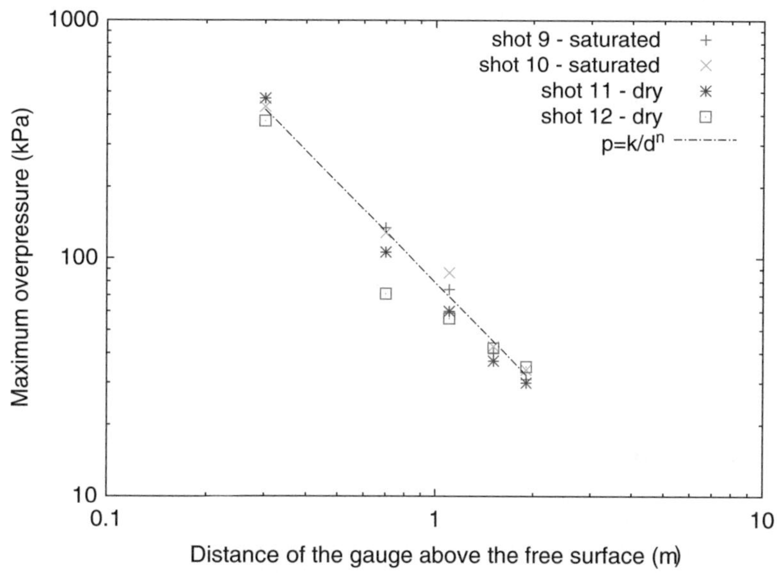

13.4.4 *Measurement of the Vertical Force at the Bottom of the Tank*

An attempt was made to measure the stresses in the soil with a force gauge placed at the bottom of the tank. Unfortunately the results are not reproducible. Two main reasons are thought to cause dispersion. First, the horizontal positioning of the gauge is difficult to control. Second, further disturbances during the tamping of the soil, even if of lower intensity than during the detonation, may perturb the measures.

13.5 Conclusion

The detonations of small explosive charges experiments have focused on the influence of water content on the aerial pressures, the stress in the soil, the size of craters, and the speed of shock wave propagation. Density and water content have been measured with traditional laboratory methods and with a nuclear gauge. The two methods provide close results. Except for crater results where a difference can be noticed between dry and saturated soils, no clear influence of the saturation degree was visible on the pressure and on the shock wave propagation. In view of the results of a second test campaign using a larger mass of explosive, this lack of sensitivity to the saturation is traced to the small range of water content covered in the present tests.

References

1. D. Bergeron, R. Walker, C. Coffey, Detonation of 100-gram anti-personnel mine surrogate charges in sand. A test case for computer code validation. *Report of Defence Research Establishment Suffield no 668* (Defence Research Establishment Suffield, Ralston, 1998), p. 212
2. J. An, C.Y. Tuan, B.A. Cheeseman, G.A. Gazonas, Simulation of soil behavior under blast loading. Int. J. Geomech. **11**(4), 323–334 (2011)
3. NATO standard AEP-55, Procedures for evaluating the protection level of armoured vehicles, Version 1, vol. 2, Edition C (2014)
4. E. Roger, Buried mines in dry and saturated soils, Ph.D. thesis, Université de Grenoble, France, 2015

Chapter 14
Similitude Analysis of Composite I-Beams with Application to Subcomponent Testing of Wind Turbine Blades

Mohamad Eydani Asl, Christopher Niezrecki, James Sherwood, and Peter Avitabile

Abstract The mechanical behavior of new materials for wind turbine blades is initially characterized by using coupon testing. If the results of the coupon testing look promising, then the materials are incorporated into a blade and certified through a full-scale blade test. The coupon testing is not always representative of performance of the new materials, and full-scale-blade testing is time consuming and very expensive—on the order of several hundred thousand dollars. To bridge the large gap between coupon testing and a full-scale test, subcomponent testing is proposed as a cost-effective alternative. To design a meaningful scaled-down subcomponent emulating the structural conditions experienced in the full-scale component, it is proposed that similitude theory can be applied to a scaled replica of the I-beam structure of a wind turbine blade involving spar caps and the shear web. In the current research, the governing equations for the bending of a simply-supported shear deformable thin-walled composite I-beam are analyzed to derive the scaling laws. Fidelity of the derived scaling laws is then examined as a model-validation criterion by mapping the maximum deflection of variant subcomponents to the full-scale composite I-beam. The combined effects of the size of the subcomponents and the ply stack-up schemes on the fidelity of the scaling laws are then investigated through complete and partial similarity conditions. According to the results, preserving the aspect ratio plays a critical role in successful prediction of the maximum bending of the full-scale I-beam. Also, subcomponents with a modified ply stack-up could be found with good accuracy in maximum bending prediction using the derived scaling laws.

Keywords Wind turbine blade • Similitude • Sub-component • Composite I-beam • Distorted layup scaling

14.1 Introduction

Wind turbine blade certification continues to become more and more challenging because of the desire for the rapid and large increase in the size of the blades. Coupon testing of the materials and full-scale blade testing are both required for certification of the blade. Coupon testing is not always representative of the material properties and overall blade strength. Therefore, a final full-scale blade test is required. Full-scale blade testing as a necessary part of the blade certification is very expensive and time consuming. If meaningful and accurate small-scale subcomponent tests could be designed, they may potentially bridge the gap between coupon and full-scale tests by expediting and facilitating the introduction of new materials and blade certification procedure.

Several studies have been pursued to design a subcomponent test for wind turbine blades. Mandell et al. [1] designed, fabricated and tested composite I-beams with shear webs and flanges representative of the cross section of a wind turbine blade. I-beams were placed into a four-point bend test, and the resulting measurements were in good agreement with the levels of strain and stiffness predicted by using finite element modeling and simple beam theory. A subcomponent test was designed for the root section of the blade fabricated by Cairns et al. [2] in which specimens were designed to replicate a single insert of a blade to hub joint and were evaluated by using a pull-out test. The primary focus of the study was on manufacturing, however, a large amount of fatigue and static strength data were generated. Mandell et al. [3] studied the sandwich panel terminations and skin-stiffener intersections of blades. The objective was to estimate the skin-stiffener fracture loads and to determine the optimum transition angle between sandwich panel and normal laminate by using tension tests.

There are a few studies focusing on the adhesive joints and bond-line durability in wind turbine blades by means of the subcomponent testing concept. The tests were designed with the intent to simulate the complex stress field which bond-lines

M.E. Asl (✉) • C. Niezrecki • J. Sherwood • P. Avitabile
Structural Dynamics and Acoustic Systems Laboratory, University of Massachusetts Lowell,
One University Avenue, Lowell, MA 01854, USA
e-mail: mohamad_eydaniasl@student.uml.edu

C. Sciammarella et al. (eds.), *Experimental and Applied Mechanics, Volume 4*,
Conference Proceedings of the Society for Experimental Mechanics Series, DOI 10.1007/978-3-319-22449-7_14

experience in the working condition of a utility-scale blade. An asymmetric three-point bending test was performed on what is called a Henkel beam by Sayer et al. [4]. The goal was to introduce a comparable amount of shear and normal stress to the bond-line to certify the adhesive performance. Later, the Henkel beam was tested for a parametric study to investigate the influence of the bond-line thickness between web and flanges on the beam flexural strength [5]. Zarouchas et al. [6] proposed a static four-point bending test on two symmetric and asymmetric composite I-beams that attempted to replicate the spar caps and shear web. However, the asymmetric design had undesirable failure modes.

The subcomponent test concept usually is referenced to a laboratory scale test rig; however a few mid-level tests were designed to emulate wind turbine blades which still could be considered subcomponent tests. Jensen [7] conducted a comprehensive study on a flap-wise loaded box girder of a wind turbine blade to study the dominant failure modes. White et al. [8] developed a dual-axis test setup on a truncated wind turbine blade to reduce the total test time for fatigue evaluation while combining resonance excitation with forced hydraulic loading.

The subcomponent tests designed to date for the wind turbine blades, were generally aimed to facilitate and expedite the material certification and to give insight into the dominant failure modes in the blade structure. Some of the previous tests were fairly simple, while others were very involved. The question is how should a subcomponent test be designed to be effectively representative of its full-scale component? It is important to know if the data from the subcomponent tests are able to accurately characterize and estimate the behavior of its corresponding component in the full-scale blade, because the subcomponent specimens are not necessarily an exact scaled-down version of the original full-scale blade. In other words, it is important to verify that the designed subcomponent is able to predict the behavior of its corresponding full-scale component and data from the subcomponent test can be used to predict or estimate the response of the full-scale component. As blades are made of composite materials, scaling is a challenge because thickness of the plies in a laminate cannot be scaled down to any desired ratio. Therefore a deliberate methodology should be considered to scale down laminated components in the process of subcomponent design. This possible connection between subcomponent (referred to as "model") and its corresponding full-scale component (referred to as "prototype", see Fig. 14.A1) must be based on the existence of the same field equations and structural parameters that govern the response of both model and prototype when subjected to different load conditions. If such structural parameters could be found, a scaled down model (subcomponent) can be built that duplicates the behavior of its prototype but on a much smaller scale. The descriptor which describes this relationship is "structural similarity".

The necessary and sufficient similarity conditions between two systems could be provided by similitude theory. Rather than dimensional analysis, similarity condition of two systems could be directly extracted from governing equations which provides more specific scaling laws. Simitses and Rezaeepazhand [9] proposed a technique to derive the scaling laws directly by means of the governing equation of a system. The response of the prototype was then predicted by projecting the response of its associate models using derived scaling laws. As a benchmark, the buckling response of an orthotropic and symmetric cross-ply laminated plate was investigated in that study. They also analyzed the vibration of scaled laminated rectangular plates in later studies [10] and investigated the effect of axial and shear loads on stability of scaled laminated rectangular plates [11, 12]. Later, this method was extensively used in their works regarding the prediction of vibration response of laminated shells [13, 14]. According to their results, the scaling laws that are obtained directly from the governing equations yield perfect accuracy for regular cross-ply laminates, while for the angle-ply laminates, the scaling laws did not show good accuracy. In the current paper which is an extension of Authors' previous work [15, 16], similitude theory is applied to governing equation of bending for a thin-walled laminated composite I-beam [17, 18] to assist in the design of meaningful scaled-down models which are representative of the spar caps and shear webs of utility-scale blades [19].

The spar caps and shear web, are the backbone of a wind turbine blade, and make an interesting case study for subcomponent test design. In this study, similitude theory is applied to the subcomponent test concept, to develop a scaled-down while representative model of the I-beam inside the utility-scale blade. Governing equations for the static deflection of a simply-supported thin-walled composite I-beam are used to derive the scaling laws. Then, models which satisfy all the scaling laws are developed as an example for complete similarity between model and prototype. Next, models which partially satisfy the derived scaling laws are developed to propose different cases for partial similarity. For both cases, accuracy and degree of similarity between developed models and prototype is investigated by predicting the maximum deflection of the prototype. Maximum deflections of the developed models for both cases are projected to that of the prototype by means of the derived scaling laws. Within this analysis, a prototype ply scheme and the resulting geometry were based on a portion of a Sandia National Laboratories 100-m long wind turbine blade near the maximum chord. Several different case studies were modeled using complete and partial similarity for an I-beam whose scaled geometries are shown in Table 14.1 and Fig. 14.A2.

14.2 Governing Equations

In this section, the governing equations for the deflection of a shear deformable thin-walled composite I-beam are introduced based on a variational formulation, and a closed-form solution is derived for a beam having a uniformly distributed loading and simple-support boundary condition. Then, scaling laws are derived, and their applicability is discussed.

14.2.1 Bending of a Thin-Walled Composite Beam Using Shear-Deformable Beam Theory

The objective is to predict the maximum deflection of the prototype which in this case is a simply-supported thin-walled composite I-beam with a layup scheme similar to a spar cap of a wind turbine blade near its maximum chord. For this study, the static bending in the absence of thermal effects for a symmetric I-beam is considered. Neglecting all coupling effects due to the symmetric geometry, the equations governing the bending with respect to x-axis are given by [17] which are well known Timoshenko beam equations:

$$(GA_x)_{com}\left(V'' + \Psi_x'\right) = q \tag{14.1a}$$

$$(EI_x)_{com}\Psi_x'' - (GA_x)_{com}\left(V' + \Psi_x\right) = 0 \tag{14.1b}$$

where Ψ_x denotes the rotation of the cross section with respect to x axis shown in Fig. 14.A2, V the displacement in y direction, q the distributed load, the prime (') is used to indicate differentiation with respect to z, $(GA_x)_{com}$ and $(EI_x)_{com}$ are shear and flexural rigidity with respect to x, respectively which could be expressed as:

$$(EI_x)_{com} = \left[A_{11}^{\alpha}\, y_{\alpha}^{(2)} - 2B_{11}^{\alpha}\, y_{\alpha} + D_{11}^{\alpha}\right]b_{\alpha} + \frac{b_3^{(3)}}{12}A_{11}^{3} \tag{14.1c}$$

$$(GA_x)_{com} = A_{55}^{\alpha}b_{\alpha} + A_{66}^{3}b_3 \tag{14.1d}$$

where A_{11}, A_{66}, A_{55}, B_{11} and D_{11} are elements of extensional, coupling and bending stiffness matrices for a composite layup. The superscript in the parenthesis () denotes the power of the exponent, and repeated index denotes summation. Index α varies from 1 to 3 where the indices 1 and 2 represent the top and bottom flanges, and 3 is for the web, respectively as shown in Fig. 14.A2 and b_{α} denotes width of the flanges and web. The closed-form solution of the maximum deflection for the uncoupled (14.1a–14.1d) may be directly calculated for the simple support boundary condition and uniform distributed load as [18]:

$$\upsilon_{max} = \frac{5ql^4}{384(EI_x)_{com}} + \frac{ql^2}{8(GA_x)_{com}} \tag{14.2}$$

In (14.2), the first term expresses the deflection by Euler-Bernoulli beam theory, and the second term is the deflection by shear deformation.

14.3 Complete Similarity

The maximum deflection of the prototype and its models (subcomponents) are all described by (14.2). To establish similitude theory and to derive the scaling laws, it is assumed that all variables of the governing equation of the model (x_m) are connected to their corresponding values in prototype (x_p) by one to one mapping. Then the scale factor may be defined as $\lambda_x = x_p/x_m$ and standard similitude procedure [9] can be applied to the (14.2) that results in the following scaling laws:

$$\lambda_l^2 = \frac{\lambda_{EI}}{\lambda_{GA}} \tag{14.3}$$

$$\lambda_\upsilon = \frac{\lambda_q \lambda_l^4}{\lambda_{EI}} \tag{14.4a}$$

$$\lambda_\upsilon = \frac{\lambda_q \lambda_l^2}{\lambda_{GA}} \tag{14.4b}$$

To achieve the complete similarity between the prototype and the model, (14.3) must be satisfied first. Then the maximum deflection scale factor (λ_υ) may be obtained exactly using either (14.4a) or (14.4b). For convenience, the flanges and web were assumed to have the same width (b_1=b_2=b_3) and same layup, then the flexural (EI_x) and shear (GA_x) rigidity may be expanded as:

$$(EI_x)_{com} = \left[2A_{11}y^2 - 4B_{11}y + 3D_{11}\right]b + \frac{b^3}{12}A_{11} \tag{14.5a}$$

$$(GA_x)_{com} = [3A_{55} + A_{66}]b \tag{14.5b}$$

The coupling stiffness (B_{11}) is zero for a symmetric layup and bending stiffness (D_{11}) is negligible compared to the other terms in (14.5a) which means that for a symmetric layup, I-beam flexural and torsional rigidities are mostly dependent on extensional elements of stiffness matrix. Based upon this assumption, similarity between two symmetric composite I-beams does not depend on the position of plies in the flanges and web laminates stack up. Also, satisfying (14.3) will simultaneously result in models which have the same aspect ratio as the prototype (λ_b=λ_l). The rigidity coefficient ratios of the different models are compared to the prototype with (14.3) using (14.5a and 14.5b). If satisfactory, then either of the scaling laws (14.4a and 14.4b) could be applied to connect the maximum deflection of the validated model to that of prototype. To satisfy (14.3), ply-level scaling method $\left[(+\Theta_n/-\Theta_n)_s, \text{ n}=1,2,3,\ldots\right]$ might be used in which the angle of plies in the layup scheme of the laminate is denoted by Θ [14]. In this approach, the total number of the plies with the same fiber angle n, is divided by an integer number resulting in models with the same extension, coupling and bending ratios as the prototype [15, 16]. Ply-level scaling if applicable, yields models which automatically satisfy (14.3).

To apply ply-level scaling and achieve complete similarity, the prototype layup must have multiple plies for each fiber direction. The prototype geometry and lay-up scheme considered in this study emulate the spar-cap flanges of the Sandia 10 MW wind turbine blade [19] near the max chord. The ply scheme used for the prototype in this analysis is based upon the ply scheme of the Sandia blade near maximum chord. This area was chosen for analysis because many failures in blades occur near maximum chord, and the area of interest is responsible for carrying most of the aerodynamic loads. The prototype is assumed to be an I-beam consisting of three identical laminated plates having an overall dimension of $29.3 \times 2.446 \times 0.136$ m^3 with layup $[0\ ]_{100}$ and ply thickness of t = 1.36 mm. Because the blade is tapered throughout its length and the geometry of the structure changes significantly, the geometry of the full-scale component (i.e. first 1/3 of the Sandia blade from its root) was averaged to simplify analysis. All computations are implemented for the glass/epoxy materials with following material characteristics [19]: E_1=41.8 GPa, E_2=E_3=14 GPa, G_{12}=G_{13}=2.63 GPa, G_{23}=1.83 GPa, υ_{12}=υ_{13}=0.28, υ_{23}=0.47. Within this analysis, it was assumed that the geometry and layup scheme of the prototype is kept fixed while comparing to the developed models with various dimensions and ply stack up.

In the complete similarity case, the maximum deflection of the prototype with simply-supported boundary conditions was compared to the models with ply-level scaled-down thickness having an aspect ratio that is the same as the prototype. Theoretical values of the models and the prototype were obtained directly by (14.2), and then predicted values were projected by applying derived response scaling laws (14.4a and 14.4b) to the theoretical values of the models. For the complete similarity case, all response scaling laws predict the deflection of the prototype with high accuracy because the complete similarity assumption in (14.3) were satisfied. For the specific I-beam design considered for the prototype, the complete similarity is obtainable.

Figure 14.1 shows the accuracy of (14.3) as a necessary assumption for the complete similarity case. The ratio parameters on the left-hand side (green circles) and right-hand side (blue stars) of the equation must be equal for complete similarity. Within the graph, complete similarity is demonstrated as the number of plies (n) in the model changes. The thicknesses of the models $[0\ ]_n$ have been scaled down using ply-level scaling and overall dimensions are varying $\left(b_p \neq b_m\right)$ while preserving

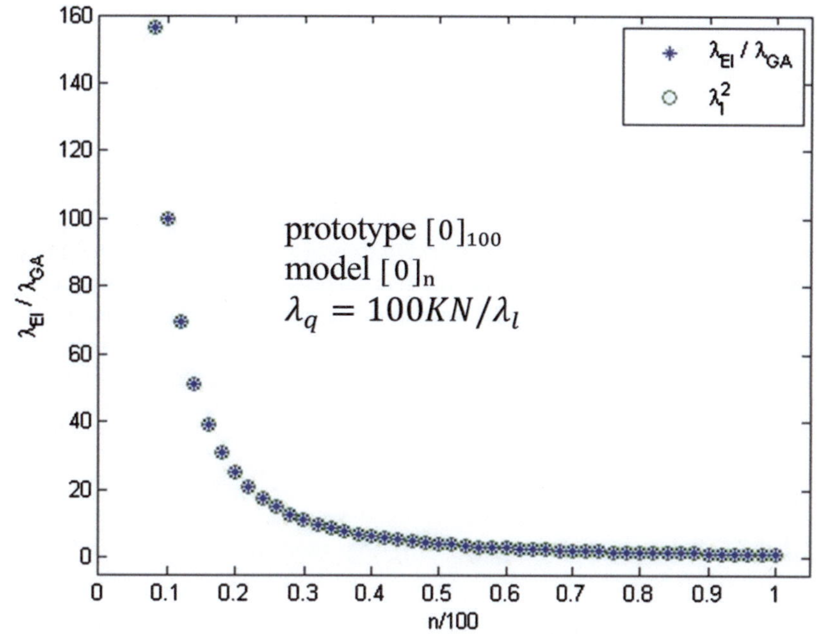

Fig. 14.1 Comparison of the terms in (14.3) as a function of the number of plies (n) in the model which demonstrates complete similarity

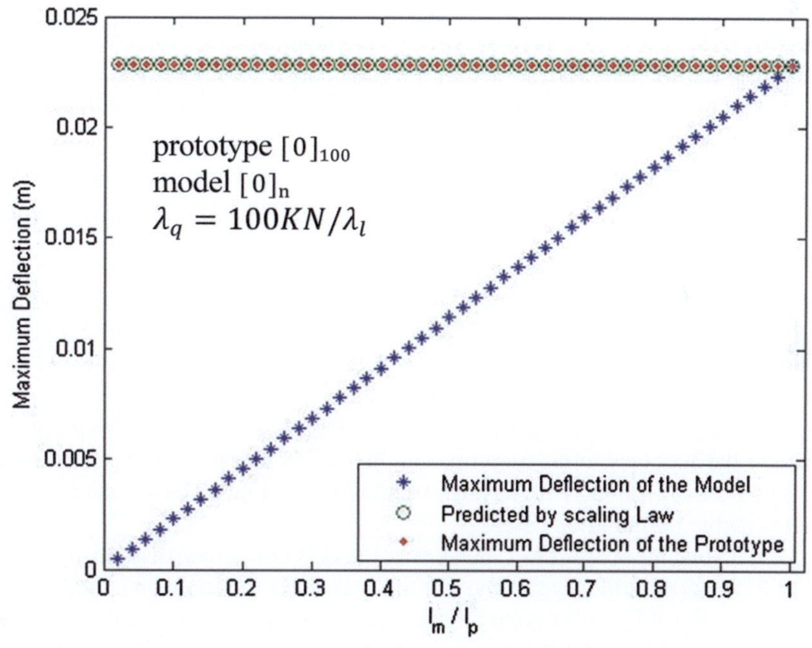

Fig. 14.2 Comparison of the predicted maximum deflection of the prototype (*green circles*) to the actual deflection (*red diamond*) by using the maximum deflection of the model (*blue star*) via the response scaling law for complete similarity

the same aspect ratio as the prototype. The scale factor refers to the ratio of length of the model to that of the prototype. The distributed load is equal to scale factor of each model. According to the results, (14.3) is valid for all models developed using complete similarity. Therefore, (14.4a and 14.4b) may be used to predict the maximum deflection of the prototype accurately. Figure 14.2 shows the accuracy of response scaling laws (14.4a and 14.4b) in predicting maximum deflection of the prototype by means of maximum deflection of the model. Maximum deflection of the models was projected to that of prototype using the derived scaling laws.

14.4 Partial Similarity

In the previous section, it was shown that complete similarity can be achieved in the special case of ply-level scaling when keeping the same aspect ratio for the prototype and its model. To extend the applicability of scaled-down models, partial similarity with distortions in ply stack up while keeping the same aspect ratio is investigated. In these cases, the necessary condition for complete similarity in (14.3) is not fully satisfied. Discrepancy in (14.3) yields inaccuracy in the response scaling laws defined in (14.4a and 14.4b). The level of inaccuracy for various distortions is investigated, and the most accurate response scaling law is selected to project the data from the model to prototype. Basically, ply-level scaling is not applicable when the laminate thickness is supposed to be scaled-down below the range in which there is no integer multiple lamina left in the stack up and changing the lamina thickness is not an option. For example, consider the prototype with the ply stack up $[\pm 45_2/0_{46}]_s$ which is more representative of a utility scaling blade layup for the spar caps, it is not possible to scale down the thickness to one quarter of the prototype thickness using ply-level scaling, because it will result in a non-integer number of plies.

14.4.1 *Distorted Layup Scaling Approach*

As the only constraint to achieve complete similarity in scaling down the prototype thickness is obtained by satisfying (14.3), any ply scheme for the model that approximately satisfies (14.3) may be considered a solution. Based on the prototype layup scheme, a permutation algorithm can be used to search for the potential model ply schemes with overall laminate thickness less than that of the prototype which approximately satisfies (14.3). For this case, it is assumed that models have the same aspect ratio (λ_b=λ_l) as the prototype but ply-level scaling is not considered, so models rigidity ratios might not be exactly equal to the square of length ratio $\left(\lambda_l^2 \neq \lambda_{EI}/\lambda_{GA}\right)$. As the prototype ply stack up $[\pm 45_2/0_{46}]_s$ is symmetric and (14.3) was derived for a symmetric layup in spar caps and the web, models with symmetric layup were searched. The error criterion to find the desired layups then was defined as:

$$e = \left[\left(\lambda_l^2 - \frac{\lambda_{EI}}{\lambda_{GA}}\right) \Big/ \lambda_l^2\right] * 100 \tag{14.6}$$

As a benchmark for the distorted layup scaling technique, models with a total of 12 layers and λ_l=8.33 were developed. The error defined by (14.6) for all possible ply schemes for the symmetric models with a total of 12 layers that only include 0, 90 and ±45 plies is shown in Fig. 14.3. The 0, 90 and ±45 plies were selected intentionally to make the designed model practical to manufacture. After discarding the unbalanced ply scheme cases among all 117 possible cases, the cases with permutation number 15 and 20 and layups $[0_5/\,90]_s$ and $[0_4/\,90_2]_s$ were found with less than 3 and 6 % error, respectively. Figure 14.4 shows the accuracy of the developed models using a distorted layup scaling technique in predicting the maximum deflection of the prototype along with response scaling law (14.4a). The selected cases 15 and 20 also show the best predictions among all possible permutation cases.

Extending the analysis to a smaller size, models with a total of eight layers and λ_l=12.5 were investigated. The error defined by (14.6) for all possible ply schemes for the symmetric models with a total of eight layers that only include 0, 90 and ±45 plies is shown in Fig. 14.5. After removing the unbalanced ply-scheme cases among the 34 possible cases, the cases with permutation number 7 and layup $[0_3/\,90]_s$ was found with less than 2 % error. Figure 14.6 shows the accuracy of all the potential models using the distorted layup scaling technique in predicting the maximum deflection of the prototype along with response scaling law (14.4a). The selected case with permutation number 7 shows the best prediction among all other potential models.

Extending the analysis to an even smaller size having a total of six layers and λ_l=16.66 were investigated. Figure 14.7 shows all possible ply schemes for the symmetric models with a total of six layers that only include 0, 90 and ±45 plies. After removing the unbalanced ply scheme cases among 12 possible cases, the cases with permutation number 4 and layup $[0_2/\,90]_s$ was found with less than 2 % error. Figure 14.8 shows the accuracy of all the potential models using the distorted layup scaling technique in predicting the maximum deflection of the prototype along with response scaling law (14.4a). The selected cases with permutation number 4 shows the best prediction among all other potential models.

Fig. 14.3 Error for all possible ply schemes for the symmetric models with a total of 12 layers that only include 0, 90 and ±45 plies

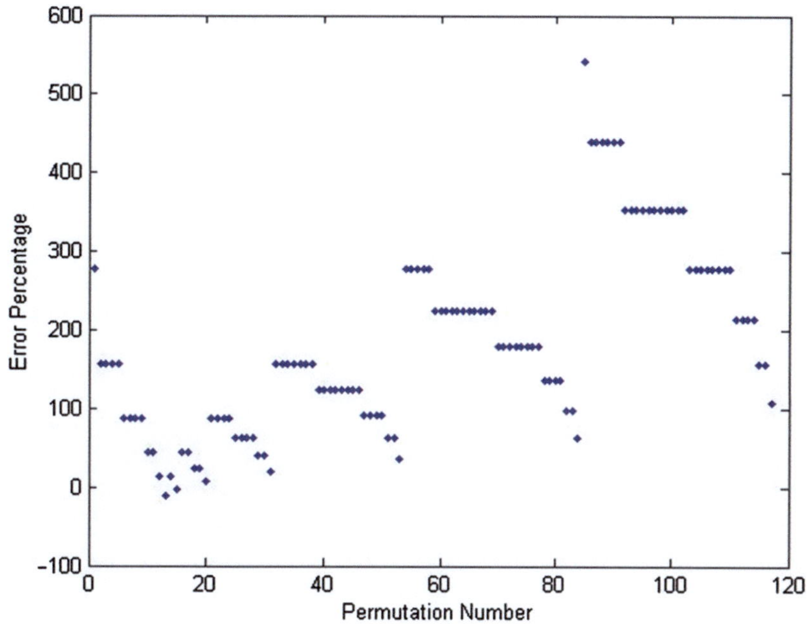

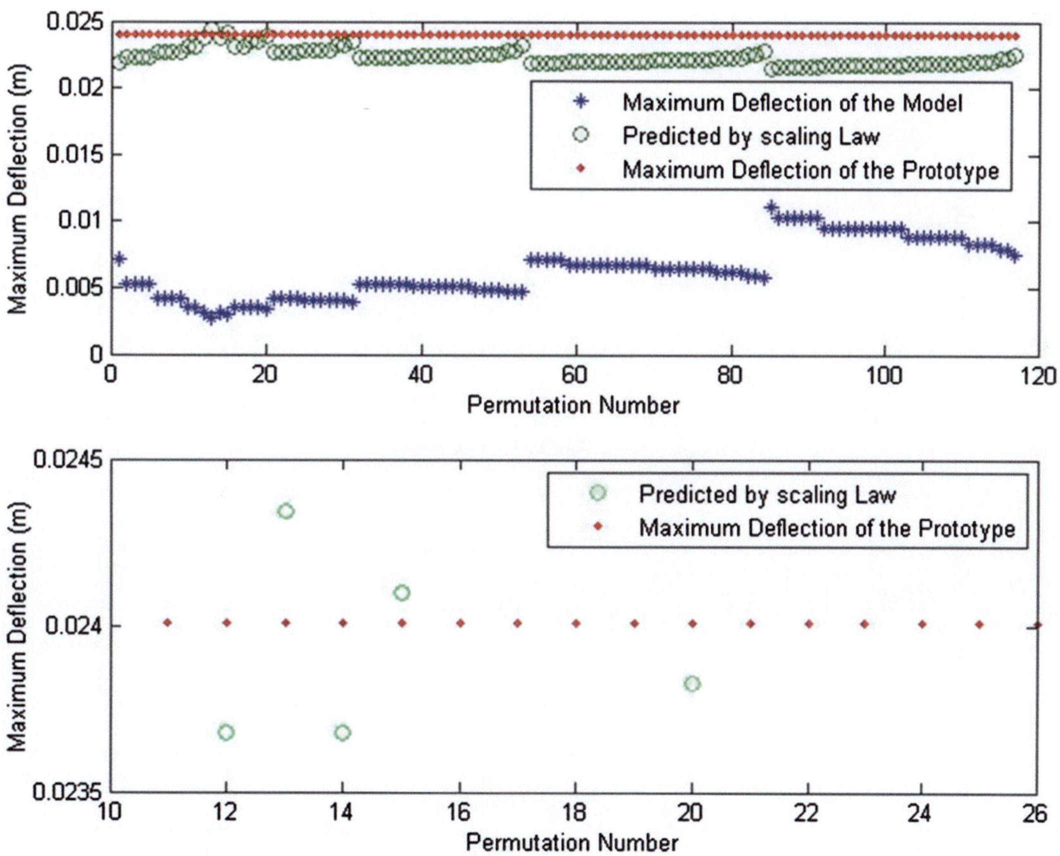

Fig. 14.4 *Top*: comparison of the predicted maximum deflection of the prototype (*green circles*) to the actual deflection (*red diamond*) by using the maximum deflection of the model (*blue star*) via the response scaling law for partial similarity (models with 12 layers); *Bottom*: zoom in for the region that has models with the least error

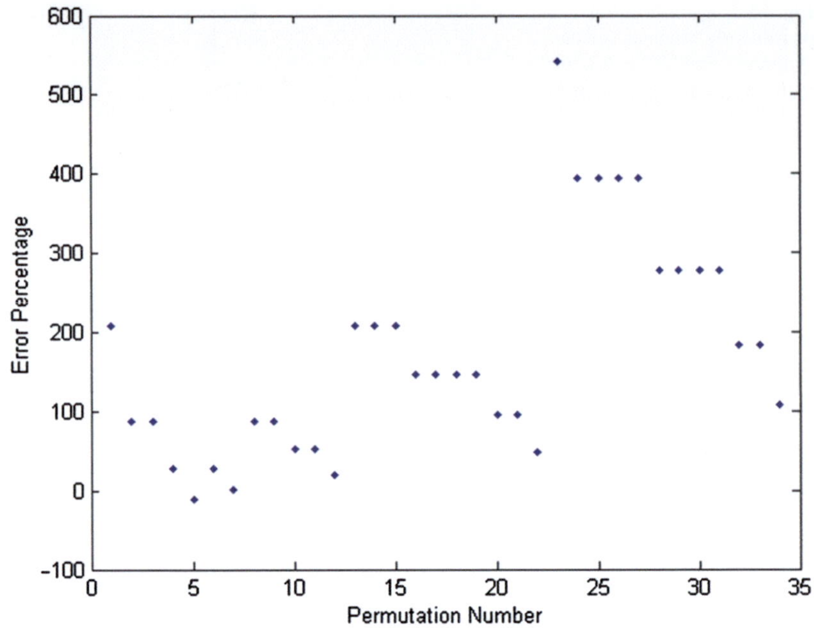

Fig. 14.5 Error for all possible ply schemes for the symmetric models with a total of eight layers that only include 0, 90 and ±45 plies

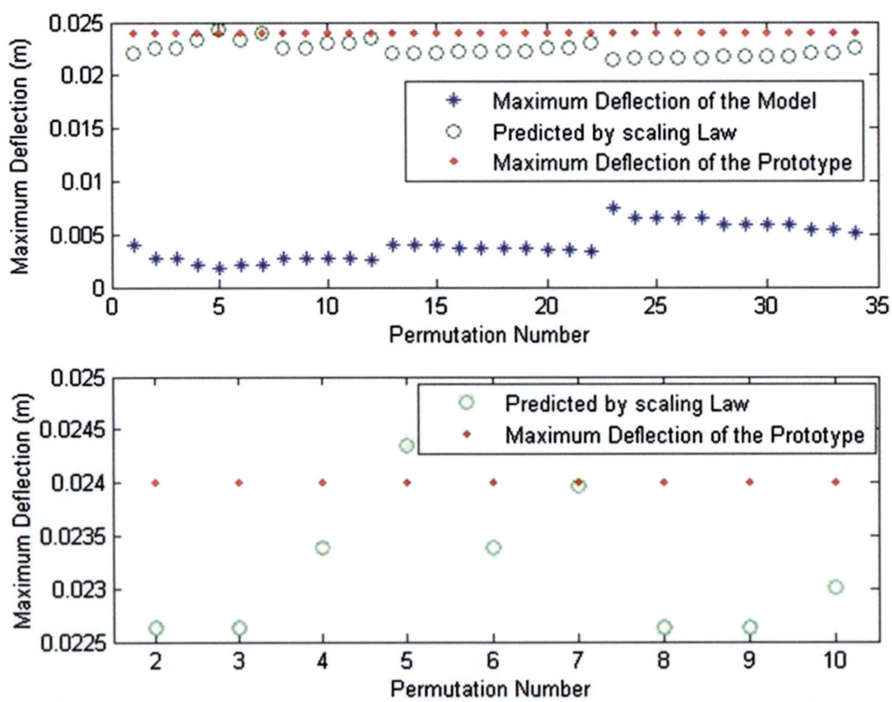

Fig. 14.6 *Top*: comparison of the predicted maximum deflection of the prototype (*green circles*) to the actual deflection (*red diamond*) by using the maximum deflection of the model (*blue star*) via the response scaling law for partial similarity (models with eight layers); *Bottom*: zoom in for the region that has models with the least error

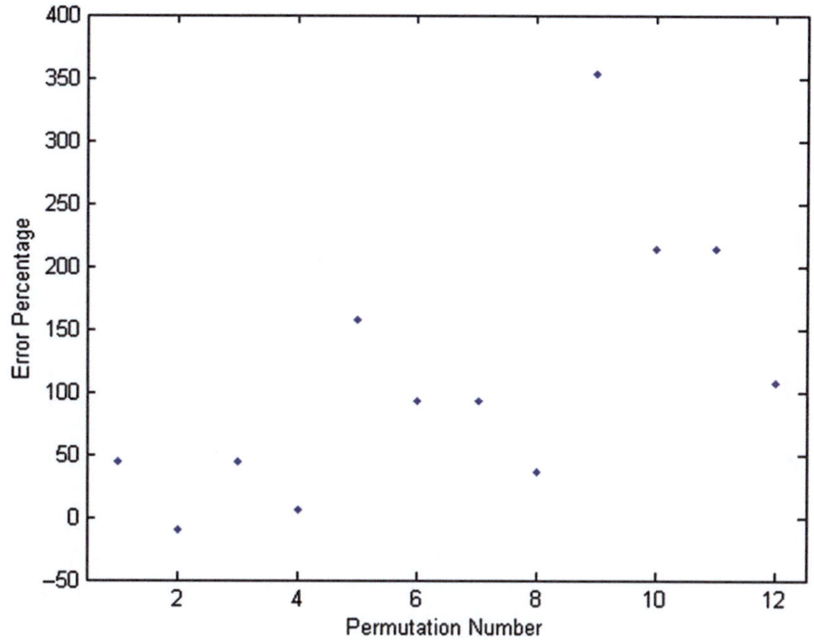

Fig. 14.7 Error for all possible ply schemes for the symmetric models with a total of six layers that only include 0, 90 and ±45 plies

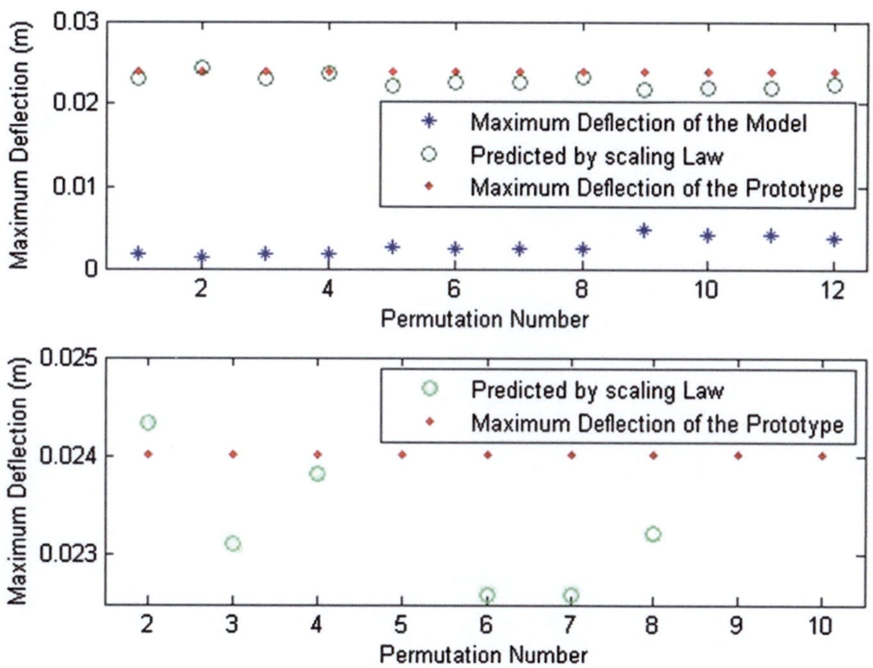

Fig. 14.8 *Top*: comparison of the predicted maximum deflection of the prototype (*green circles*) to the actual deflection (*red diamond*) by using the maximum deflection of the model (*blue star*) via the response scaling law for partial similarity (models with six layers); *Bottom*: zoom in for the region that has models with the least error

14.5 Discussion

Based on the presented results, (14.3) is always valid for ply-level scaled down models that have the same aspect ratio as the prototype and the projection of scaling laws (14.4a and 14.4b) yield excellent accuracy in predicting the maximum deflection. Applying the ply-level scaling approach to the proposed prototype with layup $[\pm45_2/0_{46}]_s$ resulted in a model with an overall thickness equal to half of the prototype thickness. To scale down the thickness of the model more than half, models with a total number of layers less than 50 should be searched that satisfy the similarity condition expressed by (14.3). Applying the distorted layup scaling approach and based on the results shown in Figs. 14.3, 14.4, 14.5, 14.6, 14.7, and 14.8, symmetric laminates including 0, 90 and ±45 plies were found as models with a total number of layers (e.g. 12, 8 and 6) that satisfy (14.3) with less than 6 % error. Selected models using this approach then were shown to have a very good ability to predict the prototype maximum deflection depending on their error in satisfying (14.3). According to Figs. 14.3, 14.5 and 14.7, by decreasing the desired total number of the layers for a model, the probability to find an accurate model which works well with the response scaling laws (14.4a and 14.4b) will decrease. This decrease implies that there is a practical limit to the amount of scaling that can be achieved using the proposed distorted layup scaling technique for laminated structures.

14.6 Conclusion

To expedite the certification and testing of new materials for wind turbine blades, a scaled-down model may be developed. To evaluate applicability of scaled-down models in testing new materials, an analytical study was conducted. Similitude theory was applied to the governing equations for bending of a thin-walled composite I-beam emulating a spar cap of a wind turbine blade. Derived scaling laws were treated as design criteria to propose scaled-down models. The effect of different design parameters on the accuracy of scaling laws was investigated and possible partial similarity scenarios were evaluated. Then, the scaling laws were evaluated for various conditions of both complete and partial similarity. The distorted layup scaling approach was developed to search for scaled-down models when ply-level scaling is not available. Applicability of this method to design the scaled-down models, for predicting the maximum bending of the prototype, was established based on an analytical investigation. According to the results, established scaling laws can be applied with excellent accuracy in complete similarity conditions and very good accuracy for partial similarity to predict the maximum bending deflection of a thin-walled composite I-beam.

Acknowledgments This material is based upon work supported by the National Science Foundation under Grant Number 1230884 (Achieving a Sustainable Energy Pathway for Wind Turbine Blade Manufacturing). Any opinions, findings, and conclusions or recommendations expressed in this material are those of the author(s) and do not necessarily reflect the views of the National Science Foundation.

Appendix

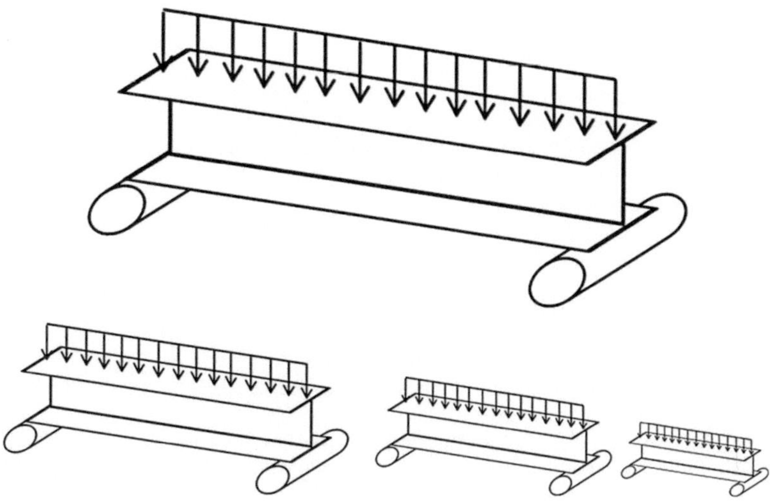

Fig. 14.A1 Prototype (*top*) and models with different scales and layups (*bottom*)

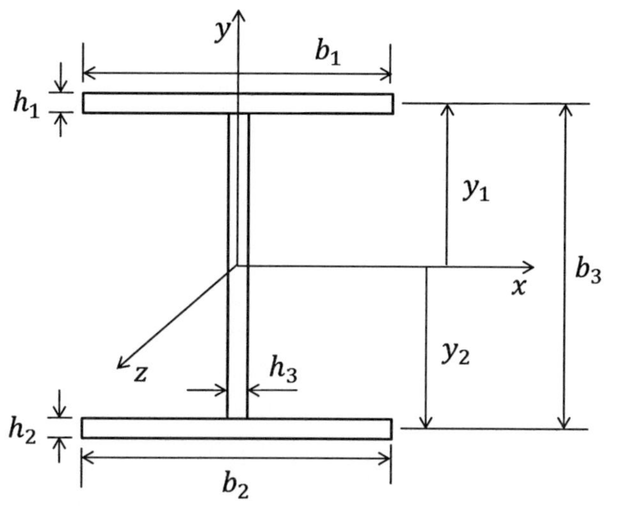

Fig. 14.A2 Geometry and coordinate of the I-beam [17]

Table 14.1 Different case studies were conducted using complete and partial similarity for the I-beam

	Total number of the plies	$b_1=b_2=b_3$ (m)	Length (m)
Prototype (complete similarity)	100	2.446	29.3
Models (complete similarity)	1–100	0.02446–2.446	0.293–29.3
Prototype (partial similarity)	100	2.446	29.3
Models (partial similarity-case 1)	12	0.29352	3.516
Models (partial similarity-case 2)	8	0.19568	2.344
Models (partial similarity-case 3)	6	0.14676	1.758

References

1. J.F. Mandell, D. Combs, D.D. Samborsky, Fatigue of fiberglass beam substructures. Wind Energy **16**, 99 (1995)
2. D.S. Cairns, J.D. Skramstad, J.F. Mandell, Evaluation of hand lay-up and resin transfer molding in composite wind turbine blade structures, in *20th 2001 ASME Wind Energy Symposium*, 2001
3. J.F. Mandell, R. Creed Jr., Q. Pan, D.W. Combs, M. Shrinivas, Fatigue of fiberglass generic materials and substructures. Wind Energy **15**, 207–213 (1994)
4. F. Sayer, N. Post, A. Van Wingerde, H. Busmann, F. Kleiner, W. Fleischmann, M. Gansow, Testing of adhesive joints in the wind industry, in *European Wind Energy Conference and Exhibition 2009*, 2009, pp. 288–315
5. F. Sayer, A. Antoniou, A. Van Wingerde, Investigation of structural bond lines in wind turbine blades by sub-component tests. Int. J. Adhes. Adhes. **37**, 129–135 (2012)
6. D.S. Zarouchas, A.A. Makris, F. Sayer, D. Van Hemelrijck, A.M. Van Wingerde, Investigations on the mechanical behavior of a wind rotor blade subcomponent. Compos. Part B **43**, 647–654 (2012)
7. F.M. Jensen, B.G. Falzon, J. Ankersen, H. Stang, Structural testing and numerical simulation of a 34 m composite wind turbine blade. Compos. Struct. **76**, 52–61 (2006)
8. D. White, W. Musial, S. Engberg, Evaluation of the B-REX fatigue testing system for multi-megawatt wind turbine blades, in *Collection of the 2005 ASME Wind Energy Symposium Technical Papers at the 43rd AIAA Aerospace Sciences Meeting and Exhibit*, pp. 52–65, 2005
9. G.J. Simitses, J. Rezaeepazhand, Structural similitude for laminated structures. Compos. Eng. **3**, 751–765 (1993)
10. J. Rezaeepazhand, G.J. Simitses, Use of scaled-down models for predicting vibration response of laminated plates. Compos. Struct. **30**, 419–426 (1995)
11. J. Rezaeepazhand, G.J. Simitses, J.H. Starnes Jr., Design of scaled down models for stability of laminated plates. AIAA J. **33**, 515–519 (1995)
12. G.J. Simitses, J. Rezaeepazhand, Structural similitude and scaling laws for buckling of cross-ply laminated plates. J. Thermoplast. Compos. Mater. **8**, 240–251 (1995)
13. J. Rezaeepazhand, G.J. Simitses, J.H. Starnes Jr., Design of scaled down models for predicting shell vibration response. J. Sound Vib. **195**, 301–311 (1996)
14. J. Rezaeepazhand, G.J. Simitses, Structural similitude for vibration response of laminated cylindrical shells with double curvature. Compos. Part B **28**, 195–200 (1997)
15. M. Eydani Asl, C. Niezrecki, J. Sherwood, P. Avitabile, Predicting the vibration response in subcomponent testing of wind turbine blades, in *Proceedings of the IMAC-XXXIII*, Orlando, 2–5 February 2015
16. M. Eydani Asl, C. Niezrecki, J. Sherwood, P. Avitabile, Application of structural similitude theory in subcomponent testing of wind turbine blades, in *Proceedings of the American Society for Composites 2014*, La Jolla, 8–10 September 2014
17. J. Lee, Flexural analysis of thin-walled composite beams using shear-deformable beam theory. Compos. Struct. **70**, 212–222 (2005)
18. J.N. Reddy, *Mechanics of laminated composite plates and shells: theory and analysis* (CRC Press, Boca Raton, 2004)
19. D.T. Griffith, T.D. Ashwill, The Sandia 100-meter all-glass baseline wind turbine blade: SNL100-00, Sandia National Laboratories, Albuquerque, Report no. SAND2011-3779, 2011

Chapter 15
Validating FSI Simulations in LS-DYNA 971 R7

Kevin A. Gardner, Jeremy D. Seidt, and Amos Gilat

Abstract The latest version of LS-DYNA includes multi-physics capabilities that allow the computational fluid dynamics and structural solvers to be combined. This presents a powerful tool for automotive engineers that can be used in designing body panels. Body panels are one of the easiest locations to eliminate weight from, however, if panels are made too thin they become susceptible to a phenomenon known as oil-canning. Oil-canning can occur under loading conditions such as those found in automotive air dryers. Oil-canning is a complex phenomenon that can result in either temporary or permanent deformation of a panel. Fluid–structure interaction simulations provide a possible design tool for predicting oil canning events, however the numerical results must be validated. An experimental program is presented where automotive roof panels are placed into a custom test rig and loaded with a high velocity air jet. Flow characterization is performed by building a pressure map for various conditions with piezo-resistive pressure transducers. The flow data is used in an LS-DYNA simulation of the experiment and the results are compared to determine the validity of the numerical simulation as a design tool.

Keywords Fluid–structure interaction • Oil-canning • Digital image correlation • Numerical methods • Automotive

15.1 Introduction

LS-DYNA 971 R7 [1] includes the capability to perform fluid–structure interaction (FSI) simulations with both the incompressible and compressible fluid dynamics solvers. This new capability is a powerful design tool that can be used by the automotive industry to help reduce the overall weight of their production vehicles and improve fuel economy. Automotive body panels are typically stamped from thin sheet steel and can represent a significant portion of the vehicle's overall weight. Weight reduction is one of the easiest methods for increasing fuel economy and thinning body panels can provide a significant weight reduction. However, if the panels are thinned too much a phenomenon known as oil-canning can arise. This phenomenon occurs when an automobile exits an automatic car wash and is subjected to high loads from the air dryers. The high velocity air dryers can cause deformation of the panels that is temporary or in some cases permanent. For small deformations the panel will simply snap back once the loading is removed, however, if the deformation is large enough permanent deformation can occur. The oil-canning phenomenon should be avoided, even if temporary, since it results in customer dissatisfaction. Results from fluid–structure interaction simulations can be used to predict and prevent oil-canning.

Many researchers have studied fluid–structure interactions in many scenarios. The deformation of composite structural plates has been investigated with underwater blast loadings using a test setup that recreates common fluid–structure interactions [2–4]. Similarly, tests on carbon panels have been performed using air blasts to determine material response and the effect of panel curvature on the fluid–structure interaction [5–7]. Fluid–structure interaction effects have been studied for steel roofing systems to minimize the uplift created by natural air currents [8].

K.A. Gardner (✉) • J.D. Seidt • A. Gilat
Department of Mechanical and Aerospace Engineering, Scott Laboratory, The Ohio State University, 201 W 19th Avenue, Columbus, OH 43210, USA
e-mail: gardner.796@osu.edu

C. Sciammarella et al. (eds.), *Experimental and Applied Mechanics, Volume 4*,
Conference Proceedings of the Society for Experimental Mechanics Series, DOI 10.1007/978-3-319-22449-7_15

15.2 Experimental

The oil-canning phenomenon has proven to be a complex phenomenon that is governed by many different properties including the geometry of the sample, the material used, the forming history of the material, the boundary conditions, and the ratio of the jet area to the panel area. In order to closely mimic the conditions observed by a production vehicle it was decided to use actual automotive roof panels from a compact sedan. The provided roof panels had a nominal thickness of 0.65 mm. The panels are clamped along their edges by matching forms that were machined from medium density polyethylene blocks. The panel and matching forms are held in place within a clamshell fixture as shown in Fig. 15.1.

The test rig consists of the clamshell which provides the clamping force along the boundary of the panels and the support stand. The panels and matching forms are placed in the clamshell and clamped in place with evenly spaced bolts. The weight of the clamshell is supported by two turnbuckles that allow for proper alignment. There are three Interface load cells placed between the clamshell and the support frame to determine the total force acting on the panel from the air jet. The load cells are able to accurately measure the force from the air jet since the entire weight of the clamshell is supported by the turnbuckles.

A 1.25 in. pipe was used to generate the high velocity air jet. Previously, an industrial air nozzle was used, however, the pipe allows for higher forces to be generated and is simpler to implement into the LS-DYNA model. The pipe generates approximately 30 lb of force when operated at 40 psi at a distance of 1 in. from the panel. Air is supplied by a 120 gal Ingersoll-Rand air compressor. The compressor is sized to provide constant supply pressure for the entire duration of the test. The supply pressure can be adjusted which allows the force applied to the panels to be controlled.

The roof panel was tested using a supply pressure of 40 psi and a distance of 1 in. from the end of the pipe to the surface of the panel. This pressure results in a peak force of approximately 30 lb as indicated by Fig. 15.2. The resultant force was determined by summing the measurements from the three load cells.

Displacement data was obtained using three-dimensional image correlation. This allows for the displacements to be measured directly from the surface of the panel. The high contrast speckle pattern was hand applied to the rear surface of the panels. The speckle pattern covered an area measuring roughly 24 in. by 18 in.. Images were recorded using a pair of Photron Fastcam SA1.1 cameras operating at 1000 frames per second with a resolution of 1024 × 1024 pixels. Correlated Solutions Vic-3D software was used to calculate the displacements on the surface of the panel. Figure 15.3 shows a typical displacement plot as determined by the DIC software. The maximum displacement occurs in the center of the panel where the air jet is focused. As shown in Fig. 15.4 the maximum displacement has a magnitude of approximately 0.7 in..

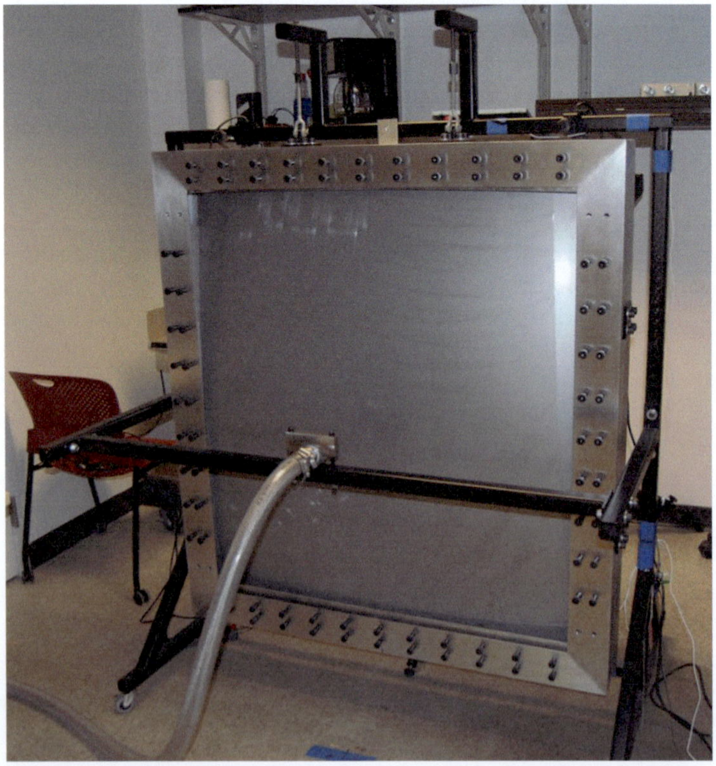

Fig. 15.1 Test rig

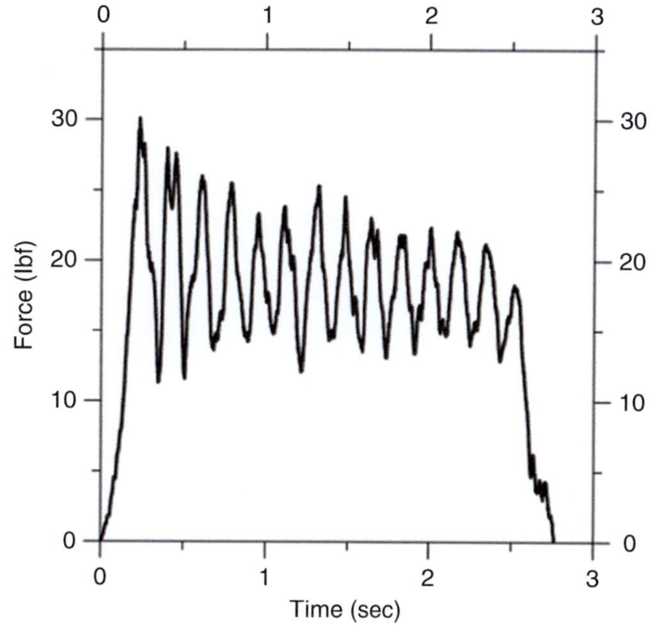

Fig. 15.2 Force vs. time, force determined from summation of three load cell values

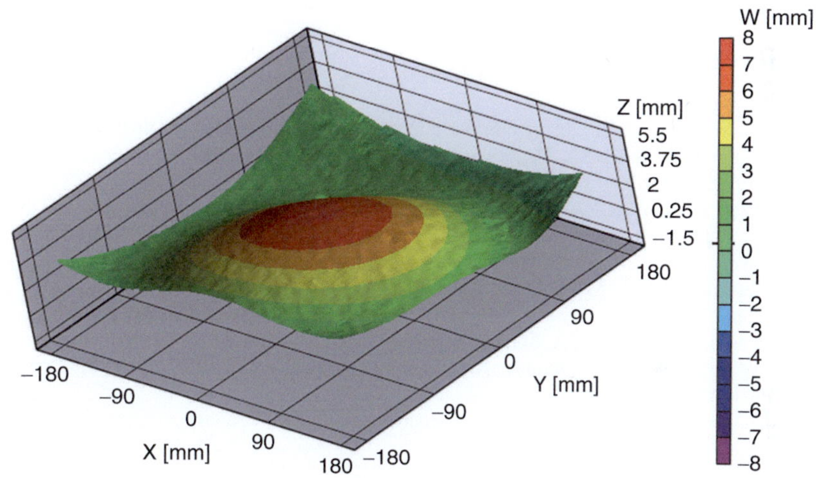

Fig. 15.3 Displacement contour from digital image correlation

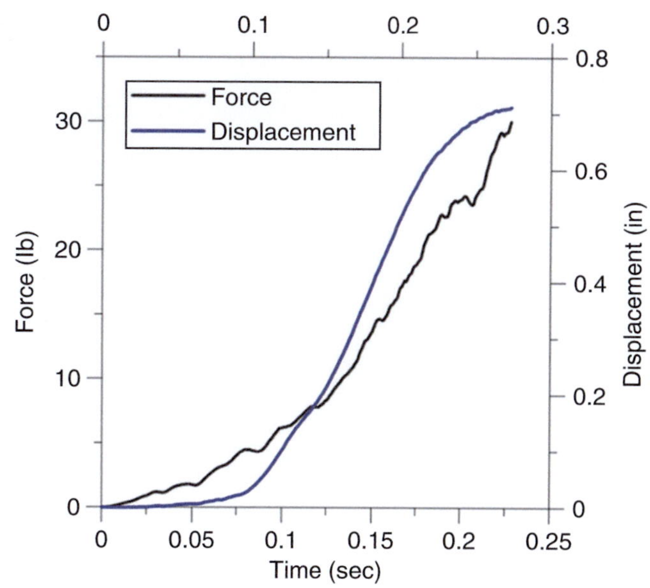

Fig. 15.4 Force vs. displacement

There is no strain observed in the panel under these test conditions which affirms the notion that the oil-canning phenomenon is typically a reversible process that does not damage the automotive panel.

15.3 Summary and Conclusions

An experimental test procedure is presented to replicate oil-canning in automotive roof panels for validation of concurrent fluid–structure simulations using the same setup. A test rig was designed that allowed tests to be performed on full size automotive roof panels with boundary conditions similar to those seen on production vehicles. Tests were performed using a 1.25 in. pipe to generate the high velocity air jet similar to those created by automotive air dryers. A total force of 30 lb was applied to the panel which resulted in a peak displacement of 0.7 in.. Full field displacement data was obtained using digital image correlation. These experimental data will be used as the validation metrics for fluid–structure interaction simulations in LS-DYNA. Validated FSI simulations will be an important tool for automotive designers to satisfy the customer while reducing overall vehicle weight.

Acknowledgements The authors would like to thank Duane Detwiler, Andy Fields, Allen Sheldon, Thomas Ramsay, and Yong-Bae Cho of Honda R&D Americas, Inc., and Ed Helwig and Emily Nutwell of LSTC for their support of this project. The authors would also like to thank Michael Tomaso of the Ohio State School of Architecture who machined the custom forms to clamp the roof panels.

References

1. F.D. Pin, Advances on the incompressible CFD solver in LS-DYNA. in 11th International LS-DYNA Users Conference (2010)
2. H.D. Espinosa, S. Lee, N. Moldovan, A novel fluid structure interaction experiment to investigate deformation of structural elements subjected to impulsive loading. Exp. Mech. **46**, 805–824 (2006)
3. X. Wei, P. Tran, A. Vaucorbeil, R.B. Ramaswamy, F. Latourte, H.D. Espinosa, Three-dimensional numerical modeling of composite panels subjected to underwater blast. J. Mech. Phys. Solids **61**, 1319–1336 (2013)
4. X. Wei, A. Vaucorbeil, P. Tran, H.D. Espinosa, A new rate-dependent unidirectional composite model—application to panels subjected to underwater blast. J. Mech. Phys. Solids **61**, 1305–1318 (2013)
5. P. Kumar, A. Shukla, Dynamic response of glass panels subjected to shock loading. J. Non-Cryst. Solids **357**, 3917–3923 (2011)
6. P. Kumar, D.S. Stargel, A. Shukla, Effect of plate curvature on blast response of carbon composite panels. Compos. Struct. **99**, 19–30 (2013)
7. E. Wang, J. Wright, A. Shukla, Analytical and experimental study on the fluid structure interaction during air blast loading. J. Appl. Phys. **110**, 114901 (2011)
8. W.X. He, H.P. Hong, Probabilistic characterization of roof panel uplift capacity under wind loading. Can. J. Civ. Eng. **39**, 1285–1296 (2012)

Chapter 16
In Vitro Experimental Study for the Determination of Cellular Characteristics of Mesenchymal Stem Cells Using a Non-uniform Deformation Field

Yasuyuki Morita, Toshihiro Sato, Sachi Watanabe, and Yang Ju

Abstract In the present study, relationship between cell orientation angle and strain value of membrane was comprehensively investigated using inhomogeneous strain field. And an axial strain threshold of cell, which corresponds to launch of cell orientation migration, was elucidated. One of the advantages in this study was that the inhomogeneous strain distribution was easily created by making a little improvement in a commonly-used uniaxial stretching device. The strains of two-dimensional stretched membrane were quantified position by position using digital image correlation (DIC) method. A 3D histogram of the cell frequency, which correlated with the cell orientation angle and normal strain of the membrane, made it possible to determine the axial strain threshold accurately. The value was 4.4 ± 0.3 %, which was reasonable compared with past study conducted by other researcher, although the past experiments were based on cyclic uniaxial stretch stimulation (homogeneous strain field). In addition, a preferential axial strain of the cell was achieved using the same technique of the determination of the axial strain threshold. This work has novel values at three points: (i) Determining axial strain threshold of the cells precisely. (ii) First suggestion of preferential axial strain of the cells. (iii) Investigating methodically cell behavior in inhomogeneous strain field.

Keywords Strain field • Inhomogeneous deformation • Digital image correlation (DIC) • Mesenchymal stem cell (MSC) • Cell orientation

16.1 Introduction

Connective tissues are always subjected to a wide range of mechanical loads which maintain the normal health and homeostasis of cells comprised in the tissues [1]. Cyclic stretch stimulation applied to tendons or ligaments, for instance, involves important effects on homeostasis and remodeling of the tissues [2]. Thus, various reactions of cells, such as orientation [3–6], morphology [7], migration [8], proliferation [9], differentiation [6, 10–12], proteins expression [10, 11], and so on, have been investigated with respect to mechanical stretch stimulation in vitro. Among them, cell orientation behavior due to mechanical stretch stimulation is familiar enough phenomenon, and one of the most classical subjects in this research field. It would be assured that Buck [13] was a precursor to study genealogically about the cell orientation response and hypothesized that the response was a stretch avoidance reaction. Buckley et al. [14] also suggested that cells aligned to minimize the strain acting on them. If their suggestions are correct, it is propounded that a certain strain threshold which launches orientation migration of cells could have to exist [4, 5]. Then, Dartsch et al. [15] determined that cells did not reorient about less than 2 % stretch amplitudes. Wang et al. [5] and Neidlinger-Wilke et al. [4] advocated that the strain threshold could be related to a longitudinal axial strain of cell, when the cell was considered as ellipsoid. And they named it axial strain threshold or axial strain limit. Wang et al. [5] explained cell orientation behavior by using their proposed cell model and the axial strain threshold based on theoretical analysis of elasticity. Neidlinger-Wilke et al. [4] determined in vitro experimentally that the axial strain thresholds of fibroblasts and osteoblasts were 4.2 ± 0.4 % (mean ± 95 % confidence limit) and 6.4 ± 0.6 %, respectively.

In this study, a two-dimensional inhomogeneous strain field of membrane was accomplished in different ways than other research groups. The prominent advantage of this technique was to easily obtain inhomogeneous strain distribution of membrane by making a little improvement in a commonly-used uniaxial stretching device. Cell orientation reaction was

Y. Morita (✉) • T. Sato • S. Watanabe • Y. Ju
Department of Mechanical Science and Engineering, Graduate School of Engineering, Nagoya University, Furo-cho, Chikusa-ku, Nagoya 464-8603, Japan
e-mail: morita@mech.nagoya-u.ac.jp

C. Sciammarella et al. (eds.), *Experimental and Applied Mechanics, Volume 4*,
Conference Proceedings of the Society for Experimental Mechanics Series, DOI 10.1007/978-3-319-22449-7_16

investigated minutely in the inhomogeneous strain field. Digital image correlation (DIC) method [16] was applied to quantify the two-dimensional strain distribution. The application of the DIC method to bioengineering field has been widely spread [17, 18], since the method does not require a complicated process of specimen preparation, and enables to quantify two-dimensional mechanical deformation of a measured object (There is no need that a measurement area must be flat.). These experimental techniques made it possible to make a correlation between the strain of membrane and cell orientation angle, and then a 3D histogram of cell frequency can be obtained. The cells were employed human bone marrow mesenchymal stem cells (hBMSCs). The 3D histogram of cell frequency and the theoretical equation of Wang et al. [5] enabled to elucidate the axial strain threshold of hBMSCs. The axial strain threshold of hBMSCs was 4.4 $\pm$ 0.3 %, which was similar value of the axial strain thresholds of other kinds of cells studied by Neidlinger-Wilke et al. [4] above mentioned. It was confident that the determination technique of the axial strain threshold in this study, which used by continuous inhomogeneous strain distribution of membrane, was more reliable and accurate that the one of Neidlinger-Wilke et al. [4], since their axial strain thresholds were determined by extrapolating experimental results of a couple of uniaxial stretch conditions. Furthermore, a preferential axial strain of the cells, which represents the mechanical circumstance that the cells want to settle down, was tried to be quantified by applying the determination method of the axial strain threshold. And the preferential axial strain of the cells was -2.0 ± 0.1 % in consequence.

16.2 Materials and Methods

16.2.1 Cell Preparation

Human bone marrow mesenchymal stem cells (hBMSCs; UE6E7T-3; Health Science Research Resources Bank, Tokyo, Japan) were maintained in low-glucose Dulbecco's modified Eagle's medium containing 10 % fetal calf serum, 0.5 % GlutaMax, and 0.05 % gentamicin (Invitrogen, Carlsbad, CA) at 37 °C under an atmosphere of 5 % CO_2 in a humidified incubator (MCO-5 AC; Sanyo, Tokyo, Japan). The hBMSCs were cultured in 25 cm^2 culture flasks (BD Biosciences, Franklin Lakes, NJ) at an initial density of 1.0×10^4 cells/cm^2 for expansion without differentiation. The medium was replaced with fresh medium every 3 days. At near-confluence, which occurred every 5–7 days, cells were detached from culture flasks with 0.25 % trypsin (Wako Pure Chemical Industries, Ltd., Osaka, Japan)/1 mM EDTA (B3232; Takara Bio, Shiga, Japan) and seeded into new culture flasks.

16.2.2 Application of Cyclic Stretch with Inhomogeneous Strain Field

A simple uniaxial stretching device (STB-140; Strex, Osaka, Japan), which were commonly used in the research field of cyclic mechanical stretch stimulation with homogeneous strain distribution [19], was employed to create inhomogeneous strain distribution. Elastic silicone elastomer chambers (STB-CH-10; Strex) used had a transparent bottom (32 $\times$ 32 mm^2) and walls that were 10 mm in height. The bottom of the chambers was undulated by the applied inhomogeneous strain field since the bottom membrane thickness of the chambers was thin (t = 100 [μm]). Therefore, we cut away the bottom membrane and adhered a stiffened thicker silicone elastomer (t = 300 [μm]) as a new bottom by using adhesive (SILPOT 184; Dow Corning Tray, Tokyo, Japan). The chamber was deformable by up to 20 % along a single direction by use of a stainless steel grappling-hook device. In this study, the upper side holes of the chamber were removed and the upper-side of the chamber was not grappled by the hooks to make non-uniform deformation (inhomogeneous strain distribution). The normal strain $\varepsilon^{mem}{}_{xx}(x, y)$ of the membrane in the stretched direction (x-axis) was dependent upon position of the membrane. A digital image correlation (DIC) method was employed to measure actual non-uniform strain distribution of the membrane.

The p-p strain distribution of the membrane obtained by the DIC method was exhibited in Fig. 16.1. The $\varepsilon^{mem}{}_{xx}$ represent the normal strain in the stretched direction (x-axis). It can be observed that the normal strain $\varepsilon^{mem}{}_{xx}$ in the stretched direction was increased from the upper side (free end) to the lower side (fixed end) of the membrane. Although the hBMSCs were seeded in the whole area of the membrane, region of interest (ROI) was set in the center area of the membrane (ROI in Fig. 16.1, 24 $\times$ 5 mm^2) since the normal strain $\varepsilon^{mem}{}_{xx}$ was changed smoothly along the orthogonal (y) direction. In addition, the center of the membrane was insusceptible to the edge effect which means that the edge area had undesirable drastic strain change and high/low cell density.

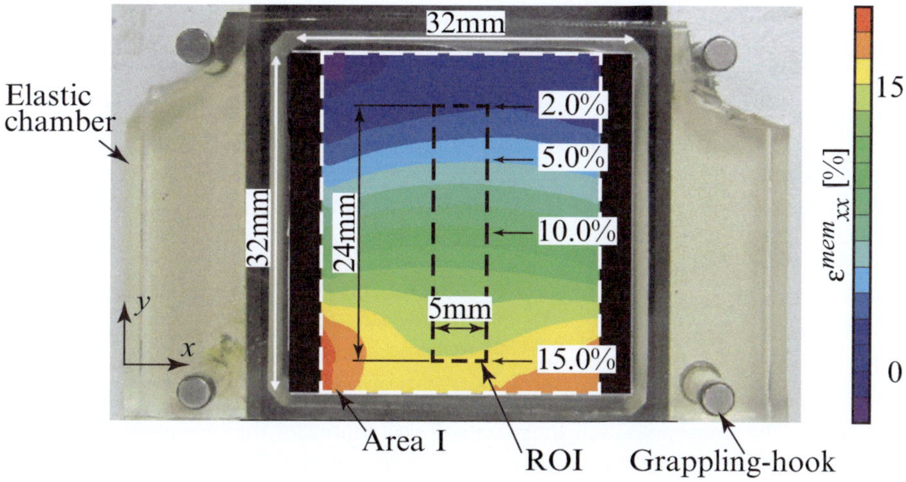

Fig. 16.1 The distribution of the normal strain $\varepsilon^{mem}{}_{xx}$ in the direction of stretch (x-axis)

16.2.3 Measurement of Cell Orientation

Cell morphology was observed using a phase-contrast microscope (IX71; Olympus, Tokyo, Japan). The orientation angle θ of each cell with respect to the stretch axis was determined. To evaluate the degree of orienting response of the cells, we constructed a 3D histogram that cell frequency was plotted against the orientation angle θ and the normal strain $\varepsilon^{mem}{}_{xx}$ of the membrane. If a cell was located in the conjugated position, i.e., an orientation angle of $-\theta$, the cell orientation was considered to be identical to a cell oriented at θ.

16.2.4 3D Histogram of the Cell Frequency

We obtained the three components of the strain, $\varepsilon^{mem}{}_{xx}$, $\varepsilon^{mem}{}_{yy}$, and $\gamma^{mem}{}_{xy}$ in respective positions of the membrane, i.e., (x, y, $\varepsilon^{mem}{}_{xx}$, $\varepsilon^{mem}{}_{yy}$, $\gamma^{mem}{}_{xy}$). The location and orientation angle of each cell were also obtained as (x, y, θ), by referring to the previous section, measurement of cell orientation. Therefore, a cell in the position (x, y) must have a property of ($\varepsilon^{mem}{}_{xx}$, $\varepsilon^{mem}{}_{yy}$, $\gamma^{mem}{}_{xy}$, θ). In this study, we focused on the effect of the normal strain $\varepsilon^{mem}{}_{xx}$ which corresponds to the stretched direction, since the values of $\varepsilon^{mem}{}_{yy}$, and $\gamma^{mem}{}_{xy}$ of the membrane were relatively small. Therefore, we investigated about a cell property of (x, y, $\varepsilon^{mem}{}_{xx}$, θ) from this section. A software of graphics (IGOR Pro 5; WaveMetrics, Portland, OR) was used to build the 3D histogram of the cell frequency with some smoothing process. The paired Student's t test was used for statistical analyses when statistical significance ($P < 0.05$) was evaluated to the samples of 0 h, which means the initial state of cells.

16.2.5 Determination of Axial Strain Threshold and Preferential Axial Strain of the Cells

Axial strain $\varepsilon^{cell}{}_{aa}$ of the cell can be determined by the following equation which Wang et al. [5] used,

$$\theta = \cos^{-1}\left(\sqrt{\frac{2(\varepsilon^{cell}{}_{aa}/\varepsilon^{mem}{}_{xx}) + 2\nu - \nu^2\varepsilon^{mem}{}_{xx}}{2(1+\nu) + (1-\nu^2)\varepsilon^{mem}{}_{xx}}}\right) \tag{16.1}$$

where ν represents Poisson's ratio of the membrane, which we measured to be $\nu = 0.43$. As the reference [5] described, this equation does not take into account the shear strain $\gamma^{mem}{}_{xy}$ of the membrane, though the normal strain $\varepsilon^{mem}{}_{yy}$ is considered. As mentioned above, the shear strain $\gamma^{mem}{}_{xy}$ was also considered as negligible small in this study, since the $\gamma^{mem}{}_{xy}$ values were less than one-tenth of the normal strain $\varepsilon^{mem}{}_{xx}$ at the respective positions of the membrane. The curve of (16.1) was fitted to the boundary line using regression analysis. The value of $\varepsilon^{cell}{}_{aa}$ at that time corresponds to the axial strain threshold $\varepsilon^{cell}{}_{aa\ thr}$ of the cells.

A preferential axial strain $\varepsilon^{cell}{}_{aa\ prf}$ of the cells can be obtained using the same technique. Maximum values of the means of the normalized cell frequency $\bar{f}$ at the each normal strain $\varepsilon^{mem}{}_{xx}$ of the membrane indicated that the cells wanted to settle down at the value of the axial strain $\varepsilon^{cell}{}_{aa}$. Therefore, the regression analysis was conducted to the maximum $\bar{f}$ at the each $\varepsilon^{mem}{}_{xx}$. The value of $\varepsilon^{cell}{}_{aa}$ at that time corresponds to the preferential axial strain $\varepsilon^{cell}{}_{aa\ prf}$ of the cells.

16.3 Results

The 3D histograms showing the mean of the normalized cell frequency $\bar{f}$ of the no (0 h)- and 48 h-stretched cells are shown in Fig. 16.2a, b, respectively. The transverse- and longitudinal-axis represent the orientation angle θ of each cell and the normal strain $\varepsilon^{mem}{}_{xx}$ of the membrane, respectively. Then, the contour colors indicate the $\bar{f}$. Separating from green of neutral color into purple of cold color and red of warm color in the $\bar{f}$ indicates that cell orientation behavior was occurred. Therefore, at first, we considered that the $\bar{f}$ distribution had to become approximately uniform 5.6 % (=Total frequency 100 %/Total angle 90°/Divided angle 5°) since the cells would exist randomly in no (0 h)-stretched experimental condition. However, non-uniform distribution of the $\bar{f}$ could be appeared as shown in Fig. 16.2a. The mean and standard deviation values are 5.6 ± 2.1 %, respectively. The standard deviation of 2.1 % should be considered the inherent dispersion of the cell in the present study.

From the 3D histogram of the cell frequency, the axial strain threshold $\varepsilon^{cell}{}_{aa\ thr}$ and preferential axial strain $\varepsilon^{cell}{}_{aa\ prf}$ of the cells were experimentally determined. When the $\varepsilon^{cell}{}_{aa\ thr}$ and $\varepsilon^{cell}{}_{aa\ prf}$ were determined, the 48 h-stretched cells (Fig. 16.2b) were used since the cell migration became steady state. The boundary of the mean of the normalized cell frequency $\bar{f} = 2.1$ (The reason of $\bar{f} = 2.1$ was described in the discussion.) was drawn by the white dots as shown in Fig. 16.3. Then, the fitted curve using (16.1) was also drawn by the red solid curve in the figure. The value of $\varepsilon^{cell}{}_{aa\ thr}$ was determined to 4.4 ± 0.3 % in consequence. On the other hand, the maximum $\bar{f}$ at the each $\varepsilon^{mem}{}_{xx}$ (in the $\varepsilon^{mem}{}_{xx}$ range which launched cell orientation behavior) were plotted as black dots in Fig. 16.3. The fitted curve using (16.1) was also drawn by the blue curve in the figure. The value of $\varepsilon^{cell}{}_{aa\ prf}$ was determined to −2.0 ± 0.1 % in consequence.

16.4 Discussion

Regarding the axial strain threshold $\varepsilon^{cell}{}_{aa\ thr}$ of cells, Neidlinger-Wilke et al. [4] determined it by extrapolating experimental results of three uniaxial stretch conditions of membrane, 4, 8, and 12 %. Their results showed that the axial strain thresholds $\varepsilon^{cell}{}_{aa\ thr}$ of fibroblasts and osteoblasts were 4.2 ± 0.4 % and 6.4 ± 0.6 %, respectively.

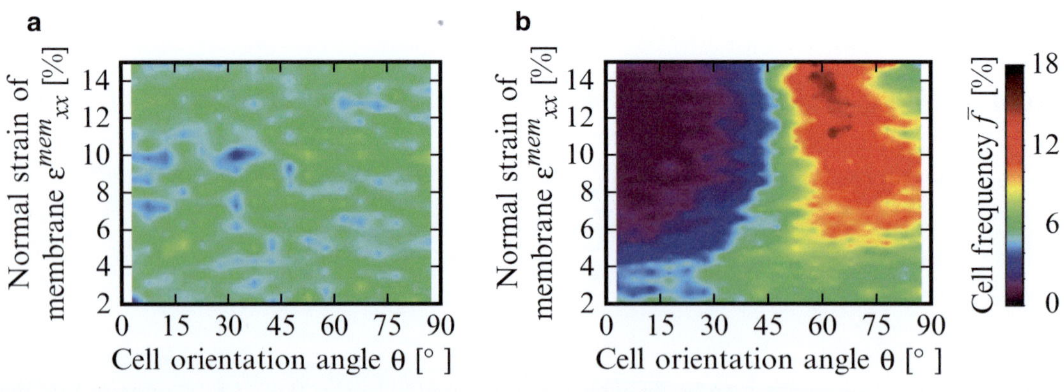

Fig. 16.2 3D histogram of the mean of the normalized cell frequency $\bar{f}$ with respect to the orientation angle θ of each cell and the normal strain $\varepsilon^{mem}{}_{xx}$ of the membrane. (**a**) Stretching for 0 h; (**b**) stretching for 48 h

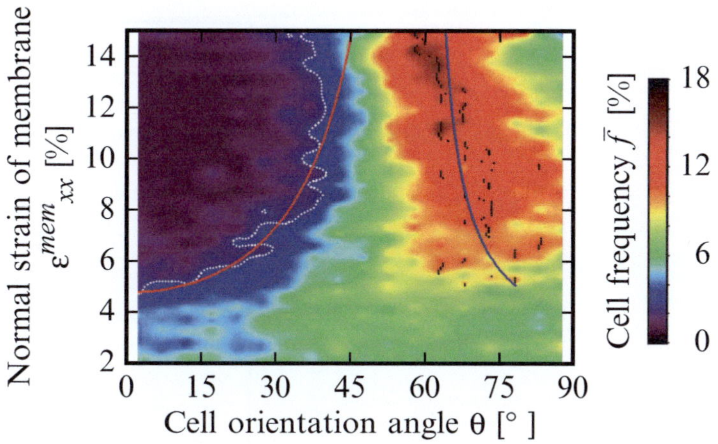

Fig. 16.3 Determination of the axial strain threshold $\varepsilon^{cell}{}_{aa\ thr}$ and the preferential axial strain $\varepsilon^{cell}{}_{aa\ prf}$ of the cells from the 3D histogram shown in Fig. 16.2b. The *white dots* and *red solid curve* represent the boundary of $\overline{f} = 2.1$ and the fitting curve of (16.1), respectively, for determination of $\varepsilon^{cell}{}_{aa\ thr}$. The *black dots* and *blue solid curve* represent the maximum $\overline{f}$ and the fitting curve of (16.1), respectively, for determination of $\varepsilon^{cell}{}_{aa\ prf}$

Determining the axial strain threshold is difficult and complicated to define the boundary between non-existing (cell frequency = 0) and existing (cell frequency > 0) regions of cells with respect to strains. It is, however, impossible to distinguish the boundary (cell frequency = 0) since the cell frequency does not become zero due to cell inherent dispersion and some experimental errors. Neidlinger-Wilke et al. [4] did not describe how to determine the boundary and draw the extrapolation lines. Yung et al. [20] pointed out that the study of Neidlinger-Wilke et al. [4] might be a small number of samples. The boundary of $\overline{f} = 0$ was not also obtained in the present study. Therefore, the standard deviation of $\overline{f}$ in no-stretched cells (Fig. 16.2a) became important to determine the axial strain threshold $\varepsilon^{cell}{}_{aa\ thr}$. The value of the standard deviation was 2.1 % as mentioned in the previous chapter, and would be able to be considered as cells inherent dispersion. Thus, the axial strain threshold $\varepsilon^{cell}{}_{aa\ thr}$ of the cells was determined by fitted (16.1) to the $\overline{f} = 2.1$. The 4.4 ± 0.3 % of the $\varepsilon^{cell}{}_{aa\ thr}$ was reasonable value compared with the values which Neidlinger-Wilke et al. [4] obtained. It was worthy to note that the axial strain threshold $\varepsilon^{cell}{}_{aa\ thr}$ of the cells was determined from innumerable continuous strain conditions, and by taking account of the inherent dispersion of the cells, under only one experiment in the present study. Therefore, it can be said that the suggesting technique is more simple, more precise, and higher reliable compared with cyclic uniaxial stretching experiments [4].

Moreover, other valuable results are able to be led from the 3D histogram of the cell frequency (Fig. 16.2). It was the same trend as well as cyclic uniaxial stimulation [3] that the cell orientation behavior became prominent in the respective normal strain values $\varepsilon^{mem}{}_{xx}$ with the duration time. It suggests that a cumulative value of p-p strain might be important factor in respect to cell orientation, since the cell orientation behavior extended to the lower normal strain values $\varepsilon^{mem}{}_{xx}$ with the duration time (Fig. 16.2b). Figure 16.4 shows the relationship between the cell orientation angle θ and the axial strain $\varepsilon^{cell}{}_{aa}$ of cells [9]. The parameter is the normal strain $\varepsilon^{mem}{}_{xx}$ of the membrane. From this figure, the absolute value of the axial strain $\varepsilon^{cell}{}_{aa}$ of cells took minimum value at almost 56.7° of the orientation angle, and then the tension and compression of the axial strain were flipped around the 56.7° by Poisson effect. The cell frequency $\overline{f}$ was higher in the compression region of the axial strain as shown in Fig. 16.2b. Therefore, Fig. 16.2 also indicated that the cells tend to tolerate and prefer moderate compressive axial strain $\varepsilon^{cell}{}_{aa} < 0$ rather than tensile axial strain $\varepsilon^{cell}{}_{aa} > 0$. Therefore, we might be able to define the preferential axial strain $\varepsilon^{cell}{}_{aa\ prf}$ of the cells in the same manner as the determination of the axial strain threshold $\varepsilon^{cell}{}_{aa\ thr}$ of the cells. Then the $\varepsilon^{cell}{}_{aa\ prf}$ determined by fitting (16.1) to the maximum $\overline{f}$ was −2.0 ± 0.1 %. The cells might definitely prefer compressive axial strain.

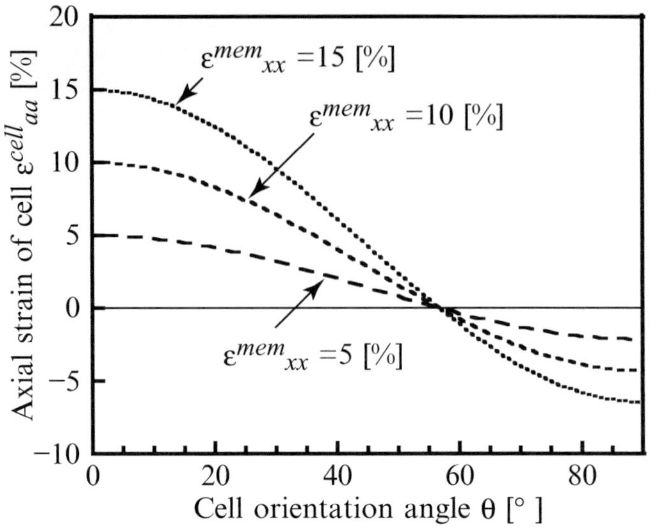

Fig. 16.4 Relationship between cell orientation angle θ and axial strain $\varepsilon^{cell}{}_{aa}$ of cell with respect to normal strain $\varepsilon^{mem}{}_{xx}$ of the membrane

16.5 Conclusions

It has advantage that our new experimental technique makes it possible to easily approach investigation of cells activities in inhomogeneous strain field by making a little improvement in a commonly-used uniaxial stretching device. And also combining the proposed device and DIC method, which is a method of deformation quantification, enable to directly correlate cell orientation angle with strains of membrane, and create 3D histogram of cell frequency in just one experiment. Furthermore, by using the 3D histogram, axial strain threshold of cells can be obtained more precise and reliable than the other researches so far. As a consequence, the axial strain threshold of hBMSCs is determined 4.4 ± 0.3 % in this study. In addition, a preferential axial strain of hBMSCs was also calculated by extending the same manners of the determination of axial strain threshold. The value of the preferential axial strain became -2.0 ± 0.1 %.

References

1. G. Giannone, M.P. Sheetz, Substrate rigidity and force define form through tyrosine phosphatase and kinase pathways. Trends Cell Biol. **16**, 213–223 (2006)
2. S.L.Y. Woo, M.A. Gomez, Y.K. Woo, W.H. Akeson, Mechanical-properties of tendons and ligaments II. The relationships of immobilization and exercise on tissue remodeling. Biorheology **19**, 397–408 (1982)
3. Y. Morita, S. Watanabe, Y. Ju, B. Xu, Determination of optimal cyclic uniaxial stretches for stem cell-to-tenocyte differentiation under a wide range of mechanical stretch conditions by evaluating gene expression and protein synthesis levels. Acta Bioeng. Biomech. **15**, 71–79 (2013)
4. C. Neidlinger-Wilke, E.S. Grood, J.H.C. Wang, R.A. Brand, L. Claes, Cell alignment is induced by cyclic changes in cell length: studies of cells grown in cyclically stretched substrates. J. Orthop. Res. **19**, 286–293 (2001)
5. H.C. Wang, W. Ip, R. Boissy, E.S. Grood, Cell orientation response to cyclically deformed substrates: experimental validation of a cell model. J. Biomech. **28**, 1543–1552 (1995)
6. L. Zhang, C.J.F. Kahn, H.Q. Chen, N. Tran, X. Wang, Effect of uniaxial stretching on rat bone mesenchymal stem cell: orientation and expressions of collagen types I and III and tenascin-C. Cell Biol. Int. **32**, 344–352 (2008)
7. S.P. Arold, J.Y. Wong, B. Suki, Design of a new stretching apparatus and the effects of cyclic strain and substratum on mouse lung epithelial-12 cells. Ann. Biomed. Eng. **35**, 1156–1164 (2007)
8. G.P. Raeber, M.P. Lutolf, J.A. Hubbell, Part II: fibroblasts preferentially migrate in the direction of principal strain. Biomech. Model. Mechan. **7**, 215–225 (2008)
9. G.G. Yang, R.C. Crawford, J.H.C. Wang, Proliferation and collagen production of human patellar tendon fibroblasts in response to cyclic uniaxial stretching in serum-free conditions. J. Biomech. **37**, 1543–1550 (2004)
10. B. Xu, G. Song, Y. Ju, Effect of focal adhesion kinase on the regulation of realignment and tenogenic differentiation of human mesenchymal stem cells by mechanical stretch. Connect. Tissue Res. **52**, 373–379 (2011)
11. B. Xu, G. Song, Y. Ju, X. Li, Y. Song, S. Watanabe, RhoA/ROCK, cytoskeletal dynamics and focal adhesion kinase are required for mechanical stretch-induced tenogenic differentiation of human mesenchymal stem cells. J. Cell. Physiol. **227**, 2722–2729 (2012)
12. L. Zhang, N. Tran, H.Q. Chen, C.J.F. Kahn, S. Marchal, F. Groubatch, X. Wang, Time-related changes in expression of collagen types I and III and of tenascin-C in rat bone mesenchymal stem cells under co-culture with ligament fibroblasts or uniaxial stretching. Cell Tissue Res. **332**, 101–109 (2008)

13. R.C. Buck, Reorientation response of cells to repeated stretch and recoil of the substratum. Exp. Cell Res. **127**, 470–474 (1980)
14. M.J. Buckley, A.J. Banes, L.G. Levin, B.E. Sumpio, M. Sato, R. Jordan, J. Gilbert, G.W. Link, R.T.S. Tay, Osteoblasts increase their rate of division and align in response to cyclic, mechanical tension in vitro. Bone Miner. **4**, 225–236 (1988)
15. P.C. Dartsch, H. Hammerle, Orientation response of arterial smooth muscle cells to mechanical stimulation. Eur. J. Cell Biol. **41**, 339–346 (1986)
16. M.A. Sutton, Digital image correlation for shape and deformation measurements, in *Springer handbook of experimental solid mechanics*, ed. by W.N. Sharpe (Springer, New York, 2008), pp. 565–600
17. Y. Morita, M. Uchino, M. Todo, Y. Matsushita, K. Arakawa, K. Koyano, Relationship between load-displacement curve and deformation distribution in porcine mandibular periodontium. J. Biomech. Sci. Eng. **4**, 336–344 (2009)
18. Y. Morita, Y. Matsushita, M. Todo, K. Koyano, Experimental study on displacement and strain distributions of bone model with dental implant. Appl. Mech. Mater. **83**, 73–77 (2011)
19. H. Inoh, N. Ishiguro, S.I. Sawazaki, H. Amma, M. Miyazu, H. Iwata, M. Sokabe, K. Naruse, Uni-axial cyclic stretch induces the activation of transcription factor nuclear factor B in human fibroblast cells. FASEB J. **16**, 405 (2002)
20. Y.C. Yung, H. Vandenburgh, D.J. Mooney, Cellular strain assessment tool (CSAT): precision-controlled cyclic uniaxial tensile loading. J. Biomech. **42**, 178–182 (2009)

Chapter 17
Development of New Method for Direct Measurement of High Strain Rate Testing Parameters

Rafid M. Kully

Abstract High strain rate testing and dynamic characterization of materials have been always serious challenges. Traditional Split Hopkinson Pressure Bar (SHPB) is being used for this purpose for the last 100 years. Traditional SHPB concept relays completely on the one dimensional stress wave propagation theory that adopts several assumptions. Besides, this method is subjected to some limitations of specimen dimensions, specimen material type, bars material, pulse shaping and others. The new developed direct measurement SHPB utilizes two force sensors to measure forces on both sides of the tested specimen thereby the stress on the specimen. Also, it utilizes two laser displacement sensors to measure the displacement on both sides of the specimen thereby measure the strain and the strain rate. This system measures the stress, the strain and the strain rate simultaneously all through the test period. The new developed method opens the door for great developments in the field of dynamic testing and characterization of all types of materials. It is a non-assumptions method independent of specimen material or dimensions.

Keywords SHPB • Direct measurement • High strain rate • Dynamic test • Displacement and force sensors

17.1 Introduction

Development of new materials and improvement of existed ones is a never ending task of nowadays technology. The industry of automobiles, aircrafts, aerospace, marines and other mobile and stationary applications are moving in wide steps. That increased the demand for more development in the mechanical and physical properties of the new and existed material. These developments require vast acknowledgment in materials behavior under deferent types of loading. Dynamic loading is the most important as it involves the effects of the strain rate on the material behavior. In particular, high strain rate response of the material carries a significant importance in its characterization. It determines if the material meets the durability and safety requirements of an application. For instance, it evaluates the energy absorption capability of the material to endure special conditions as in car crash, bird impact on blades of turbojet engines, hail damage and others. An accurate dynamic characterization of material depends completely on a reliable dynamic testing.

Traditionally, high strain rate (dynamic) tests of materials are conducted using test apparatus known as Split Hopkinson Pressure Bar (SHPB). The apparatus was first developed by Bertram Hopkinson to measure stress wave propagation in metal bars a 100 years ago [1]. Twenty five years later the system was modified by Kolinsky to measure dynamic response of materials [2]. Basically, when testing with SHPB, a test specimen is sandwiched between two bars and a strain/stress pulse propagates in the bars and the specimen. The stress, strain and strain rate of the specimen are calculated from the strains in the bars (more details in the next sections). This method adopts the one dimensional stress wave propagation theory with several assumptions like: The one dimensional elastic stress wave propagates in the bars without dispersion, no friction between the specimen and the bars, the specimen is under uniaxial stress conditions, the specimen is under dynamic equilibrium all through the test (no stress propagation in the specimen), bars material is homogenous, bars deform elastically, bars are physically straight and bars and specimen are perfectly aligned [3, 4]. Achieving these assumptions is not guaranteed during a dynamic test and it imposes some limitation. For instance, the specimen should be relatively thin to go with the assumption of dynamic equilibrium and lubricated to go with the assumption of uniaxial stress state.

To overcome the limitations and to avoid errors that could result by the assumptions; many have attempted to acquire direct measurement during dynamic tests as they are accurate and reliable. In the literature, several studies were found about using force sensor to directly measure the force acting on the specimen instead of strain measurement on transmission bar.

R.M. Kully (✉)
97th Air Mobility Wing, Air Education and Training Command, United States Air Force, 97th CES, Bldg 358, Room 1082, 401 L Ave., Altus AFB, Altus, OK 73523, USA
e-mail: rafid.kully@us.af.mil

C. Sciammarella et al. (eds.), *Experimental and Applied Mechanics, Volume 4*,
Conference Proceedings of the Society for Experimental Mechanics Series, DOI 10.1007/978-3-319-22449-7_17

Different brands of force sensors were used in different studies. However, in most cases the sensors contained piezoelectric crystals (quartz transducers) for their extremely high natural frequency and excellent linear behavior over a wide amplitude range [5–13]. On the other hand, few alternative attempts to measure dynamic strain history were found in the literature. One approach is using the high speed camera to capture the specimen edge deformation during the dynamic test and conduct some analysis to calculate the strain [6, 14–16]. Other approach is to measure the deformation of the specimen by tracking the amount of light passing around the specimen. The light is generated by laser line sensor (LLS) and transmitted through the opening around the specimen to a photodiode of a collecting lens. The amount of light reaching the photodiode diminishes as the specimen deforms. With some calculations the dynamic strain can be calculated [11, 17].

This paper discusses a new approach of measuring the displacement history of the specimen's two edges simultaneously, thereby determining the dynamic strain and strain rate directly. Two laser displacement sensors are to be employed for this purpose. Also, two force sensors are to be used to measure the forces thereby the stress on the specimen. The dynamic results acquired from the direct measurement are to be compared with results of the conventional SHPB. The methodology followed in this paper is to validate the available conventional SHPB by testing polycarbonate and nylon 6/6 specimens and compare the test results to published results. Then modify the validated apparatus to acquire direct measurement as well as conventional measurement simultaneously. Use the modified SHPB apparatus to test the same type specimens again to compare the results and to validate the direct measurement.

17.2 Test Materials and Specimens

Two types of materials were selected to conduct the dynamic tests for this study, polycarbonate and nylon 6/6. The dynamic test results for these two materials are available in the literature. That makes the selection practical for comparison of test results and validation of test instrumentations and analysis. The polycarbonate of 1.27 g/cm^3 density and 77 MPa static yielding strength and the nylon 6/6 of 1.16 g/cm^3 density and 55 MPa static yielding strength were provided by McMaster-Carr [18]. Polycarbonate and nylon 6/6 specimens of 6.35 mm diameter and 6.35 mm length were prepared. Both materials were provided as 6.35 mm diameter rods. Specimen preparation process included cutting 7 mm length specimen and inserting it in 6.35 mm diameter hole in a 6.35 mm thick aluminum plate. The two ends of the specimen were sanded down to the plate level. The resulted specimen length was exactly 6.35 mm and the two faces of specimen were perfectly parallel to reduce the misalignment with the test apparatus during the test.

17.3 Conventional SHPB Testing

SHPB apparatus appears in Fig. 17.1 consists of two bars known as incidence bar and transmission bar and a third projectile bar driven by the pressure of a gas gun, known as strike bar. When testing with SHPB a test specimen is sandwiched between incidence bar and transmission bar. A stress/strain compression wave is generated by the impact of the strike bar on the impact end of the incidence bar. The stress/strain pulse propagates through the incidence bar toward the specimen, called incident pulse, $\varepsilon_i(t)$. When the pulse reaches the specimen-incidence bar interference, part of the pulse passes through the specimen to the transmission bar, called transmitted pulse, $\varepsilon_t(t)$. The other part of the incident pulse reflects back to the incidence bar as a tensile pulse, called reflected pulse, $\varepsilon_r(t)$. Pulse reflection occurs because of the impedance deference of the specimen and bars materials. The incident and reflected pulses are measured by a strain gauge

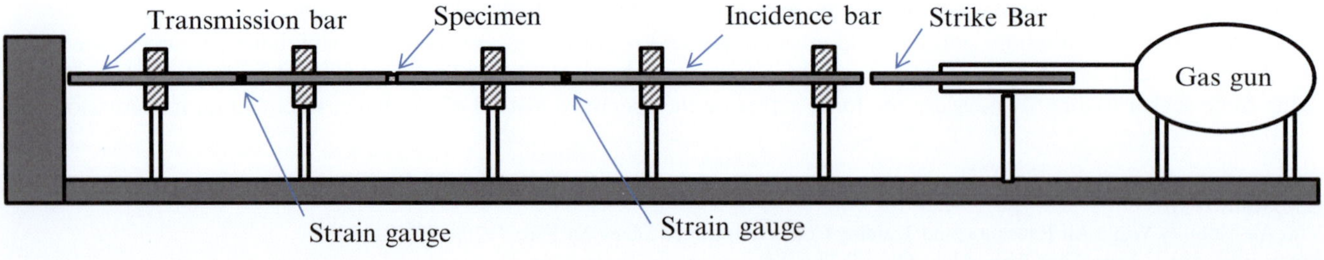

Fig. 17.1 Schematic of conventional SHPB

installed on the incidence bar and the transmitted pulse is measured by a strain gauge installed on the transmission bar. The two gauges are at the same distance for the specimen (0.9 m). During the test, the specimen undergoes deformation until it reaches the dynamic limit. The concept of the one dimensional stress wave propagation theory is used to calculate the strain rate in the specimen from the reflected wave signal, $\varepsilon_r(t)$ as:

$$\dot{\varepsilon}_s(t) = \frac{2c_b\varepsilon_r(t)}{l} \tag{17.1}$$

Where c_b is the speed of sound in the bars and l is the specimen length. The strain of the specimen is calculated by integrating the strain rate with respect to time as:

$$\varepsilon_s(t) = \frac{2c_b}{l}\int_0^t \varepsilon_r(t)dt \tag{17.2}$$

The stress of the specimen is calculated from transmitted wave signal, $\varepsilon_t(t)$ as:

$$\sigma_s(t) = \frac{A_bE_b}{A_s}\varepsilon_t(t) \tag{17.3}$$

Where A_b is the cross-section area of incidence/transmission bar, A_s is the cross-section area of the specimen and E_b is the elastic modulus of bar material. The stress strain response of the specimen is obtained by (17.2) and (17.3). This method of computation is given in number of text books and references on impact; like [10, 19, 20]. SHPB used in this experiment has its three bars made of aluminum 7075-T6. Strike, incidence and transmissions bars are all of 19 mm diameter and lengths of 0.76 m, 3.66 m and 1.83 m, respectively.

Polycarbonate and nylon 6/6 specimens were tested at 1200/s and 1250/s strain rates, respectively. Test results were compared to data in literature for validation. Polycarbonate test result were compared to results of Salisbury [21] and nylon 6/6 results were compared to result of Chou et al. [8]. Comparisons showed very well agreement between experimental and published results, which validated our used SHPB (See Fig. 17.2). More details about the apparatus and validation process are given in reference [22].

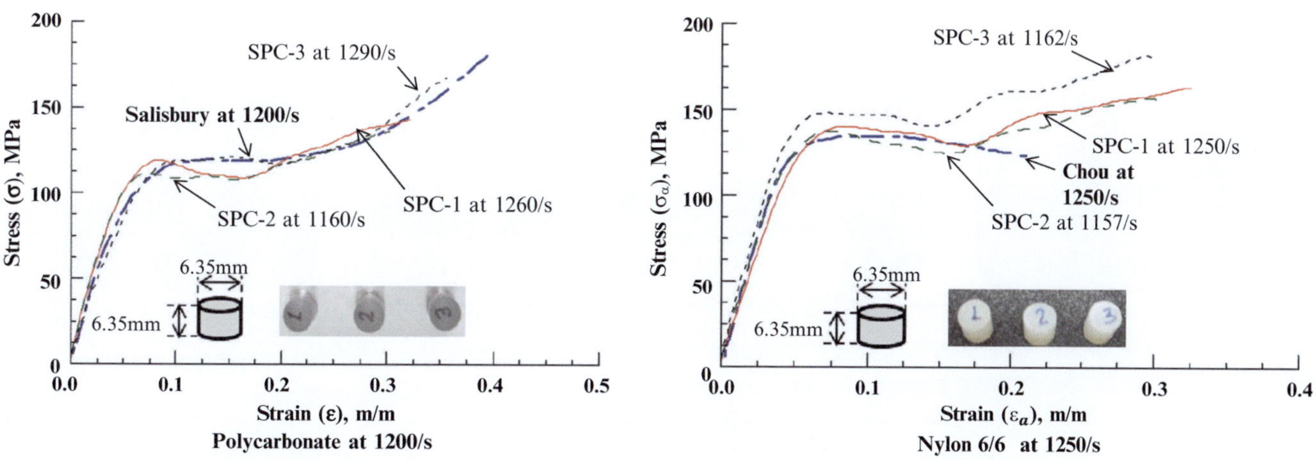

Fig. 17.2 Comparison between experimental and published test results for conventional SHPB

17.4 New Direct Measurement SHPB Testing

17.4.1 Direct Measurement SHPB Apparatus

The development of direct measurement SHPB apparatus focused on two major parts; direct measurement of the force acting on the specimen and direct measurement of specimen's both sides displacement during the test. For force measurement, two piezoelectric force sensors model 1051v6 provided by Dytran, Inc., were used. They were of 5000 lbf (22.24 kN) compression range and of 75 kHz natural frequency (high natural frequency). That made them in the safe range of measuring the low frequency forces (1.2–1.3 kHz) presented in our tests. The body of the sensor is made of stainless steel with total impedance close to the bars impedance so it doesn't disperse the pulse. A photograph and more details on the sensor can be found in reference [23]. Sensors were already calibrated by manufacture upon delivery. The factory calibration certificate indicated +2 % error of each sensor that needs to be accounted for when taken measurement. To confirm the accuracy of the sensors another calibration process was conducted in the laboratory. The two sensors were calibrated against each other's and against calibrated MTS machine. The measurements of the two sensors matched each other's exactly and both were off by +2 % from the MTS machine measurement. That process verified the accuracy of the sensors. For displacements measurements, two laser sensors model LK-H057 provided by Keyence Corporation were used. The sensor offers high repeatability of 0.005 μm, high accuracy of ±0.02 % and high sampling rate of 392 kHz. A photograph and more details of the sensor can be found in reference [24]. The sensor measures the instant distance between its front edge and its target (any opaque object, preferred white). Reference distance between the sensor and the target is 50 mm and measured rage is ±10 mm. Then the total measured displacement range is 20 mm. Sensors were calibrated against calibrated digital caliber and found to be very accurate.

After calibrating the sensors, they were used to modify the conventional SHPB to the new direct measurement SHPB. Note that the conventional setup was kept while the direct measurement setup was added to SHPB apparatus. That was for the sake of comparison. Figure 17.3a, b show the schematic and the photographs of direct measurement SHPB apparatus

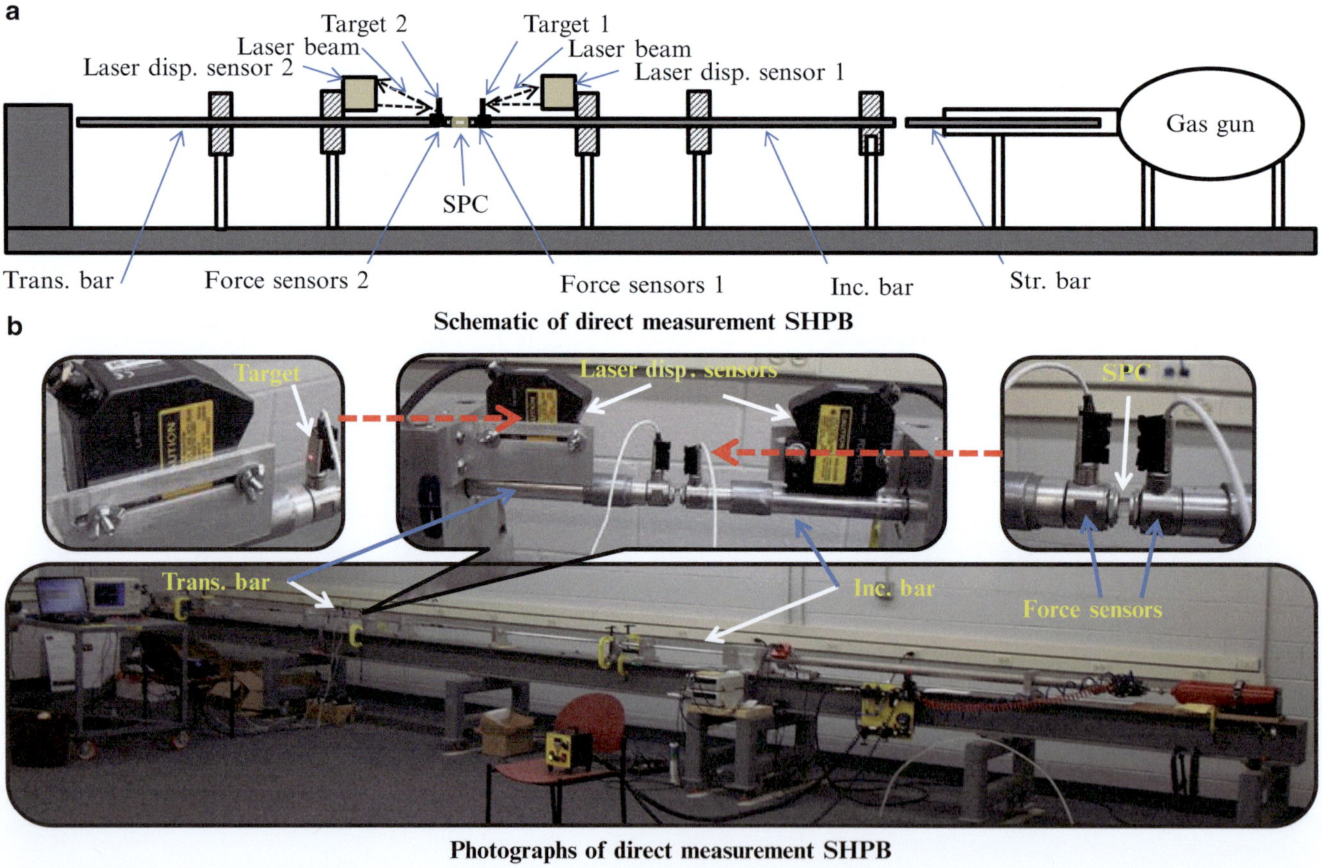

Fig. 17.3 Direct measurement SHPB

(with only the related direct measurement parts), respectively. The two force sensors were mounted on the incidence and transmission bars. The laser displacement sensors were mounted on the bearing stands. The laser beam of the first sensor pointed to a small target fixed on the incidence bar beyond the first force sensor. Then the very tiny displacement of the force sensor wouldn't be sensed by the laser sensor. Also, the laser beam of the second laser sensor was pointed to another small target mounted on the transmission bar in front of the second force sensor for the same reason. The sensors were connected to their auxiliary electronic devices for power supply and data acquisition. The outputs of all sensors were conncected to and displayed by the main oscilloscope. Also, the outputs of the incidence and transmission strain gauges were connected to the same oscilloscope to acquire direct measurement and conventional measurements at the same time.

17.4.2 Testing

Starting with the polycarbonate, a specimen was placed between the incidence and transmission bars (force sensor 1 and 2). The center line of the specimen was aligned with the centerline of the bars and the force sensors. Bars' locations were adjusted so that the laser beams of the displacement sensors are pointed to the targets on the bars and the distance between each sensor and its target was 50 mm. the readings of the laser sensors were zeroed out on the computer by the LK-G5000 viewer software. The gas gun pressure was set to 97 kPa (according to calibrating process for strain rate versus gas gun pressure) and the oscilloscope was turned on trigger position. The system was fired, strike bar impacted incidence bar and data was acquired for outputs from load and displacement sensors as well as the strain gauges at a 0.5 MHz sampling rate. Three polycarbonate specimens were tested in the same manner. Same steps were repeated to test three nylon 6/6 specimens, but, at 103 kPa gas gun pressure this time (according to calibrating process as well).

17.4.3 Test Results

A typical response of all measured parameters during the polycarbonate test appears in Fig. 17.4. It is a screenshot of the oscilloscope focused on initial important part of the impulses. The figure shows that the incidence force sensor pulse led the

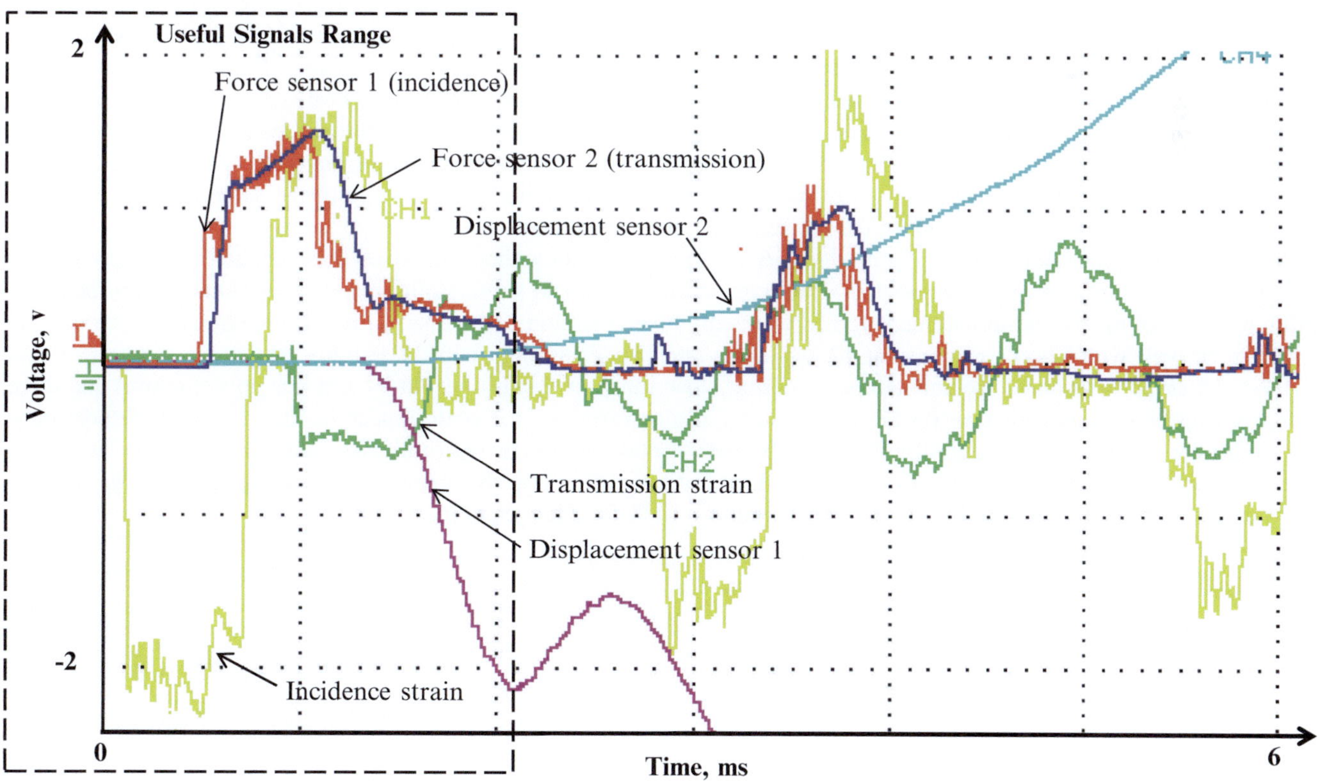

Fig. 17.4 Oscilloscope screenshot for typical direct and conventional measurements of polycarbonate dynamic testing

transmission force sensor pulse by 50 μs. After 50 μs the two pulses overlapped each other's to indicate that the specimen has reached its dynamic limit (dynamic equilibrium). The two pulses had similar pattern and amplitude. Therefore, only one pulse (transmitted pulse) was used in further analysis to calculate the force acting on the specimen during the test. The incident pulse had more noise than the transmitted pulse, but, they were very similar when smoothed. The figure also shows that the deformation of the specimen lagged the transmitted force pulse by 800 μs. The deformation started after the pulse has almost passed completely through the specimen. In addition, the figure shows that almost all the deformation occurred on the incidence side of the specimen and propagated to the other side of the specimen that only moved slightly during the deformation process. Note that the signal of the displacement sensor 1 was of a negative sign as the target was moving away from the sensor. But, as deformation the signal was corrected to positive sign.

The conventional responses were acquired and reduced as explained before. The direct measurement responses were calculated as following: The strain at each sampling was calculated from the displacements of the specimen's two edges as;

$$\varepsilon_s(t) = \frac{d_1(t) - d_2(t)}{l} \tag{17.4}$$

Where d_1(t) and d_2(t) are the displacement of the incidence side and transmission side of the specimen, respectively and l is the original length of the specimen, which is 6.35 mm in this test. The strain rate was calculated from the strain and sampling time as;

$$\dot{\varepsilon}_s(t) = \frac{\varepsilon_s(t)}{t} \tag{17.5}$$

Where t is the sampling time, which is 2 μs in this test. The stress acting on the specimen was calculated from the transmitted force pulse as;

$$\sigma_s(t) = \frac{F_t(t)}{A_s} \tag{17.6}$$

Where F_t (t) is the force of the transmission force sensor (force sensor 2) and A_s is the cross section aria of the specimen, which is 31.67 mm in this test. The stress strain response was derived from the stress versus time response and the strain versus time response by synchronizing the two responses. Figure 17.5a–d show the comparison between the conventional and direct measurement responses of the strain rate, strain, stress and stress strain, respectively. Figure 17.5d shows that the direct measurement stress strain response agreed very well with the conventional one (1180/s conventional strain rate). Both responses agreed very well with Salisbury's results at 1200/s strain rate of a conventional measurement as well [21]. That validates the direct measurement method to determine the stress strain response of materials. However, Fig. 17.5a shows that the direct measurement stain rate was higher than the conventional strain rate of the same test. As the direct measurement strain rate was 1460/s since the conventional one was 1180/s.

Nylon 6/6 specimen test results appear in Fig. 17.6a–d. The figures show the comparison between the direct measurement and conventional measurement responses of strain rate, strain, stress and stress strain, respectively. Once again the stress strain response of the direct measurement agreed well with the conventional response (1140/s conventional strain rate) and both agreed well with Chou's response at 1250/s strain rate [8]. That confirmed the validity of the direct measurement method. Three polycarbonate specimens were tested at close range of strain rates (at 97 kPa gas gun pressure) and three nylon 6/6 specimens as well (at 103 kPa). The stress strain responses for both appear in Fig. 17.7a, b, respectively. The figures indicate good repeatability of the tests, show good agreement with published data and validate the direct measurement. However, the average direct measurement strain rate of the three specimens was higher than the average conventional strain rate by 13 % for polycarbonate, and by 14 % for the nylon 6/6.

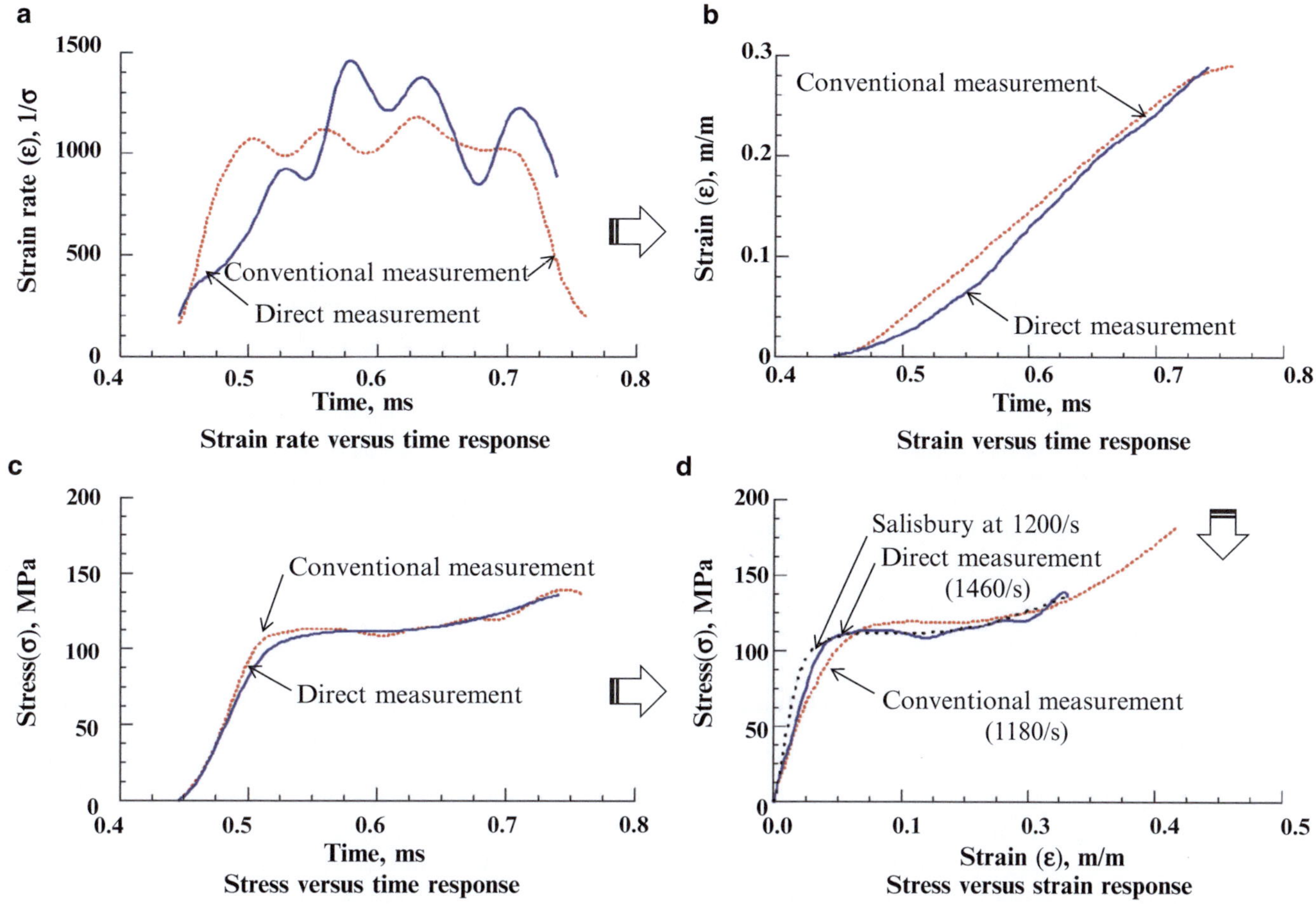

Fig. 17.5 Comparison of direct and conventional measurement responses of polycarbonate

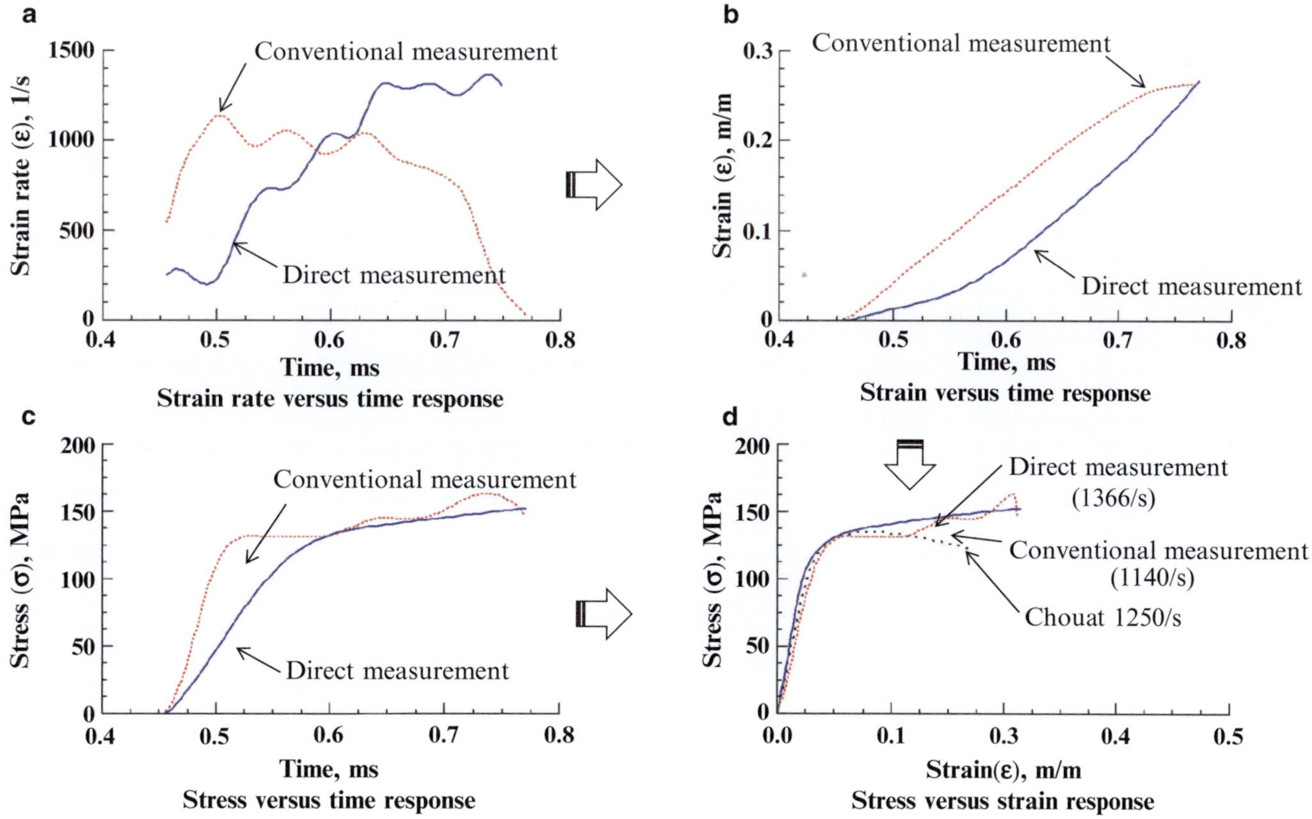

Fig. 17.6 Comparison of direct and conventional measurement responses of polycarbonate

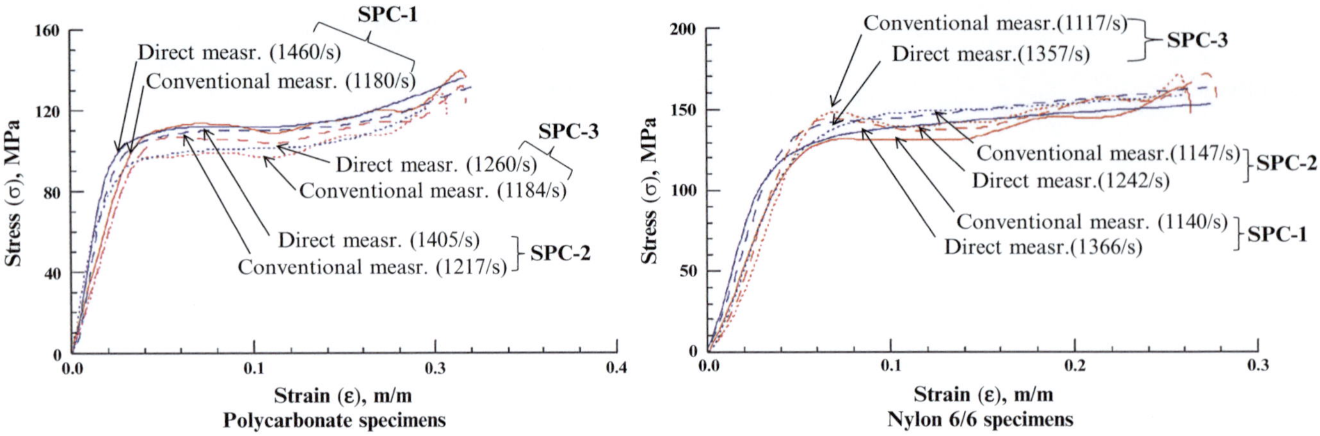

Fig. 17.7 Comparison between direct and conventional measurements of dynamic test

17.5 Conclusion

New direct measurement SHPB was developed by utilizing two force sensors to measure the forces acting on both side of the specimen thereby determine the stress of the specimen. Also, utilizing two laser displacement sensors to measure the displacement on both sides of the specimen thereby determine the stain and the strain rate of the specimen. Three polycarbonate specimens and three nylon 6/6 specimen all of 6.35 mm diameter and length were tested by the new apparatus. Results showed very good agreement between the new direct measurement method and the conventional method for the same test. Also, it showed good agreement with test results available in the literature. These agreements validate and confirmed the fidelity of the new direct measurement SHPB apparatus. Test result also showed that the strain rate measured by the direct measurement method is about 14 % higher than the strain rate measured by the conventional method for the same test.

References

1. B. Hopkinson, A method of measuring the pressure produced in the detonation of high explosives or by the impact of bullets. Proc. R. Soc. London Ser. A **89**(612), 411–413 (1914)
2. H. Kolsky, An investigation of the mechanical properties of materials at very high rates of loading. Proc. Phys. Soc. Sect. B **62**(11), 676 (1949)
3. K.T. Ramesh, High rates and impact experiments, in *Springer handbook of experimental solid mechanics*, ed. by W.N. Sharpe (Springer, Berlin, 2008), pp. 929–960
4. W.W. Chen, B. Song, *Split Hopkinson (Kolsky) bar: design, testing and applications* (Springer, New York, 2010)
5. I.-G. Kim, H.-Y. Lee, J.-W. Kim, Impact damage detection in composite laminates using PVDF and PZT sensor signals. J. Intell. Mater. Syst. Struct. **16**(11–12), 1007–1013 (2005)
6. B. Song, W. Chen, One-dimensional dynamic compressive behavior of EPDM rubber. J. Eng. Mater. Technol. **125**(3), 294–301 (2003)
7. D. Van Nuffel et al., Calibration of dynamic piezoelectric force transducers using the hopkinson bar technique. in 15th International Conference on Experimental Mechanics (ICEM15-2012). INEGI-Instituto de Engenharia Mecânica e Gestão Industrial (2012)
8. S. Chou, K. Robertson, J. Rainey, The effect of strain rate and heat developed during deformation on the stress-strain curve of plastics. Exp. Mech. **13**(10), 422–432 (1973)
9. W. Chen, F. Lu, N. Winfree, High-strain-rate compressive behavior of a rigid polyurethane foam with various densities. Exp. Mech. **42**(1), 65–73 (2002)
10. B. Song, W. Chen, Split Hopkinson pressure bar techniques for characterizing soft materials. Lat. Am. J. Solids Struct. **2**(2), 113–152 (2005)
11. M. Trexler et al., Verification and implementation of a modified split Hopkinson pressure bar technique for characterizing biological tissue and soft biosimulant materials under dynamic shear loading. J. Mech. Behav. Biomed. Mater. **4**(8), 1920–1928 (2011)
12. W. Chen, F. Lu, B. Zhou, A quartz-crystal-embedded split Hopkinson pressure bar for soft materials. Exp. Mech. **40**(1), 1–6 (2000)
13. W. Chen et al., Dynamic compression testing of soft materials. J. Appl. Mech. **69**(3), 214–223 (2002)
14. A. Cole, J.F. Quinlan, F. Zandman, The use of high-speed photography and photoelastic coatings for the determination of dynamic strains. in Proceedings of 5th International Congress on High-Speed Photography, vol. 250 (1962), p. 261
15. A.E. Abrantes, J.A. Yamamuro, Effect of strain rates in cohesion less soil, in *Constitutive modeling of geomaterials*, ed. by H.I. Ling et al. (CRC, Boca Raton, 2003), pp. 188–194

16. C. Siviour, S. Grantham, High resolution optical measurements of specimen deformation in the split Hopkinson pressure bar. Imaging Sci. J. **57**(6), 333–343 (2009)
17. K.T. Ramesh, N. Kelkar, Technique for the continuous measurement of projectile velocities in plate impact experiments. Rev. Sci. Instrum. **66**(4), 3034–3036 (1995)
18. McMaster-Carr, http://www.mcmaster.com/
19. M.A. Meyers, *Dynamic behavior of materials* (Wiley, New York, 1994)
20. W. Chen, Testing conditions on Kolsky bar, in *Materials under extreme loadings: application to penetration and impact*, ed. by E. Buzaud, I.R. Ionescu, G.Z. Voyiadjis (Wiley, New York, 2013), pp. 131–144
21. C. Salisbury, Spectral analysis of wave propagation through a polymeric Hopkinson bar. Master's thesis. Department of Mechanical Engineering, University of Waterloo, Waterloo (2001)
22. R.M. Kully, Dynamic constitutive equation for a syntactic foam under multi-axial stress state. Dissertation. Mechanical Engineering, North Carolina A&T State University (2014)
23. Incorporated, D.I., Model 1051 V6, IEPE Force Sensor
24. Keyence Corporation, Ultra high-speed/high-accuracy laser displacement sensore LK-H057 (Concord, MA)

Chapter 18
Simultaneous Measurement of Polymerization Stress Evolution, Conversion Kinetics, and Exotherm in Real-Time

Zhengzhi Wang, Forrest A. Landis, Anthony A.M. Giuseppetti, Sheng Lin-Gibson, and Martin Y.M. Chiang

Abstract Photopolymerized composites are used in a broad range of applications with their performance largely directed by reaction kinetics and contraction accompanying polymerization. Herein, we demonstrate an instrument capable of simultaneously collecting multiple kinetics parameters for a wide range of photopolymerizable systems: degree of conversion (DC), reaction exotherm, and polymerization stress (PS). Our system consists of a cantilever beam-based instrument (tensometer) that has been optimized to capture a large range of stresses. The sample configuration allows the tensometer being coupled to a fast near infrared (NIR) spectrometer collecting spectra in transmission. Using our instrument design, simultaneous measurements of PS, DC, and exotherm are performed, for the first time, on a commercial composite with ~80 % (by mass) silica particle fillers. This new system is expected to provide new insights into kinetics and property relationships for photopolymerized composites.

Keywords Polymerization stress • Tensometer • Infrared spectrometer • Photopolymerization • Composites

18.1 Introduction

Photopolymerized composites are widely used in applications ranging from electrical and optical materials, structural materials, coatings and adhesives, to various biomedical applications such as dental restoration [1]. As monomers polymerize to form covalent bonds during the curing process, the volume of the resulting polymers can contract by significant amount [2]. Associated with the shrinkage, enormous polymerization stress (PS) can develop for materials that shrink by as little as 1 % owing to the high modulus of these materials and can reduce the overall material performance. For example, in polymeric dental composites, considerable PS can develop at the tooth-composite interface that may lead to debonding, further tooth decay at the interface, and/or cracking of the tooth [3]. In coatings and adhesives, stress generated may lead to surface cracking and debonding. One of the common features in photopolymerization processes is that gelation can take place very fast, usually within seconds. Therefore, it is important to accurately follow the curing kinetics of such fast reactions for a better assessment of the performance of monomers, as well as understanding and controlling the curing process. The object of this study is to introduce new instrumentation that can continuously monitor the development of PS and curing kinetics in real-time for polymer composites with wide range of filler contents during photo-cure processing.

PS development is dictated by complex interplay among several factors, including curing kinetics (normally characterized by the degree of conversion, DC), development of resin shrinkage and modulus associated with DC, sample dimensions, and external constraint [4]. Several instruments have been introduced in the literature to characterize PS for photo-polymerized materials [5–7]. To understand the reaction kinetics and PS development, it is necessary to measure these properties simultaneously. To this end, non-destructive infrared spectroscopy technique has been utilized to couple with PS measurement instrument for the in situ measurement of DC. However, due to the limits in the instrument design and the resolution of the devices adopted, the simultaneous measurement of PS and DC reported is limited by either the special sample requirement (e.g., only for coatings or thin films [8]) or the range of composites that can be tested (e.g., only for

Z. Wang • S. Lin-Gibson • M.Y.M. Chiang (✉)
Biosystems and Biomaterials Division, National Institute of Standards and Technology, Gaithersburg, MD 20899, USA
e-mail: martin.chiang@nist.gov

F.A. Landis
Department of Chemistry, Penn State University, York Campus, York, PA 17403, USA

A.A.M. Giuseppetti
American Dental Association Foundation, Dr. Anthony Volpe Research Center, National Institute of Standards and Technology, Gaithersburg, MD 20899, USA

C. Sciammarella et al. (eds.), *Experimental and Applied Mechanics, Volume 4*,
Conference Proceedings of the Society for Experimental Mechanics Series, DOI 10.1007/978-3-319-22449-7_18

composites with filler content up to 30 % by mass [9]). Many practical applications use a significantly higher amount of fillers (as high as 80 % by mass) to improve the mechanical, tribological, and thermal properties of the materials.

In this study, a tensometer constructed according to our design criteria [10] was coupled to a high-speed near infrared (NIR) spectrometer to enable real-time, simultaneous measurement of PS and DC for a wide range of polymer composites. Since temperature rise during photo-polymerization is usually a critical parameter in determining the applicability of the material, and reaction exotherm is also associated with the curing kinetics, the evolution of exothermic temperature is measured in parallel. Due to the existence of a vast number of monomer systems, studies related to methacrylate-based composites were chosen here. Our results indicate that the simultaneous measurement can be performed on a model highly-filled composite (≈80 % by mass) at a high acquisition rate in excess of 10 data points per second, which gives sufficient data to accurately evaluate curing kinetics from the inception of polymerization.

18.2 Methods[1]

In this investigation, a commercial dental composite (TPH Spectra™, Dentsply-Caulk, Milford, DE) was used as the testing material to demonstrate the capabilities of the newly developed system. TPH is a blue-light activated, urethane modified Bis-GMA/TEGDMA (50:50 mass ratio) based composite, filled primarily with barium boron aluminum silicate glass particles at 78 % by mass (≈57 % by volume).

A cantilever beam-based instrument (tensometer) presented in our previous study [10] has been further improved (described below) and a dispersive NIR spectrometer has been coupled to the instrument as shown in Fig. 18.1. Using this device, an uncured composite (or resin) specimen was bonded with two quartz rods, the lower rod was fixed and the upper rod attached to a cantilever beam. Upon photo-polymerization, the composite shrinkage stress induced a deflection in the calibrated cantilever beam, which was recorded by a capacitive displacement sensor (resolution of ≈20 nm, CPL 190, Lion Precision, St. Paul, MN) mounted at the free end of the beam (Fig. 18.1C). The PS was then calculated through the beam formula:

$$\sigma = \frac{F}{A} = \frac{6\delta EI}{\pi r^2 a^2 (3l - a)} \tag{18.1}$$

where σ is the PS and F is the force exerted by the sample shrinkage (see schematic in Fig. 18.1B); A and r are the cross-sectional area and the radius of the sample, respectively; δ is the beam deflection at the free end; E and I ($=wh^3/12$) are the Young's modulus and the moment of inertia of the beam, respectively; h and w are the height and width of the beam cross-section, respectively; l and a are the length of the beam and the distance between the sample position and the clamped edge of the beam, respectively.

The remote NIR data acquisition was realized by configuring two optical-fiber cables (1 mm diameter, Ocean Optics, Inc., Dunedin, FL) onto the sides of the sample, with one cable connected to the IR light source (LS-1, Ocean Optics) and the other to the NIR spectrometer (NIRQuest512-2.2, Ocean Optics). The NIRQuest is a dispersive infrared spectrometer, which provides a scan rate of up to 1000 Hz over the spectral range of 900–2200 nm (4500–11,100 cm^{-1}). Simultaneous measurement of PS and DC were carried out using the tensometer and the NIR spectrometer as described above. The diameter and height of the composite sample were 2.5 and 2 mm, respectively. A non-tacky Teflon sleeve (inner diameter: 2.5 mm, wall thickness: 0.4 mm) with an injection hole and a smaller air-venting hole was used to encase the rods and composite sample. The NIR spectra were collected continuously at an integration time of 85 ms. This optimal integration time was determined based on a series of test runs, which showed spectra of sufficient resolution and minimized noise. For the calculation of DC, the absorption peak of the methacrylate functional group centered at 6165 cm^{-1} was used. The area under this peak was determined by subtracting a linear baseline across the edges of the peak for each spectrum obtained. The DC was calculated by taking the peak area of the sample prior to the start of irradiation ($Area_{monomer}$) and at each point during the polymerization process ($Area_{polymer}$) based on the following formula [11]:

[1] Certain commercial materials and equipment are identified in this manuscript in order to specify adequately the experimental and analysis procedures. In no case does such identification imply recommendation or endorsement by the National Institute of Standards and Technology (NIST) nor does it imply that they are necessarily the best available for the purpose.

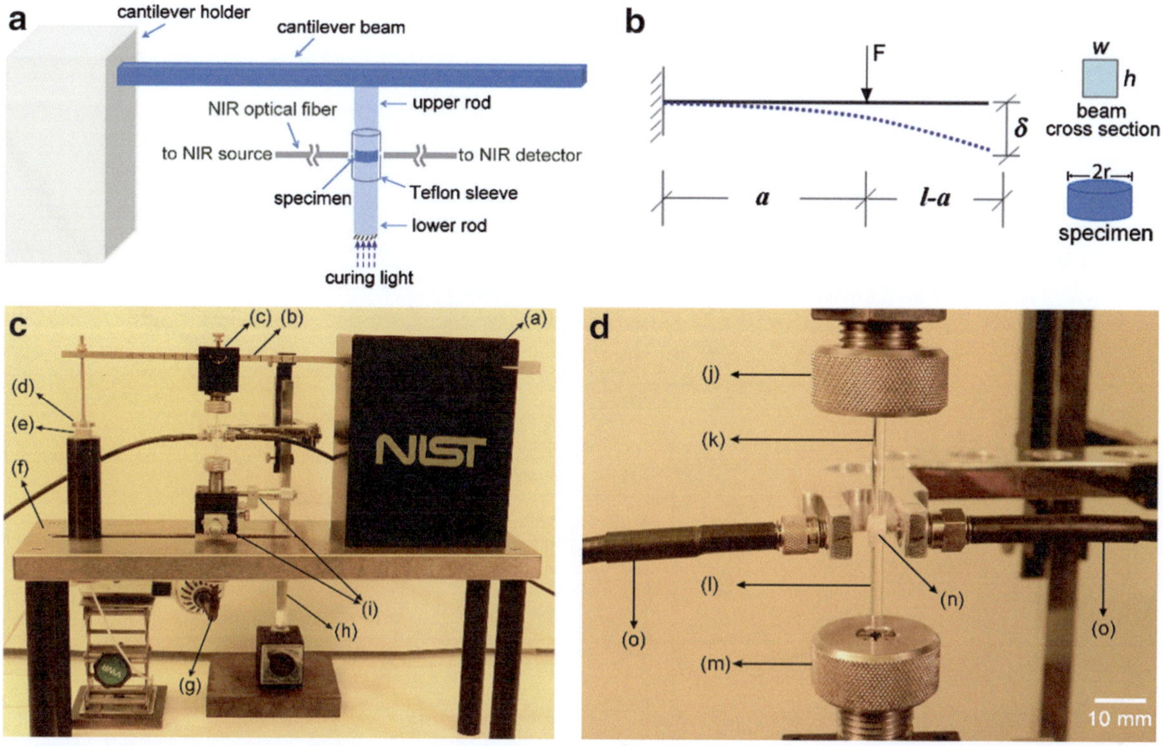

Fig. 18.1 Schematic diagrams (**A, B**) and photographs (**C, D**) of the NIST-developed tensometer instrument for the simultaneous measurement of polymerization stress, curing kinetics, and exotherm temperature. The photograph is shown from the overall view (**C**) and close view of the sample mounting region (**D**): (a) cantilever-beam holder; (b) cantilever beam (stainless steel, 6.36 mm width, 6.35 mm height, 24.5 cm length with effective beam length ranging from 6 to 20 cm); (c) upper collet holder; (d) flat metal target; (e) probe of the displacement sensor; (f) base stand; (g) curing light guide; (h) NIR cable holder; (i) X-Y stages for rods alignment; (j) upper collet; (k) upper quartz rod; (l) lower quartz rod; (m) lower collet; (n) sample encased by polytetrafluoroethylene sleeve; (o) NIR optical fibers

$$DC = \left(1 - \frac{Area_{polymer}}{Area_{monomer}}\right) \times 100\ \% \qquad (18.2)$$

Temperature rise during the exothermic polymerization was also measured in parallel using a T-type microprobe thermocouple (0.1 mm diameter, Physitemp Instruments, Clifton, NJ) inserted into the center of the specimen (not shown in Fig. 18.1). The synchronized PS/DC/temperature data were collected continuously for 30 min with 10 Hz sampling temporal resolution controlled by a LabView program (National Instruments, Austin, TX). Experiments were conducted under a yellow light environment to minimize premature photopolymerization and at room temperature.

18.3 Results

As shown in Fig. 18.2a, both the DC and PS data were simultaneously collected during polymerization. Using the current system, we were able to collect abundant information to monitor DC and PS development upon irradiation, particularly at the early stage of the polymerization (inset). These data were significantly more comprehensive compared to previous studies [8, 9, 11, 12]. The inset plot shows that the development of PS started $\approx$5 s, which occurred at a later time compared to the onset of the DC ($<$2 s). This behavior has previously been reported for similar materials [9, 12, 13] and is believed that the onset of stress development occurs near the gel point. Additionally, we observed a small negative PS within the first few seconds, which was most likely due to the thermal expansion caused by the irradiation.

The key advantage for simultaneous PS and DC measurement is that these properties can be correlated directly (Fig. 18.2b), thus providing higher measurement confidence and more in-depth understanding of the network formation process. For the material studied, no obvious stress buildup was observed until DC reached $\approx$ 20 %. If this point is taken as

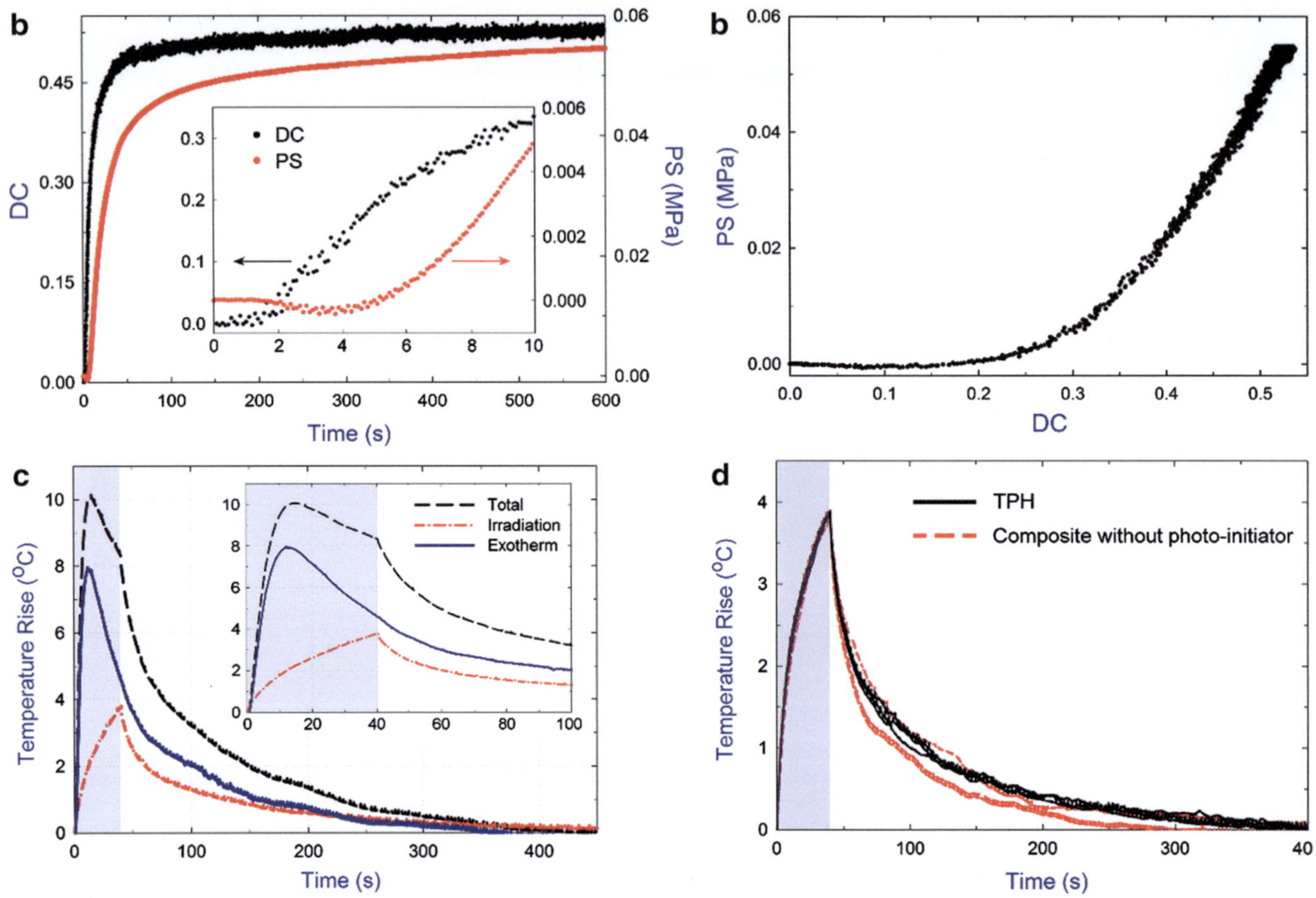

Fig. 18.2 Typical results obtained from the simultaneous measurement test of a TPH sample. The development of polymerization stress (PS) and degree of conversion (DC) in the first 10 min after irradiation (**a**). The inset shows the data in the first 10 s. The PS development as a function of DC (**b**). Superposition of the temperature rise during first (Total) and second (Irradiation) irradiation allows the calculation of reaction exotherm (inset shows the data in the first 100 s) (**c**). Triplicate measurements of temperature rise for the cured TPH and the prepared composite paste without photo-initiator under the same irradiation conditions (**d**). Shaded regions indicate irradiation time (i.e., 40 s with an effective intensity of 500 mW/cm^2)

the gel point, the observed gelation conversion is in large disagreement with those reported in the literature, where the typical values are well below 10 % for the dimethacrylate-based systems [14]. The obvious discrepancy could be explained by the special sample configuration used in our tensometer system, which facilitates extensive material flow (stress relief) from the unbonded surfaces during the pre-gel and early post-gel stages and delays the PS buildup. Therefore, since PS is an engineering property and its development depends on many internal and external factors, the onset of PS development is not necessarily representative of the sample gel point. Upon an increase in the PS, a significant amount of stress developed (Fig. 18.2b) due to increases in the sample elastic modulus during the vitrification stage (i.e., polymer network transforms from rubbery state to glass state) [15].

Free radical photo-polymerizations are generally exothermic reactions. The actual temperature rise during photo-polymerization is the result of both reaction exotherm and heat generated by light irradiation. Contribution from the later factor can be determined by applying a second irradiation that has the same irradiation conditions as the first one on the post-cured sample. Figure 18.2c shows real-time temperature rises from the combined contribution, contribution of light irradiation only, and their differences corresponding to reaction exotherm. Measurement of reaction exotherm in this way depends on two assumptions: the polymerization reaction must be completed prior to the second irradiation; and similar thermal conductivities and diffusivities through the uncured paste and cured solid must exist. The first assumption was validated by that the monitoring of DC as a function of time (Fig. 18.2a) plateaued after the first few minutes after irradiation, indicating that minimal post-polymerization was occurring afterwards. The second assumption was validated by using a prepared composite with the same formulation as the model composite without a photo-initiator. The same irradiation protocol was applied on the prepared paste and the changes in temperature of the composite were monitored. The results (Fig. 18.2d) show the temperature increase of the uncured and cured composites due to the irradiation have a nearly identical profile. This indicates that, for the material components studied, distributions of the light absorption in the composite are little changed from monomer state to polymer state.

18.4 Conclusion

A cantilever-beam based instrument, incorporated with an in-situ, high-speed NIR spectrometer, has been developed to continuously monitor the polymerization stress, exotherm, and curing kinetics in real time for polymeric composites with high filler content. This development enables us to accurately follow the kinetics of fast photopolymerization and to unravel the complex interrelationship among curing kinetics, rate of reaction, and the polymerization stress, such that a better understanding and control of the curing process can be achieved.

Acknowledgement Financial support by Interagency Agreement between the National Institute of Dental and Craniofacial Research and the National Institute of Standards and Technology, NIH/NIDCR [Y1-DE-7005-01].

References

1. K.T. Nguyen, J.L. West, Photopolymerizable hydrogels for tissue engineering applications. Biomaterials **23**(22), 4307–4314 (2002)
2. C.Y. Chen, C.K. Huang, S.P. Lin, J.L. Han, K.H. Hsieh, C.P. Lin, Low-shrinkage visible-light curable urethane-modified epoxy acrylate/SiO2 composites as dental restorative materials. Compos. Sci. Technol. **68**(13), 2811–2817 (2008)
3. J.R. Condon, J.L. Ferracane, Polymerization contraction stress of commercial composites. J. Dent. Res. **77**, 639 (1998)
4. D.C. Watts, J.D. Satterthwaite, Axial shrinkage-stress depends upon both C-factor and composite mass. Dent. Mater. **24**(1), 1–8 (2008)
5. J.R. Condon, J.L. Ferracane, Reduced polymerization stress through non-bonded nanofiller particles. Biomaterials **23**(18), 3807–3815 (2002)
6. D.C. Watts, A.S. Marouf, A.M. Al-Hindi, Photo-polymerization shrinkage-stress kinetics in resin-composites: methods development. Dent. Mater. **19**(1), 1–11 (2003)
7. H. Lu, J.W. Stansbury, S.H. Dickens, F.C. Eichmiller, C.N. Bowman, Probing the origins and control of shrinkage stress in dental resin-composites: I. Shrinkage stress characterization technique. J. Mater. Sci. Mater. Med. **15**(10), 1097–1103 (2004)
8. A.A. Stolov, T. Xie, J. Penelle, S.L. Hsu, Simultaneous measurement of polymerization kinetics and stress development in radiation-cured coatings: a new experimental approach and relationship between the degree of conversion and stress. Macromolecules **33**(19), 6970–6976 (2000)
9. H. Lu, J.W. Stansbury, S.H. Dickens, F.C. Eichmiller, C.N. Bowman, Probing the origins and control of shrinkage stress in dental resin composites. II. Novel method of simultaneous measurement of polymerization shrinkage stress and conversion. J. Biomed. Mater. Res. B **71** (1), 206–213 (2004)
10. C.J. Kleverlaan, A.J. Feilzer, Polymerization shrinkage and contraction stress of dental resin composites. Dent. Mater. **21**(12), 1150–1157 (2005)
11. M.Y.M. Chiang, A.A.M. Giuseppetti, J. Qian, J.P. Dunkers, J.M. Antonucci, G.E. Schumacher, S. Lin-Gibson, Analysis of a cantilever-beam based instrument for evaluating the development of polymerization stresses. Dent. Mater. **27**(9), 899–905 (2011)
12. L.C. Yamasaki, A.G. De Vito Moraes, M. Barros, S. Lewis, C. Francci, J.W. Stansbury, C.S. Pfeifer, Polymerization development of "low-shrink" resin composites: reaction kinetics, polymerization stress and quality of network. Dent. Mater. **29**(9), 169–179 (2013)
13. J.W. Stansbury, M. Trujillo-Lemon, H. Lu, X.Z. Ding, Y. Lin, J.H. Ge, Conversion-dependent shrinkage stress and strain in dental resins and composites. Dent. Mater. **21**(1), 56–67 (2005)
14. R.R. Braga, J.L. Ferracane, Contraction stress related to degree of conversion and reaction kinetics. J. Dent. Res. **81**(2), 114–118 (2002)
15. C.S. Pfeifer, N.D. Wilson, Z.R. Shelton, J.W. Stansbury, Delayed gelation through chain-transfer reactions: mechanism for stress reduction in methacrylate networks. Polymer **52**(15), 3295–3303 (2011)

Chapter 19
Symmetric and Asymmetric Double Cantilever Beam Methods for Interfacial Adhesion Strength Measurement in Electronic Packaging

Tsgereda Alazar, Santosh Sankarasubramanian, Sivakumar Yagnamurthy, Kyle Yazzie, Pilin Liu, and Pramod Malatkar

Abstract This paper discusses Double Cantilever Beam (DCB) test methods that were developed for characterizing adhesion strength of several critical interfaces in advanced microelectronic packaging. Those interfaces include silicon-epoxy underfill and solder resist-epoxy underfill. A unique sample preparation technique was developed for DCB testing of each interface in order to avoid the testing challenges specific to that interface—for example, silicon cracking and voiding in silicon-underfill samples and cracking of solder resist films in solder resist-underfill samples. Additionally, asymmetric DCB samples (i.e., different cantilever beam thickness on top compared to the bottom) were found to be more effective in maintaining the crack at the interface of interest and in reducing the occurrence of cohesive cracking when compared to symmetric DCB samples. Several case studies using DCB for material selection and assembly process optimization are also discussed. Furthermore, fractography results from SEM examination of the fractured surfaces are also presented for better understanding of the failure mode.

Keywords Double cantilever beam • Interfacial adhesion strength • Asymmetric DCB • Silicon-underfill adhesion • Solder resist-underfill adhesion

19.1 Introduction

Advanced microelectronic packages utilize a multitude of materials with dramatically different mechanical properties. Delamination occurring at the interfaces between these materials, due to poor adhesion and/or moisture exposure, is a key failure mode affecting the thermomechanical reliability of the package. Thus, the adhesion strength of these interfaces is a critical mechanical property governing reliability performance of microelectronic packages.

Figure 19.1 shows the schematic of a flip chip microelectronic package. The silicon die is attached to a substrate through Copper—solder interconnects with a capillary underfill (CUF) flown in between the interconnects for package rigidity. The many interfaces present in this package, such as silicon die sidewall-capillary underfill (CUF) and solder resist (SR)-underfill, are susceptible to delamination during package assembly and usage due to thermomechanical stresses, exacerbated by the presence of moisture. The lack of an accurate adhesion strength metrology has made it challenging to predict reliability performance of new materials and processes. Without a good metrology, material down-selection and assembly process optimization require expensive and time consuming assembly builds, followed by reliability testing which lead to long development times.

Several methods exist to ascertain adhesion strength of interfaces. They can be broadly divided into those that provide only qualitative comparison of adhesion strength, and those that provide quantitative data. Quantitative adhesion testing has the benefits of deeper insight into failure modes, and enabling faster decision making in material and process development.

Some of the most common adhesion test methods are: pull test, scratch test, blister test, button shear test, peel test, and beam bending test [1]. Each method has its advantages and disadvantages with respect to difficulty of sample preparation or data analysis. For example, while pull tests offer simple sample preparation, this strength-of-materials type test can be

---

T. Alazar • S. Sankarasubramanian (✉) • S. Yagnamurthy • K. Yazzie • P. Liu • P. Malatkar
Intel® Corporation, 5000 West Chandler Blvd., Chandler, AZ 85226, USA
e-mail: santosh.sankarasubramanian@intel.com

C. Sciammarella et al. (eds.), *Experimental and Applied Mechanics, Volume 4*,
Conference Proceedings of the Society for Experimental Mechanics Series, DOI 10.1007/978-3-319-22449-7_19

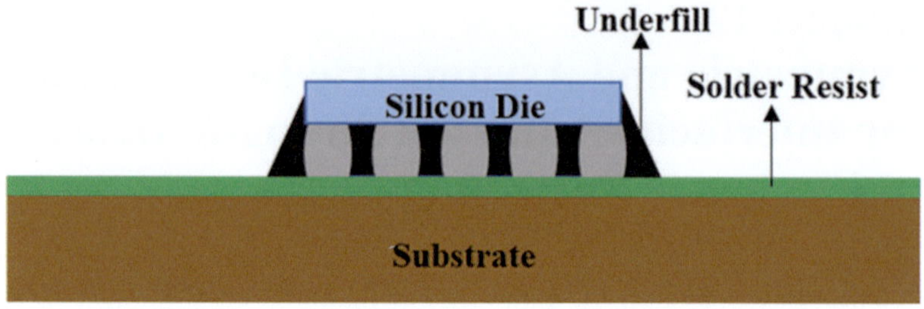

Fig. 19.1 Schematic cross-section of a flip chip microelectronic package

sensitive to defects or misalignment [2]. Scratch testing also offers simple sample preparation, but this method lends itself more to testing hard or brittle thin films that can be easily deposited on test coupons. Blister tests have the advantage of being able to yield quantitative adhesion data, but sample preparation is quite difficult. Button shear has the advantage of ease of preparation, but results are qualitative and typically suffer from significant variation. Peel testing has the advantage of simpler sample preparation, but traditionally adhesion data has only been semi-quantitative. Finally, beam bending methods, such as double cantilever beam (DCB), utilize fracture mechanics to extract quantitative adhesion data with less variation compared to button shear.

Another consideration in method selection is the mode mixity of each adhesion test method. An interface can be subjected to Mode I (tensile), Mode II (shear), Mode III or a combination [1, 3]. Unlike in a homogeneous material, a crack at a bimaterial interface always sees a combination of these modes irrespective of the external loading. The ratio of these modes is referred to as the mode mixity, and it influences the measured adhesion strength [4]. When the mode-mixity value is low, the crack sees predominantly Mode-I loading and when the mode-mixity value is high it sees Mode-II loading. Typically, an interface will fail more easily in Mode I, so a test designed for Mode I will be more sensitive. Button shear and blister tests are examples where the external loading is predominantly mixed mode which strongly influences the mode-mixity of the crack and so the data generated from these tests shows large variation.

In DCB, the interface of interest is held between two elastic substrates which are then pulled apart, in tension, to slowly grow a pre-crack previously introduced at the interface. From the resulting load–displacement data, the critical Mode I strain energy release rate (G_{IC}) is calculated at multiple locations to characterize the adhesion strength as a function of crack length. Strain energy release rate is the energy dissipated per unit area of the newly created surface during fracture. Since the interface toughness tested for at low mode-mixity is lower than at other mode-mixity factors typically experienced in real units, the DCB test provides data which envelopes (in a lower-bound sense) the loading seen in a real package. Another salient feature of DCB is the ease of producing a mixed-mode loading condition during the test by using cantilever beams of unequal thickness. This test is called Asymmetric DCB. For these reasons, the DCB test was chosen as the metrology to tackle the critical package technology developments outlined herein.

The DCB test requires one material that can be cut into cantilever beams and then be bonded together using the other interface material, which should be some sort of epoxy. Materials which are plated or deposited cannot be tested using the DCB test method.

Many key challenges had to be overcome to successfully develop a mature DCB metrology. The challenges varied based on the interface being studied. In the case of silicon-underfill, silicon coupons were prone to premature failure (silicon cracking) due to edge defects. This was addressed with a unique sample geometry that created crack growth away from sample edges. Also, a void free CUF layer was formed after much trial-and-error, by building a semi-automated CUF dispense station with temperature control. Dispense and cure profiles were optimized using a thermal IR camera to visualize void formation. Solder resist films are brittle; to make SR-CUF samples without damaging the film, solder resist film was laminated to a block of underfill. Voids were eliminated by pre-heating the underfill block.

19.2 DCB Sample Preparation

This section highlights the DCB sample preparation techniques that were developed for each interface of interest. As will be explained, each interface had its own set of challenges and hence required different sample preparation techniques.

19.2.1 Silicon-CUF DCB Sample Preparation

Blank silicon wafers were cut into 12 mm by 50 mm rectangular pieces using a dicing saw. The silicon beams were cleaned with acetone, isopropyl alcohol (IPA), and then dried with nitrogen to remove organic residues. One side of the silicon surface was masked with polytetrafluoroethylene tape and a thin palladium (or gold) film was sputter-coated onto it to create a pre-crack. The masking tape was peeled off and the exposed silicon was again cleaned with IPA. This exposed region was the region of interest for measuring G_C between silicon and CUF. The palladium/gold-coated silicon beam was firmly placed on top of another piece of clean silicon with a 50 μm polyimide film at either end as spacers. Using a custom built apparatus equipped with a syringe and needle assembly, underfill was carefully dispensed into the 50 μm gap with the help of capillary action, as shown in Fig. 19.2.

Since silicon is transparent to near infrared, a high sensitivity infrared camera was used to monitor the CUF flow front. This enabled the identification of voids in the underfill since they could be clearly observed in the IR image. Specimens with identifiable voids were discarded as previous testing had shown that crack growth in such specimens was neither purely adhesive nor purely cohesive. The good specimens were then placed in an oven to cure the CUF material. After the curing process, the IR camera was once again used to check the quality of the interface as there is a potential for voids to form during the curing process. Loading tabs were then bonded to the DCB specimens to enable testing of these specimens in Mode I. A schematic of the final sample is shown in Fig. 19.3a and side-view of an actual sample is shown in Fig. 19.3b.

For testing the silicon-CUF interface, both symmetric and asymmetric DCB samples were created. For symmetric DCB, both pieces of silicon had the same thickness, around 800 μm (referred to as "full thickness"). For asymmetric DCB, the piece of silicon on which the pre-crack was sputtered was picked from a thinned silicon wafer while the other piece was maintained at the full thickness of 800 μm. As will be discussed in Sect. 19.4.1, a nominal thickness of 300 μm for the thinner silicon was found to be ideal for obtaining adhesive failures without silicon cracking.

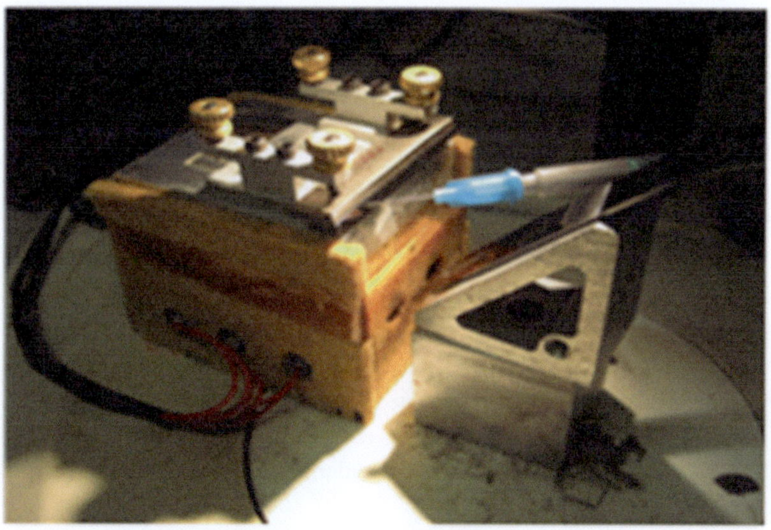

Fig. 19.2 Custom-made fixture allowed CUF to be carefully dispensed in controlled amounts. The fixture was heated so that capillary action aided in uniform CUF dispense

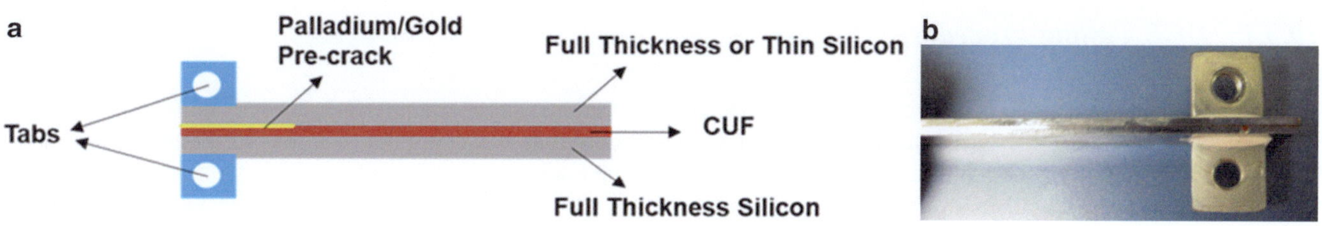

Fig. 19.3 (**a**) Schematic of the final silicon-underfill DCB sample, (**b**) Side view of final sample

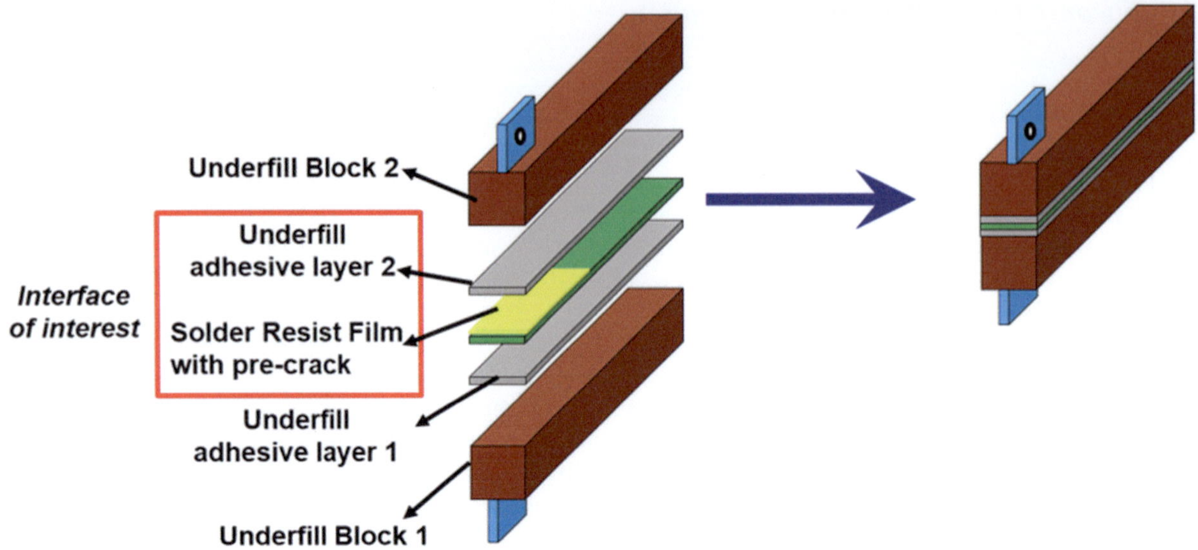

Fig. 19.4 Stack-up of DCB sample used to test solder resist-underfill adhesion

19.2.2 Solder Resist-CUF DCB Sample Preparation

The solder resist used for DCB measurements was a brittle, thin film, about 20 μm thick which made it very challenging to handle without breaking. As a result, the film was attached to an elastic coupon in order to be able to create a sample where the solder resist-underfill interface could be pulled apart without the cracking the resist. In this study, fully cured underfill material itself was used to make the elastic substrate, henceforth referred to as an underfill block, and the same underfill material in uncured form was used as an adhesive to attach the solder resist film to the underfill block.

The solder resist film was attached to an underfill block of dimensions 50 mm × 15 mm × 5 mm after which the sample was heated in an oven to cure the underfill adhesive. Once this lamination process was completed, a gold pre-crack layer was sputtered on to the top of the solder resist to about 15 mm length. This created one half of the DCB sample. This half was attached to a second underfill block using uncured underfill as the adhesive once again. A 50 μm polyimide film was used a spacer between the blocks. Additionally, the second block was oversized by about 5 mm on all sides compared to the first block so as to provide sufficient area for excess underfill adhesive to overflow.

This sandwich was clamped together and once again heated in an oven to cure the second layer of underfill adhesive. After cure, the sandwiched blocks were cut into samples 5 mm wide and tabs attached at the top and bottom. Finally, a blade was used to cut an additional mechanical notch on top of the pre-crack to facilitate the initial crack propagation. Figure 19.4 shows the stackup of the final DCB sample used to measure the solder resist to underfill adhesion. The final dimensions of the top and bottom underfill block were 50 mm × 5 mm × 5 mm. In this figure, the interface of interest is between underfill adhesive layer 2 and the side of the solder resist film containing the gold pre-crack.

19.3 Overview of DCB Testing and Calculations

19.3.1 DCB Testing Procedure

The DCB samples were tested using a commercially available load frame to evaluate interface strength, consisting of a low-noise load cell and a screw-driven actuator. This assembly was connected to a computer to record load and displacement. The samples were held on the load frame using clevis grips and the load tabs.

During the test, the specimens were displaced at a rate of 5 μm/s which was slow enough to be considered to be quasi-static. The initial delamination typically occurred at the side of the interface containing the pre-crack. The specimens were loaded at constant rate until local crack propagation was observed resulting in a sharp drop in the load–displacement curve.

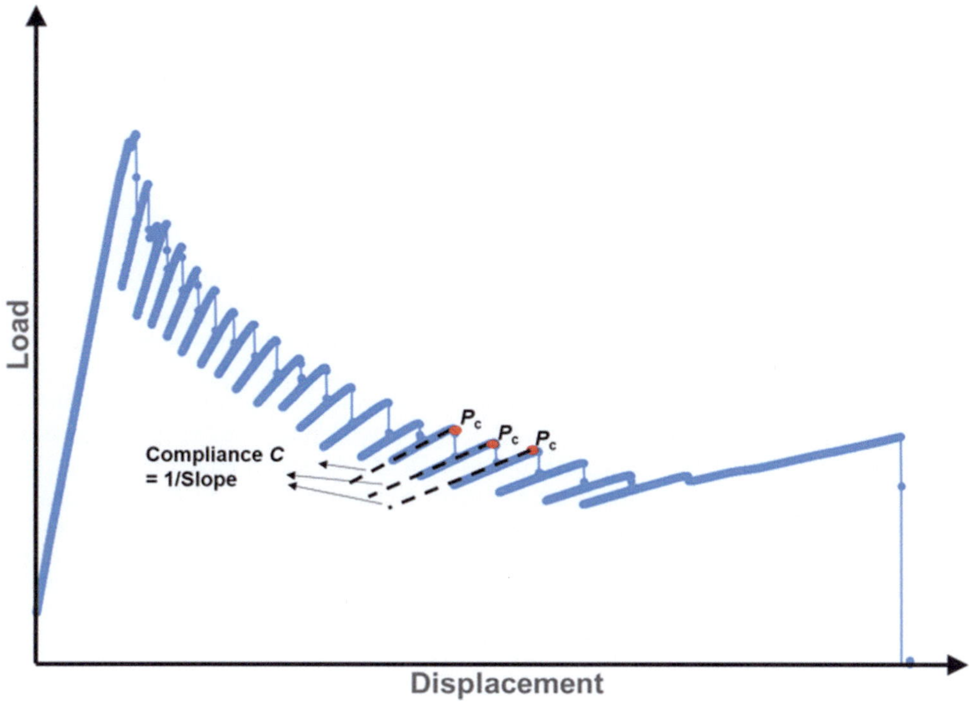

Fig. 19.5 Load–displacement curve from a typical DCB test illustrating the compliance C and critical load P_c values

At this point, the specimen was unloaded and then reloaded until the next local crack propagation was observed. This load-unload-reload procedure was repeated until complete failure was observed. A typical load–displacement curve recorded during DCB testing is shown in Fig. 19.5.

19.3.2 Critical Strain Energy Release Rate Calculations for DCB

Once the test was completed, the critical strain energy release rate (G_C) was calculated at each instance of local crack propagation. In other words, multiple values of G_C can be calculated for each sample. The calculation procedure is described below:

1. Calculate compliance C and critical load (P_c) values from the load–displacement data at each instance of crack propagation, as shown in Fig. 19.5.

2. For each value of C and P_c, calculate the crack length (a) based on the following equations. The compliance, C, and crack length, a, at each crack growth location are related according to [5]

$$C = \sum_{i=top,\, bottom} \left\{ \frac{1}{EI_i\lambda_i^3} \left[\frac{\lambda_i^3 (2a^3 - 2d^3)}{3} + 2\lambda_i a + 2\lambda_i^2 a^2 + 1 \right] + \left(\frac{3(a-d)}{2bGh_i} \right) \right\} \tag{19.1}$$

Here, d is half the length of the loading tabs. λ_i is defined for each half of the DCB sample as:

$$\lambda_i = \sqrt[4]{\frac{k_i}{4EI_i}} \tag{19.2}$$

where E is the elastic modulus of the material of the underfill blocks, I_i is the moment of inertia for each half of the DCB sample,

$$I_i = \frac{bh_i^3}{12} \tag{19.3}$$

and k_i is the foundation modulus of each half of the DCB sample

$$k_i = \frac{2Eb}{h_i} \tag{19.4}$$

Here b is the width of the DCB sample and h_i is the thickness of each half of the DCB sample. G is the shear modulus of the material of the underfill block. An iterative procedure was implemented and a computer program developed to solve for the crack length a from (19.1). It must be noted the above equations for calculating crack length are applicable to both symmetric DCB ($h_{top} = h_{bottom}$) as well as to asymmetric DCB ($h_{top} \neq h_{bottom}$).

3. Once the crack length is known, multiple analytical solutions are available in literature for symmetric DCB [5–7] to calculate critical strain energy release rate (G_C) for each crack length (a)—critical load (P_c) pair. These analytical solutions gave almost identical results for G_C (within 0.2 %). Asymmetric DCB produces a mixed mode loading condition at the interface (Mode I + Mode II) and analytical solutions are available to calculate both total G_C ($G_{C,\ tot}$) as well as the G_C for Mode 1 (G_{IC}) [7]. For isotropic materials, the total and Mode I G_C at each crack location are given by [7]:

$$G_{C,\ tot} = \frac{6P_c^2a^2}{Eb^2h^3}\left[1 + \left(\frac{h}{H}\right)^3\right]\left(1 + \frac{0.667Bh}{a}\right)^2 \tag{19.5}$$

$$G_{IC} = \frac{6P_c^2a^2}{Eb^2h^3}\left[1 + \left(\frac{h}{H}\right)^3\right]\sin^2\phi\left(1 + \frac{0.667B_1h}{a}\right)^2 \tag{19.6}$$

where h and H are the smaller and larger of the two beam thicknesses while B, B_1 and ϕ are given by

$$B = 1.120 - 0.695\left(\frac{h}{H} - 0.585\right)^2 \tag{19.7}$$

$$B_1 = 1 + 0.546\left(1 - \frac{h}{H}\right) \tag{19.8}$$

$$\phi = \left[0.574 + 0.033\left(\frac{h}{H}\right) + 0.805\left(\frac{h}{H}\right)^2 - 0.413\left(\frac{h}{H}\right)^3\right]\left(\frac{\pi}{2}\right) \tag{19.9}$$

It was verified for symmetric DCB that both $G_{C,tot}$ and G_{IC} from (19.5) and (19.6) above gave identical results.

After testing, the failed samples were inspected to identify the failure mode. For each interface that was tested, the observed failure modes were either (i) adhesive (at the interface of interest), (ii) cohesive (within the adhesive), (iii) mixed (combination of adhesive and cohesive), and (iv) brittle (typically observed only for silicon-based DCB samples as silicon cracking). Only samples that showed adhesive failures (failure mode (i) above) were considered as good yielding samples and considered for further analysis. DCB yield was defined as the percentage of samples showing adhesive failures.

19.4 DCB Applications: Results and Discussion

This section discusses some of the applications of the DCB samples created using the sample preparation methods described in Sect. 19.2 and tested using the procedure described in Sect. 19.3. Unless mentioned otherwise, symmetric DCB samples were used for testing. The strain energy release rate (G_C) values presented here were normalized relative to the lowest G_C value for that study. Finally, as mentioned in Sect. 19.3.2, multiple values of G_C can be calculated for each DCB sample.

19.4.1 Effect of Flux Materials Used in Chip-Attach on Silicon-Capillary Underfill Adhesion

In the chip-attach process in flip-chip microelectronic packaging, the silicon chip (die) is attached to a substrate by forming joints between solder balls on the substrate and copper bumps on the chip. The chip-attach process uses flux materials to clean oxides on the solder balls in order to enable electrically functional joint formation between the solder balls and the copper bumps. After chip attach, an underfill material is flown through these joints by capillary action, in order to stabilize the joints. The flux materials used in chip-attach tend to leave behind a residue, which if not cleaned thoroughly could impact adhesion between the capillary underfill material and silicon sidewall.

The DCB metrology was used to study the potential degradation in adhesion between silicon and underfill as a result of flux residue. Two sets of silicon-CUF DCB specimens were produced, both consisting of a silicon-CUF-silicon stack, using the sample preparation procedure described in Sect. 19.2.1. In one set, prior to creating the DCB specimen, the silicon piece with the palladium/gold pre-crack was subjected to a simulated chip-attach in order to contaminate the side containing the pre-crack with flux residue. For the control set, no contamination was done. A schematic of the two sets of samples is shown in Fig. 19.6.

Samples created using this method was used to test the effect of two different flux materials, designated Flux A and Flux B, on silicon-CUF adhesion, with a no-flux leg as control. The results from the DCB testing are shown in Fig. 19.7. Multiple samples were tested for each case and the results shown in the figure only show samples that failed adhesively. It is clear from the results that flux contamination can reduce the silicon to underfill adhesion by more than 50 % which has significant implications for the reliability of these packages.

A smaller difference can be seen between the two fluxes. For this experiment, the amount of flux residue that was transferred to the silicon during the simulated chip-attach was not quantified. Hence, it was not possible to conclude whether the difference in adhesion between the two flux materials was due to inherent chemical differences between the two fluxes or due to different amounts of flux residue being transferred to the silicon—the latter could be a function of the flow behavior of the flux.

Fig. 19.6 Schematic of DCB samples for testing adhesion of uncontaminated and flux contaminated silicon to underfill

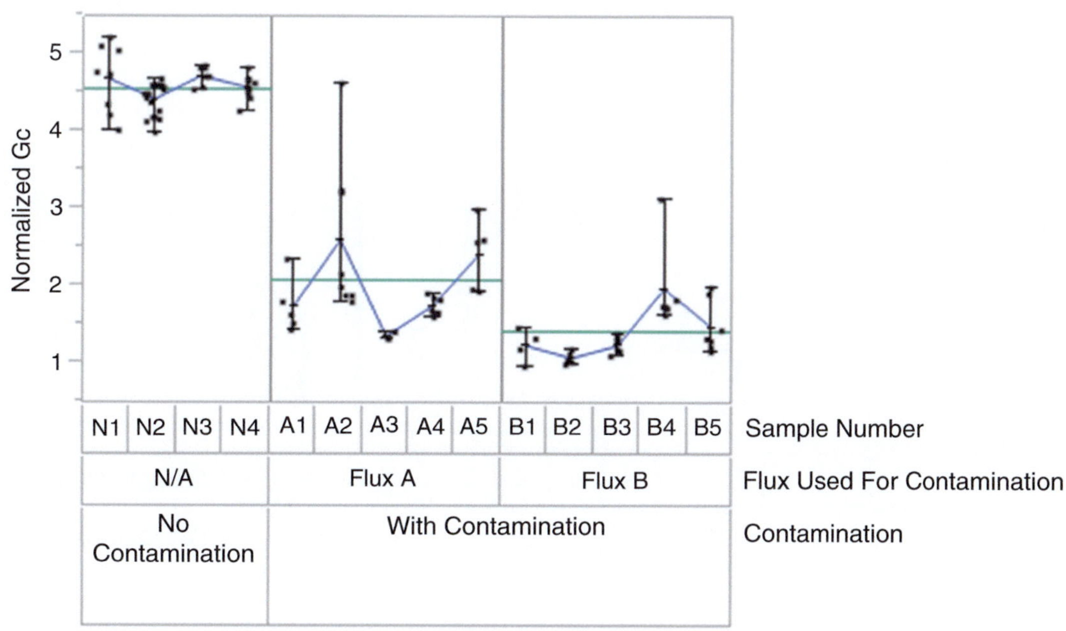

Fig. 19.7 DCB adhesion results comparing the effect of flux contamination on silicon-underfill adhesion

With "DCB yield" defined as the percentage of samples that exhibited pure adhesive failures, typical yields were in the range of 50 % for the symmetric DCB samples above, with cohesive failures inside the underfill the main source of "yield loss". Asymmetric DCB samples were found to have improved yield i.e., a higher percentage of samples showing adhesive failures, when the pre-crack was applied on the thinner silicon. Two thicknesses were attempted for the thin silicon piece in the asymmetric DCB sample, 300 and 500 μm, with the other half of the DCB sample being a full-thickness (800 μm thick) piece of silicon. It was found that the yield with asymmetric DCB samples containing 500 μm silicon was not significantly better compared to symmetric DCB samples (around 50 % adhesive). With 300 μm silicon though, yields improved to the 70–80 % range. It is suspected that using silicon thinner than 300 μm is likely to cause brittle silicon failures.

As explained in Sect. 19.3, with asymmetric DCB, the interface is subjected to mixed-mode loading (combination of Mode I + Mode II). Analytical models in existing literature [7] were used to separate out the critical strain energy release rate values for Mode I and Mode II with the Mode I value (G_{IC}) then used to quantify the adhesion strength of the interface for reasons described in Sect. 19.1. A full characterization of the asymmetric DCB test is in progress for the silicon-CUF interface with particular focus on the repeatability of the G_{IC} value for different silicon thicknesses as well as understanding the thickness window available where adhesive failures dominate i.e., high DCB yields exist. It is expected that at the lower thickness end of this window, brittle silicon cracking would dominate, while cohesive failures will dominate the upper end.

19.4.2 Adhesion of Solder Resist to Capillary Underfill Adhesion

The DCB metrology was used to measure the solder resist to underfill adhesion for two different solder resist materials (labeled SR1 and SR2 here) to two different underfill materials (labeled UF1 and UF2). The results are shown in Fig. 19.8. It is clear that SR2 had 50 % lower adhesion to underfill materials compared to SR1. The adhesion strength did not have a significant dependence on the underfill material. It is clear that using SR2 as the solder resist material would pose a higher risk for package reliability compared to SR1. Typical DCB yields for these samples were in the 70–80 % range.

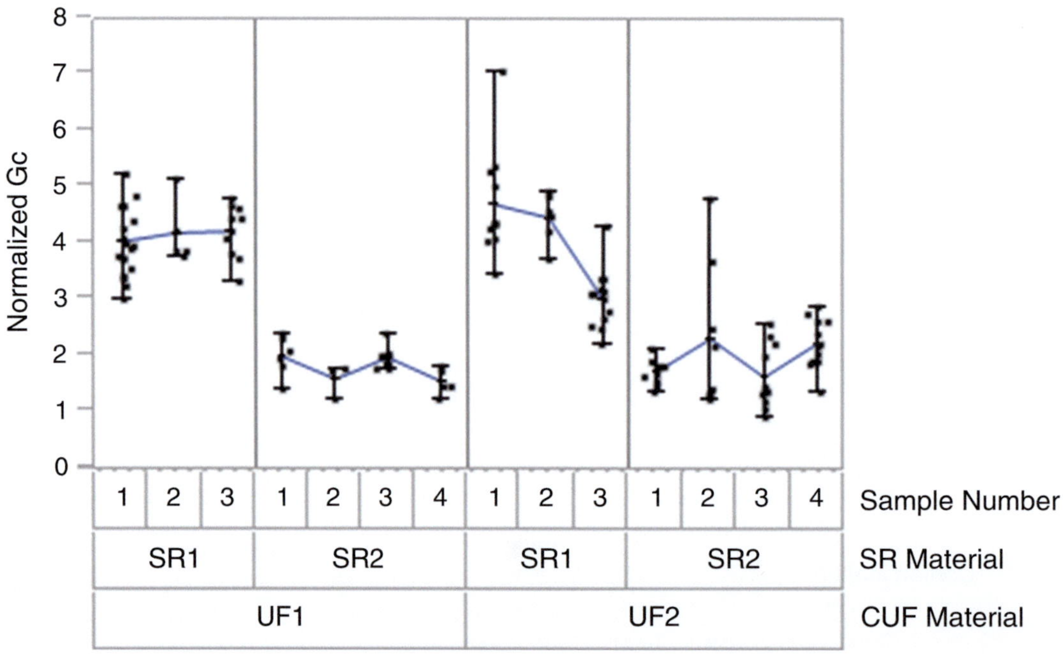

Fig. 19.8 DCB adhesion results comparing solder resist-underfill adhesion for two different solder resist and underfill materials

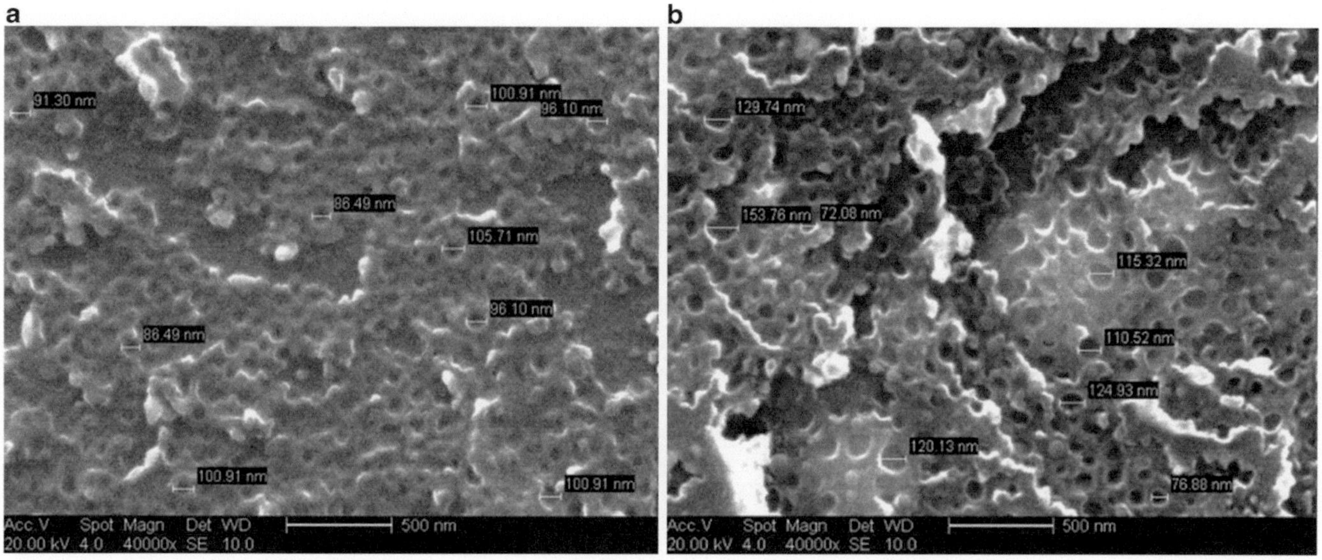

Fig. 19.9 (**a**) Top down view of DCB fracture surface on silicon side, (**b**) Fracture surface on underfill side

19.4.3 DCB Test Fractography

The fracture surfaces from the silicon-underfill DCB test were examined with SEM to understand the failure better. Figure 19.9 shows SEM images of a failed sample. Figure 19.9a shows the typical top down view from the silicon side of the fracture surface. A very thin layer of underfill residual can be seen on top of silicon, indicating the failure was not purely adhesive between silicon and underfill, although the crack path was very close to silicon/underfill interface. Figure 19.9b is the fracture surface on the underfill side. The roughness of the surface was consistent with the silicon side observation that the crack was not purely separated between silicon and underfill, but was through underfill close to silicon. In short, while the failure is mainly adhesive, there was a thin film of underfill residue on the silicon side of the fracture surface, indicating that purely adhesive failure is not common. It also noted that small dimples with about 100 nm in diameter can be observed on both side of the fracture surfaces, which could be from the outgassing of CUF during curing.

19.5 Conclusions

With the adoption of new technologies into microelectronic packages and the continuous quest for low-cost and better packaging materials and processes, study of adhesion strength between newer interfaces has become important. Poor interfacial adhesion means poor package yield during the assembly process and under real world use conditions. This paper discussed the DCB adhesion metrology that was successfully developed and applied to interfaces like silicon-underfill and solder Resist-underfill. In actual microelectronic packages, the mode of delamination at an interface can be mixed. However, DCB, having low mode mixity, provides the lowest value of interfacial toughness giving experimental sensitivity (by eliminating the impact of mixed-mode loading) necessary to detect process and material impacts on adhesion unlike in blister or button shear tests. Asymmetric DCB was found to improve the adhesive yield for silicon-underfill DCB testing.

Acknowledgments The authors would like to acknowledge the outstanding contributions from the following persons, all from Intel® Corporation: Liwei Wang, Nachiket Raravikar, Deepak Arora, Joe Bautista, Arjun Krishnan, Nisha Ananthakrishnan, Akshay Mathkar, Frank Prieto, Dave McCoy, Don Erickson, Alan Overson, Bob Dalpe, Rodney Wells, Patrick Nardi, Beverly Canham, Dilan Seneviratne, Bharat Penmecha, Edvin Cetegen, and Jon Atkins.

References

1. R. Lacombe, *Adhesion measurement methods: theory and practice* (CRC, Boca Raton, 2006)
2. R. Pearson, Adhesion fundamentals for microelectronic packaging. in IMAPS Device Packaging 2014 Professional Development Course (Fountain Hills, AZ, 2014)
3. L.F.M. da Silva, D.A. Dillard, B.R.K. Blackman, R.D. Adams, *Testing adhesive joints* (Wiley-VCH, Weinhem, 2012)
4. J.W. Hutchinson, Z. Suo, Mixed mode cracking in layered materials. Adv. Appl. Mech. **29**, 63–191 (1991)
5. X. Dai, M.V. Brillhar, P.S. Ho, Adhesion measurement for electronic packaging applications using double cantilever beam method. IEEE Trans. Comp. Packag. Technol. **23**(1), 101–116 (2000)
6. M.F. Kanninen, Augmented double cantilever beam model for studying crack-propagation and arrest. Int. J. Fract. **9**, 83–92 (1973)
7. G. Bao, S. Ho, Z. Suo, B. Fan, The role of material orthotropy in fracture specimens of composites. Int. J. Solids Struct. **29**(9), 1105–1116 (1992)

Chapter 20
Experimental Study of Mechanical Performance in Friction Stir Welded Dissimilar Titanium Alloys

Neha Kulkarni and M. Ramulu

Abstract Friction Stir Welding (FSW) is a solid state joining process widely used in aerospace industry for near net shape manufacturing. In the past, FSW has been well established with similar titanium alloys. The process is mainly optimized for titanium alloy, Ti-6Al-4V, to produce defect free joints of it. The current study evaluates FSW with dissimilar titanium alloys that include, Ti-6Al-4V, Ti-6242, and Ti-54MFG to broaden the application of the process with titanium alloys. FSW of dissimilar titanium joints was studied by varying its process conditions, namely, rotation speed and feed rate during the process. The experimental study of effect of process parameters on friction stir welded dissimilar titanium joints was conducted to investigate the mechanical performance of joints. The mechanical performance of joints showed that the modulus of elasticity close to 120 GPa and average ultimate strength of 920 MPa was obtained. The optimum values of rotation speed and feed rate were identified for joining dissimilar titanium alloys.

Keywords FSW • Dissimilar Ti-alloys • Rotation speed • Feed rate • Mechanical strength

20.1 Introduction

Friction stir welding is a solid state welding technique introduced in 1991 at TWI in UK [1]. The adoption of FSW replaces the use of riveting in joining the structures, therefore provides material and cost savings as compared to the mechanically fastened structures. In addition to this, FSW offers several advantages over fusion welding with higher welding rates and lower joining temperature that have less impact on the joint efficiency. The problems that occur in the solidification of liquid phase are not present in friction stir welding. The process does not require consumables to produce weld, there are no toxic fumes or splatter; it is a process that can be automated and thus makes it viable for industrial use. FSW can join both similar as well as dissimilar metals. In FSW, two sheets or plates are tightly abutted, and a tool is run through them generates frictional heat between the joining metal and the tool (as shown in Fig. 20.1). The heat produces plastic deformation in the material near the tool; as a result the joint is formed without the melting of materials. The two important parameters that influence the friction stir welding process are the feed rate and the rotation speed of the tool. The other variables that control the process include tooling design, material, thermal management, plunge depth, etc.

The joining of materials with friction stir welding with the aluminum alloys is well established now in aerospace structures. FSW with dissimilar aluminum alloys and that of aluminum with titanium has been investigated by some of the researchers. Chen et al. [2] lap joined dissimilar alloys of Al-Si and titanium using friction stir welding; and achieved maximum failure load of joints about sixty two percent of Al-Si alloy base metal. Aonuma [3, 4] investigated the butt joining of titanium and Mg-Al-Zn alloy and of aluminum alloys 2024 and 7075 to titanium alloys by friction stir welding. The study showed that the thickness of Ti-Al intermetallic compound affects the tensile strength of FSW butt joints.

FSW can be combined with superplastic forming to produce near-net shape manufacturing of large titanium parts [5, 6]. Titanium alloys have become strategically important in aerospace industry due to its high strength, low weight, high operating temperature and corrosion resistance. Ti-6Al-4V is a workhorse of aerospace industry, but has limitation of working temperature. The other titanium alloys such as Ti-6242 and Ti-54MFG offer advantages of high temperature sustainability and better machinability over Ti-6Al-4V. FSW is widely utilized to produce defect free joints of Ti-6Al-4V. In the previous study, FSW have also been demonstrated successfully with titanium alloys other than Ti-64 standard grain such as Ti-64 fine grain, Ti-54 fine grain, Ti-6242 standard grain, Ti-6242 fine grain [7]. However till date very few studies have been done on joining of dissimilar titanium alloys by friction stir welding. The joining of Ti-6Al-4V with the other titanium alloys will improve the adaptability of titanium in various industrial and aerospace applications. It will also further result in

N. Kulkarni (✉) • M. Ramulu
Department of Mechanical Engineering, University of Washington, 352600 Seattle, WA 98195, USA
e-mail: nbhagat@u.washington.edu

C. Sciammarella et al. (eds.), *Experimental and Applied Mechanics, Volume 4*,
Conference Proceedings of the Society for Experimental Mechanics Series, DOI 10.1007/978-3-319-22449-7_20

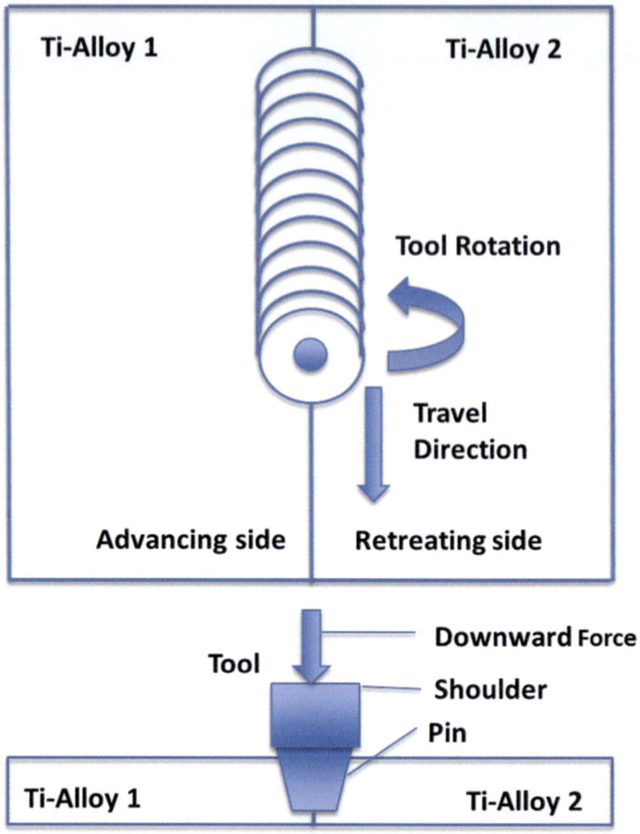

Fig. 20.1 Friction stir welding process

saving energy as they are lighter in weight and have high strength to weight ratio. Therefore, in current paper friction stir welding is investigated with dissimilar titanium alloys. The mechanical behavior of the titanium joints is examined to characterize the strength of the joining process.

20.2 Specimen Preparation

The friction stir welding of dissimilar titanium alloys was studied on three titanium alloys that included Ti-5Al-4V-0.6Mo-0.4Fe Fine Grain (Ti-54MFG), and Ti-6Al-4V Standard Grain (Ti-64SG) and Ti-6Al-2Sn-4Zr-2Mo Standard Grain (Ti-6242SG). Ti-64SG and Ti-54MFG are alpha-beta titanium alloys, while Ti-6242SG is a near-alpha titanium alloy. The composition of titanium alloys is shown in Table 20.1. The alpha stabilizer, aluminum, is present in all the three alloys; and Ti-64SG and Ti-6242SG has equal proportion of it. The beta stabilizers in these alloys are vanadium and molybdenum, with later being stronger than the former element. Ti-6242SG also has additional neutral elements, tin and zirconium, for beta stability. The fine grain of Ti-54 M was joined to standard grains of Ti-64 and of Ti-6242; and standard grains of Ti-64SG and Ti-6242SG were welded to obtain dissimilar friction stir welded joints. During friction stir welding of dissimilar titanium alloys, in Ti-64SG/Ti-54MFG, and Ti-6242/Ti-54MFG titanium alloy Ti-64SG and Ti-6242 SG were kept on advancing side and Ti-54MFG was on the retreading side. In case of Ti-6242SG/Ti-64SG, titanium alloy Ti-6242SG was on advancing while Ti-64SG was on retreading side. Table 20.2 provides the mechanical properties of the parent titanium alloys.

The optimum process conditions are necessary in order to produce good quality welded joints. Higher spindle speeds overheat and over soften the weld material resulting in unconsolidated joint formation. On the contrary slow spindle speeds with high travel rate will not produce sufficient heat to promote material flow during the process resulting in extremely high loads, shoulder disengagement and tool wear. Even the low spindle speeds with slow travel rate result in lack of penetration and to produce sufficient material flow can result in tool wear [9]. Edward's [10] has established the optimum process window of friction stir welding of Ti-6Al-4V. The conditions found in this study were used to produce the friction stir welded joints of all the titanium alloys. The samples were prepared in the Boeing welding facility by varying the rotation

Table 20.1 Composition of titanium alloys

		Element composition (%)							
Alloy	Type	Al	Fe	Mo	O	Sn	Ti	V	Zr
Ti-54 M	$\alpha + \beta$	5	0.4	0.6			90	4	
Ti-64	$\alpha + \beta$	6	0.25 (max)		0.2 (max)		90	4	
Ti-6242	Near α	6		2		2	88		4

Table 20.2 Mechanical properties of parent Ti-alloys [8]

Properties	Ti-54 M	Ti-64SG	Ti-6242SG
Yield strength (MPa)	889	880	895
Ultimate strength (MPa)	972	950	1000
% Elongation	16.5	14	12

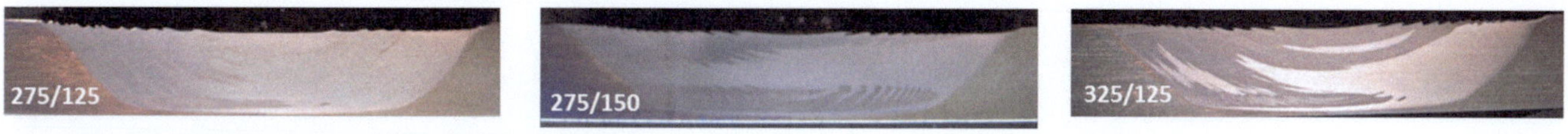

Fig. 20.2 Macrograph of FSW Ti-62424SG/Ti-64SG at different rotation speed and feed rates (RPM/mm min^{-1})

speed from 225 to 325 RPM and increasing the feed rate from 100 to 150 mm min^{-1} during the welding process. A total number of five combinations of welded samples were obtained by changing the rotation speed and the feed rate: 225 RPM/125 mm min^{-1}, 275 RPM/100 mm min^{-1}, 275 RPM/125 mm min^{-1}, 275 RPM/150 mm min^{-1}, and 325 RPM/125 mm min^{-1}. Figure 20.2 shows typical macrograph of friction stir welded joint between Ti-6242SG and Ti-64SG. During the welding process, the material moves from retreating side to advancing side in the welded zone. The change in feed rate and rotation speed shows different mixing patterns in the welded zone of dissimilar alloys. Note that these experimental conditions produced defect free butt joints with dissimilar titanium alloys. In Ti-6242SG/Ti-64SG, the material flows from Ti-64SG (retreating side) to Ti-6242SG (advancing side). In case of constant spindle speed at 275 RPM, the increase in feed rate from 125 to 150 mm min^{-1} moved less amount of material from Ti-64SG towards Ti-6242SG side. For constant feed rate at 125 mm min^{-1}, higher amount of material passed from the retreating side at 275 RPM than at 325 RPM.

20.3 Experimental Method

The mechanical behavior of friction stir welded joints was evaluated by performing tensile test. The tensile specimens were machined from the friction stir welded plates with abrasive jet machine such that the welded zone was perpendicular to the loading direction of the coupon as shown in Fig. 20.3a. This facilitated estimation of transverse tensile properties of the weldments. Four tensile coupons were cut per plate. The dimension of the tensile test specimen per ASTM E-8 is shown in Fig. 20.3b. The tensile test was performed on 250 kN capacity 5585 Instron Load Frame under displacement control at a constant cross head extension speed of 1.27 mm min^{-1}. Two specimens of all the welded titanium joints were tested till the failure. The load data were monitored by a load cell built into the load frame. The axial strain data was examined by a video extensometer. The gage length of 50 mm was monitored to collect the axial strain data. Through a data acquisition system, plots of engineering stress vs. engineering strain were produced in addition to calculating elastic modulus; yield strength at 0.2 % strain tensile strength, and elongation. Figure 20.3c shows the experimental set up used during tensile testing. There was not much of a scatter present in the stress–strain data of two samples of each specimen; therefore the testing was limited to two tensile coupons.

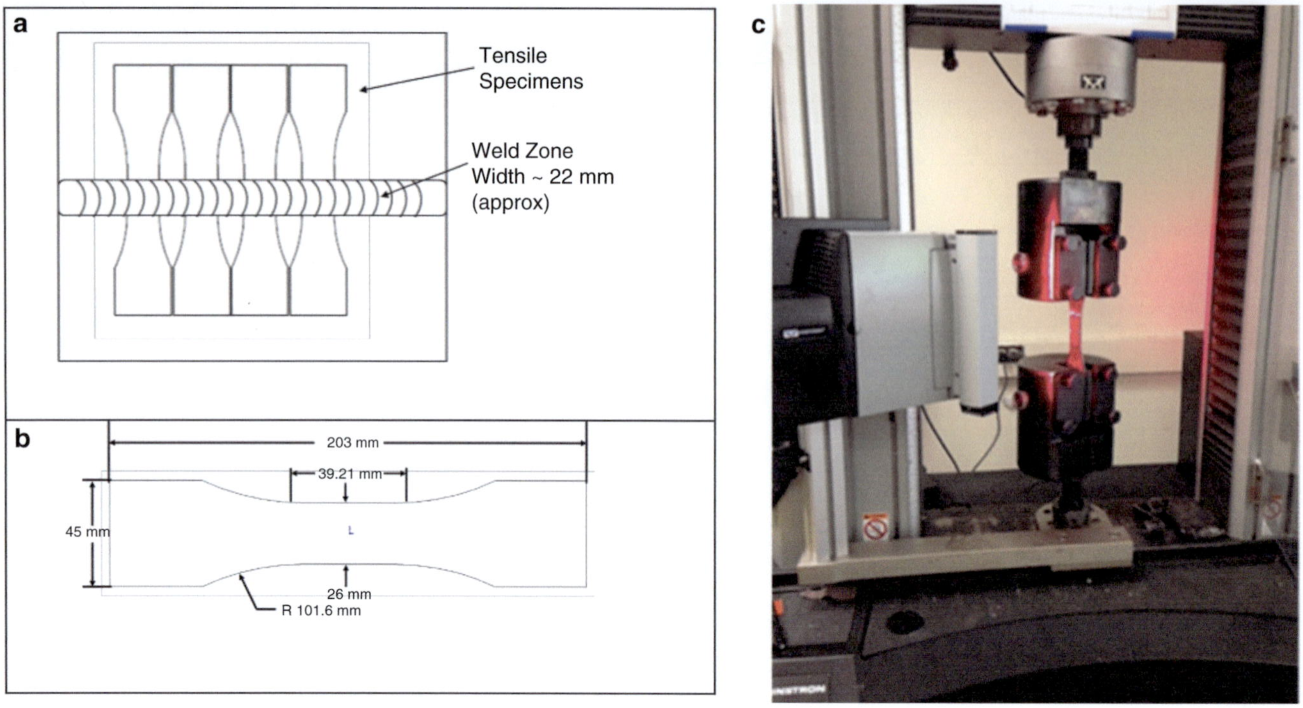

Fig. 20.3 Tensile test experimental set up

20.4 Results and Discussions

The mechanical behavior was assessed from the stress strain curve in the friction stir welded dissimilar titanium alloys. Figure 20.4 shows stress strain for dissimilar titanium alloys Ti-64SG/Ti-54MFG, Ti-6242SG/Ti-54MFG, and Ti-6242SG/Ti-64SG at different rotation speeds and feed rates. The joint strength of dissimilar friction stir weldment was 90 % and above of the parent alloys. The stress strain curve shows that the elastic modulus did not affect with the change in process parameters. The average elastic modulus for dissimilar alloys Ti-64SG/Ti-54MFG, Ti-6242SG/Ti-54MFG, and Ti-6242SG/Ti-64SG was 94.8 GPa, 129.4 GPa, and 127.1 GPa respectively. Ti-6242SG/Ti-54MFG showed highest elastic modulus of 158.6 GPa at 225 RPM/125 mm min^{-1}. The impact of feed rate and rotation speed was evident on ultimate strength and ductility than the yield strength. In case of dissimilar alloys with Ti-54MFG, the mechanical properties were higher with Ti-6242SG than with Ti-64SG. Both the ultimate strength and yield strength were higher in friction stir welded Ti-6242SG/Ti-54MFG alloy. In terms of ductility, highest value was reached in Ti-6242SG/Ti64SG at 275 RPM/125 mm min^{-1}. In the rest of the cases, the elongation was less than 10 %.

The influence of feed rate on the strength of the joint was examined by changing the feed rate from 100 to 150 mm min^{-1} maintaining the rotation speed constant at 275 RPM as shown in Fig. 20.5. The increase in feed rate caused the increase in modulus of elasticity in the alloys joined with Ti-54MFG, while decreased in Ti-6242SG/Ti-64SG. The other mechanical properties, the yield strength measured at 0.02 %, the ultimate strength, and the elongation lowered with increase in the feed rate in Ti-64SG/Ti-54MFG. In case of Ti-6242SG/Ti-54MFG, the yield strength increased followed by a decrease; while the ultimate strength slightly increased with the first step of increase in the speed, followed by a decrease from the feed rate of 125 to 150 mm min^{-1}. The ductility of Ti-6242SG/Ti-54MFG decreased with the increase in feed rate. The highest value of ductility was reached at the feed rate of 125 mm min^{-1} in Ti-6242SG/Ti-64SG welded joint. The decrease in strength as a result of increase in feed rate was less than 14 and 16 % for yield strength and ultimate strength respectively. During the friction stir welding, when the rotation speed was held constant, the feed rate of 125 mm min^{-1} recorded the maximum strength for the joint. In terms of dissimilar joined alloys, Ti-6242SG/Ti-54MFG showed higher strength properties than the other titanium joined alloys.

The effect of rotation speed on the mechanical behavior of friction stir welded joints was observed by increasing the rotation speed from 225 to 325 RPM keeping the feed rate constant at 125 mm min^{-1} as shown in Fig. 20.6. The increase in

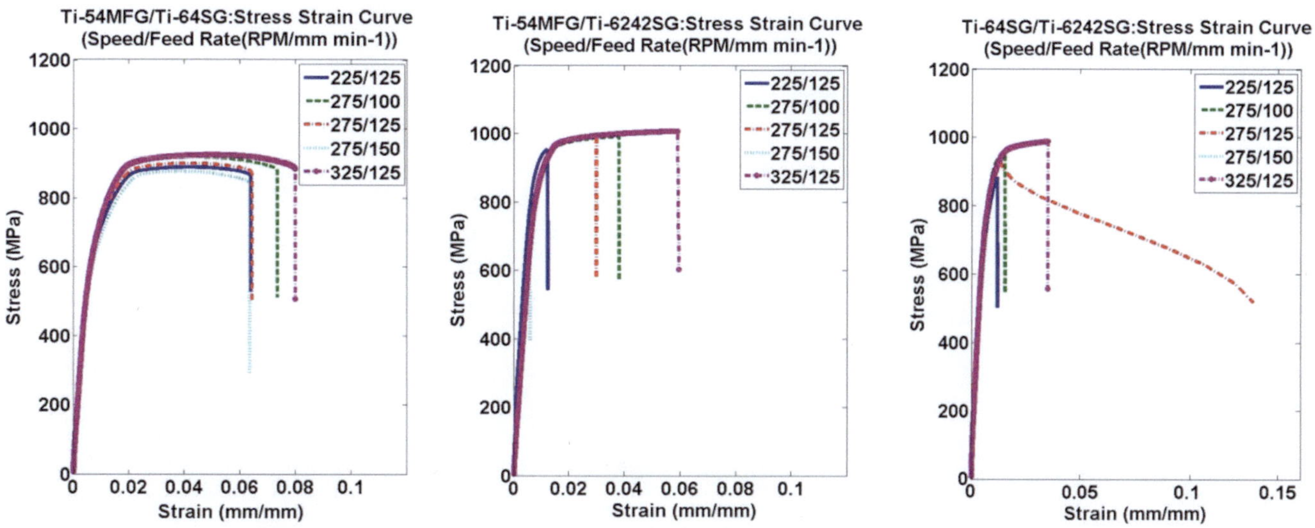

Fig. 20.4 Stress strain in Ti-54MFG/Ti-64SG, Ti-54MFG/Ti-64SG, Ti-54MFG/Ti-6242SG, and Ti-64SG/Ti-6242SG

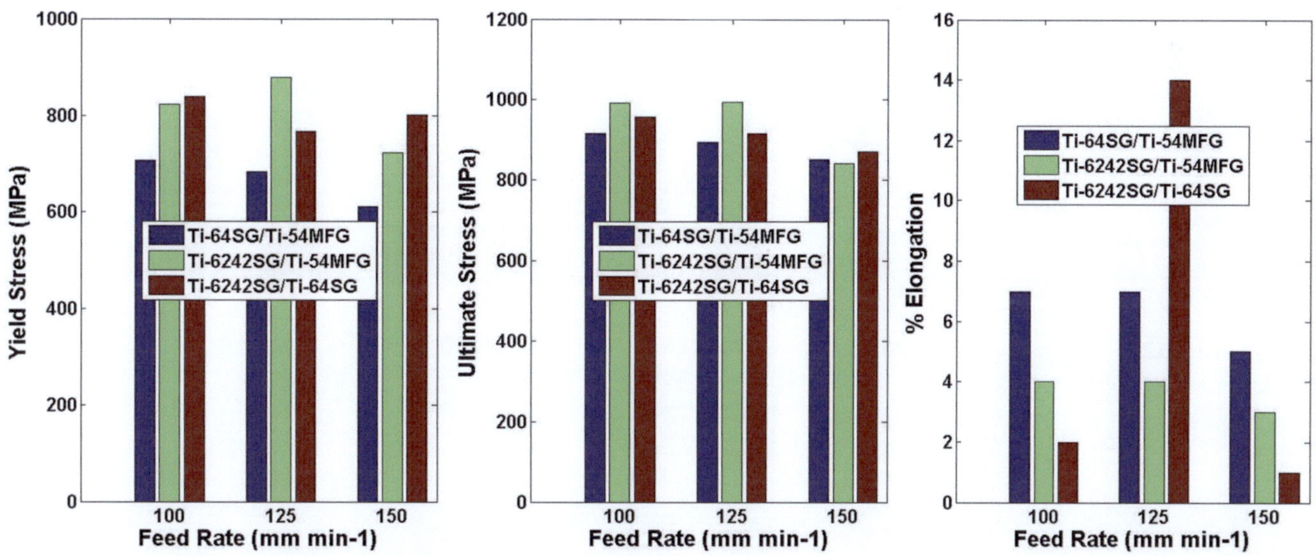

Fig. 20.5 Effect of feed rate on mechanical strength and ductility of joints in dissimilar joined Ti-alloys

rotation speed resulted in increase in yield strength; ultimate strength and the elongation in dissimilar alloy Ti-64SG/Ti-54MFG and Ti-6242SG/Ti54MFG. In Ti-6242SG/Ti64SG, the increase in rotation speed from 225 to 275 RPM decreased the yield strength, followed by the increase from 275 to 325 RPM. However the ultimate strength in this case increased with the increase in rotation speed. The ductility in case of Ti-6242SG/Ti-64SG increased with the rotation speed 225 to 275 RPM where it recorded 14 %. In the process when the feed rate was constant, maximum properties in terms of strength were observed at the rotation speed 325 RPM for all three of the welded alloys, while the maximum ductility was reached with rotation speed 275 RPM. Ti-6242SG/Ti-54MFG alloy showed higher strength than the other dissimilar welded titanium alloys. The rise of maximum 7 % in yield strength and of 10 % in ultimate strength was observed in the dissimilar alloys as an effect of increase in rotation speed. Therefore, the effect of rotation speed was more evident than the feed rate on the friction stir welded dissimilar titanium alloys.

Friction stir welded of dissimilar Ti-alloys were subjected to the tensile tests till the fracture. Figure 20.7 shows the fracture occurred in the friction stir welded dissimilar Ti-alloys. Many of dissimilar welded titanium alloys showed ductile type of failure. However in some of the welded alloys brittle phenomenon was also observed. In the ductile failure, two patterns

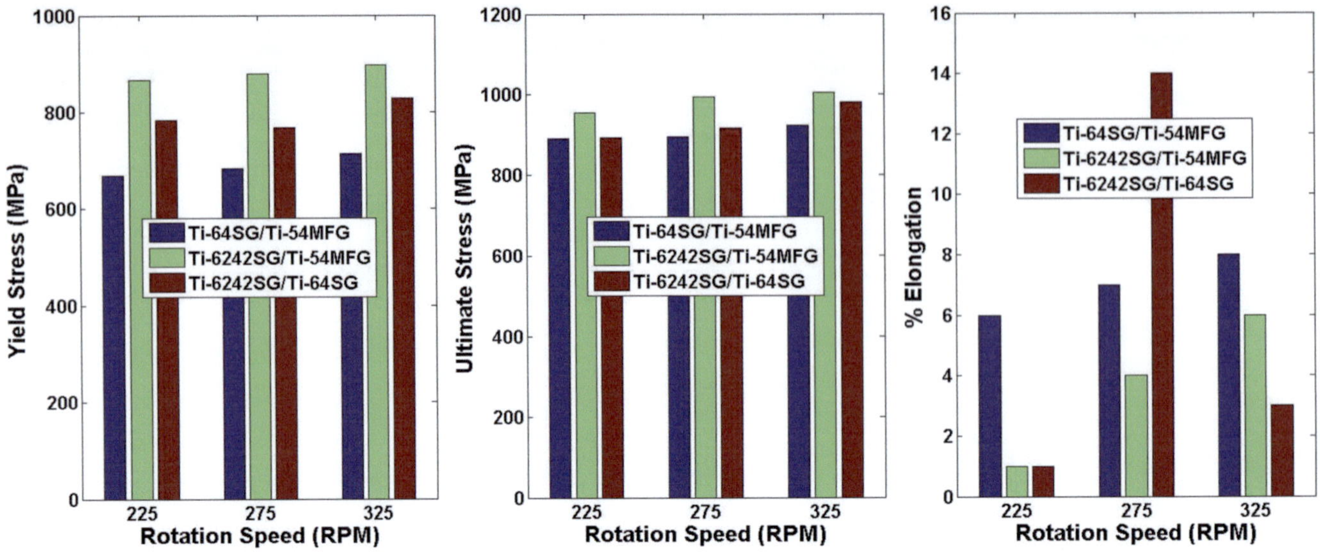

Fig. 20.6 Effect of rotation speed on mechanical strength and ductility of joints in dissimilar joined Ti-alloys

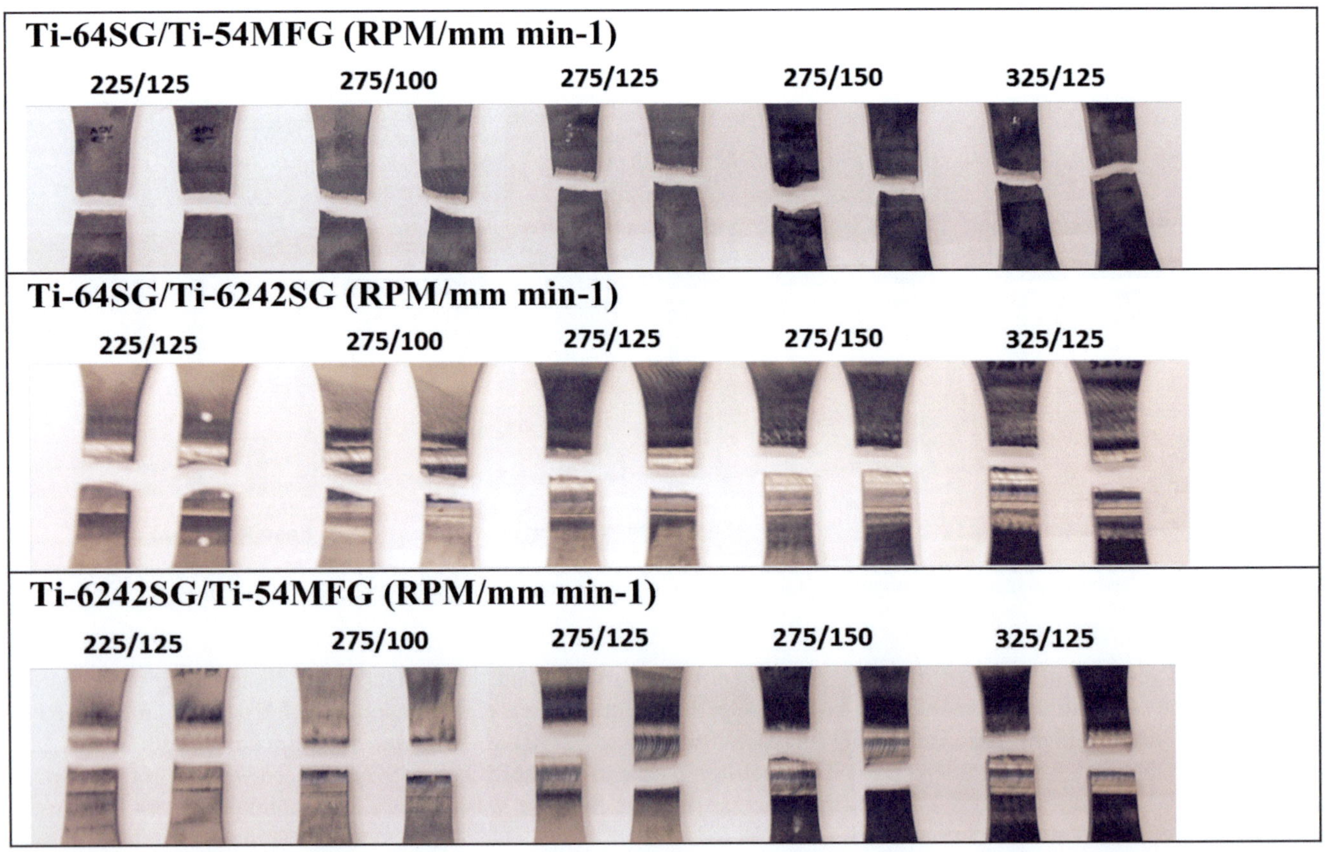

Fig. 20.7 Fracture in Ti-54MFG/Ti-64SG, Ti-54MFG/Ti-64SG, Ti-54MFG/Ti-6242SG, and Ti-64SG/Ti-6242SG

of failure: mixed mode and shear type was seen following the fracture. In most of the cases, the fracture was observed in the welded region; however there were joints wherein the fracture occurred near the parent material edges. Some cases of edge fracture happened in the advancing side, while some it occurred in the retreating side. The shear type of ductile failure was observed in Ti-64SG/Ti-54MFG at 275 RPM/100 mm min^{-1}. In Ti-6242SG/Ti-64SG, the failure occurred at the advancing side.

In Ti6242SG/Ti-54MFG, fractured specimens showed failure both at the advancing and the retreating side. The brittle fracture cases included Ti-6242SG/Ti-54MFG-225 RPM/125 mm min^{-1}, 275 RPM/100 mm min^{-1}, and Ti-6242SG/Ti-64SG-225 RPM/125 mm min^{-1}, 275 RPM/150 mm min^{-1}. These joints have recorded low values of ductility in the material behavior test.

20.5 Conclusion

Friction stir welding was demonstrated with dissimilar titanium alloys at different rotation speeds and feed rates. The joining process produced defect free butt joints with dissimilar titanium alloys. The process showed a very good strength as close as 90 % with respect to the parent alloys. The rotation speed of the welding tool showed more influence on the strength of the joints than its feed rate. The optimum feed rate and rotation speed were 125 mm min^{-1} and 325 RPM respectively. Ti-54MFG/Ti-6242SG showed higher mechanical strength than Ti-6242SG/Ti-64SG and Ti-54MFG/Ti-64SG. The maximum ductility 14 % was recorded for Ti-6242SG/Ti-64SG at 275 RPM/125 mm min^{-1}. Most of the cases showed fracture in the welded region, thus the process showed good joining capability.

Acknowledgments The authors acknowledge The Boeing Company for the supply of materials, welds and financial support. We sincerely thank Dr. D. Sanders for his encouragement throughout this research project.

References

1. W. Thomas, E. Nicholas, J. Needham, M. Murch, P. Templesmith, C. Dawes, Great Britain Patent 9125978.8 (1991)
2. D.G. Sanders, P.D. Edwards, M. Ramulu, G. Grant, Optimization of friction stir welding process for superplastic forming and improved surface texture for titanium aerospace structures. Key Eng. Mater. **433**, 153–167 (2010)
3. D.G. Sanders, M. Ramulu, P.D. Edwards, Superplastic forming of friction stir welds in Titanium alloy 6Al-4V: preliminary results. Mat.-wiss. u. Werkstofftech **39**(4–5), 353–357 (2008)
4. Y.C. Chen, K. Nakata, Microstructural characterization and mechanical properties in friction stir welding of aluminum and titanium dissimilar alloys. Mater. Design **30**, 469–474 (2009)
5. M. Aonuma, K. Nakata, Effect of alloying elements on interface microstructure of Mg–Al–Zn magnesium alloys and titanium joint by friction stir welding. Mater. Sci. Eng. **B161**, 46–49 (2009)
6. M. Aonuma, K. Nakata, Dissimilar metal joining of 2024 and 7075 aluminium alloys to titanium alloys by friction stir welding. Mater. Trans. **52**(5), 948–952 (2011)
7. N. Kulkarni, M. Ramulu, Experimental and numerical analysis of mechanical behavior in friction stir welded different titanium alloys. in Proceedings of the ASME 2014 International Mechanical Engineering Congress & Exposition (2014)
8. http://www.matweb.com/Search/MaterialGroupSearch.aspx?GroupID
9. P.D. Edwards, M. Ramulu, Effect of process conditions on superplastic forming behavior in titanium 6Al-4V friction stir welds. Sci. Technol. Welding Joining **14**(7), 669 (2009)
10. P.D. Edwards, M. Ramulu, Identification of process parameters for friction stir welding Ti-6Al-4V. J. Eng. Mater. Technol. **132**, 0310061–0310061 (2010)

Chapter 21
Testing a Multi-bay Box Subjected to Combined Loads

Marshall Rouse and Dawn Jegley

Abstract The COmbined Loads Test System (COLTS) facility at NASA Langley Research Center provides a test capability to help develop validated structures technologies. The test machine was design to accommodate a range of fuselage structures and wing sections and subject them to both quasistatic and cyclic loading conditions. The COLTS facility is capable of testing fuselage barrels up to 4.6 m in diameter and 13.7 m long with combined mechanical, internal pressure, and thermal loads. The COLTS facility is currently being prepared to conduct a combined mechanical and pressure loading for a multi-bay pressure box to experimentally verify the structural performance of a composite structure which is 9.1 m long and representative of a section of a hybrid wing body fuselage section in support of the Environmentally Responsible Aviation Project at NASA. This paper describes development of the multi-bay pressure box test using the COLTS facility. The multi-bay test article will be subjected to mechanical loads and internal pressure loads up to design ultimate load. Mechanical and pressure loads will be applied independently in some tests and simultaneously in others.

Keywords Composite • Combined loading • Hybrid wing body • PRSEUS • Rod-stiffened

21.1 Introduction

The ability to subject aerospace structures to combined loads representative of actual operating conditions is an important aspect of aerospace structural design. The standard aerospace industry practice is to design a structure that supports the various loading conditions and to verify the design by structural testing. A test fixture is usually designed to subject the structure to a subset of the loading conditions. The test fixture is often limited to a particular design or single configuration. However, the development of structures technology to be used for future aerospace design requires that full-scale and sub-scale structural concepts be tested with combinations of mechanical and internal pressure loads. Combined loads testing requires a combined loads test machine with the flexibility to accommodate these structures. NASA Langley Research Center has developed a COmbined Loads Test System (COLTS) Facility that consists of a multi-actuator test machine capable of applying combined loads and internal pressure loading to validate structures technology [1]. Structural tests have been conducted using the COLTS facility to support airframe and space structures technology development.

The subject of this paper is the preparation for conducting a sequence of combined mechanical and pressure loadings to a large-scale multi-bay pressure box to experimentally verify the structural performance of a composite structure which is 9.1 m long. This test article is representative of a section of a Hybrid Wing Body (HWB) center fuselage section and supports NASA's Environmentally Responsible Aviation (ERA) Project. In this paper, first, the COLTS facility is described and the goals of the project are presented. Subsequently, preparations and plans for testing this structure are described.

21.2 Combined Loads Test System

COLTS was designed to accommodate a range of fuselage structures and wing sections and subject them to both quasistatic and cyclic loading conditions. COLTS is capable of testing fuselage barrels up to 4.6 m in diameter and 13.7 m long with combined mechanical, internal pressure, and thermal loads. A schematic of COLTS configured with a cylindrical shell

M. Rouse (✉) • D. Jegley
NASA Langley Research Center, Mail Stop 190, Hampton, VA 23681-2199, USA
e-mail: marshal.rouse@nasa.gov; dawn.c.jegley@nasa.gov

C. Sciammarella et al. (eds.), *Experimental and Applied Mechanics, Volume 4*,
Conference Proceedings of the Society for Experimental Mechanics Series, DOI 10.1007/978-3-319-22449-7_21

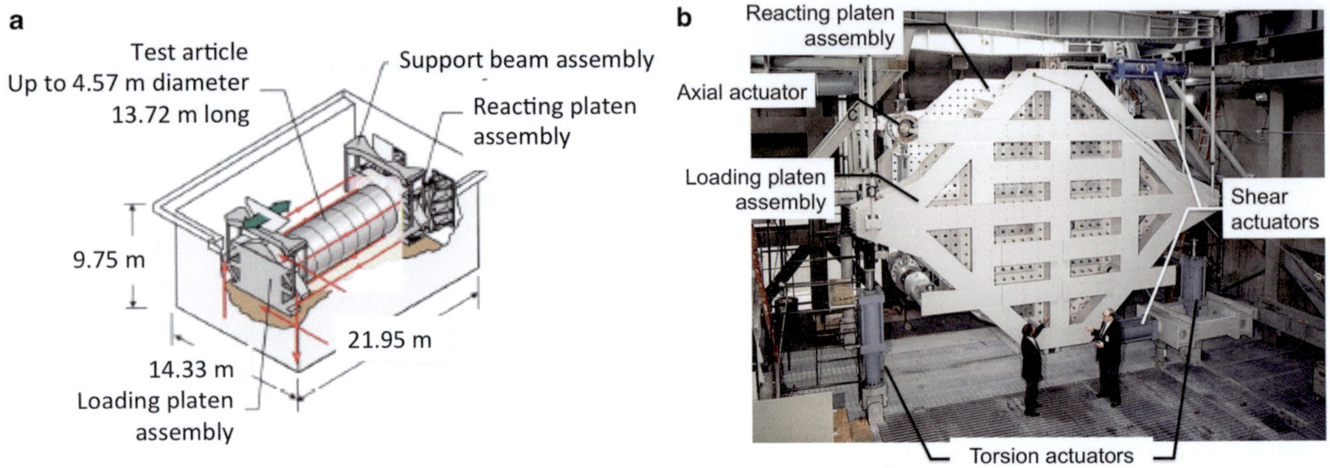

Fig. 21.1 Description of COmbined Loads Test System. (**a**) COLTS test chamber. (**b**) COLTS facility with platen at 3-m position

Fig. 21.2 Mini-COLTS

mounted between the test machine platens is illustrated in Fig. 21.1a, and a photograph of the facility is shown in Fig. 21.1b. The loading platen is suspended from a gantry that can traverse forward or backward to accommodate test articles of different lengths. Compression and bending loads are applied to the test articles by six 2001.7-kN hydraulic actuators located perpendicular to and between the loading and reacting platens along the x (longitudinal) axis. Torsion loads are applied to the test articles by two 1334.5-kN hydraulic actuators located vertically and spaced 7.0 m apart. Shear loads are applied to the test articles by two 1334.5-kN hydraulic actuators located horizontally and spaced 5.5 m apart. The test machine is located in a steel reinforced concrete pit that is approximately 9.8 m deep, 14.3 m wide, and 22.0 m long. This arrangement ensures that pressurized structural testing can be performed safely. The length of the axial actuators is extended by means of tubular extensions to connect the loading platen with the reacting platen so that test specimens that are longer than 3.05 m can be accommodated.

For the development of tests to be conducted in COLTS, an approximately 10 %-scale model of COLTS, called Mini-COLTS, is utilized. Compression, bending, shear, and torsion loads are applied to subscale test models by six 8896-N hydraulic actuators located perpendicular to and between the loading and reacting platens along the x (longitudinal) axis. Torsion loads are applied to the test articles by two 8896-N hydraulic actuators located vertically and spaced 81.6 cm inches apart. Shear loads are applied to the test articles by two 8896-N hydraulic actuators located horizontally and spaced 57.15 cm apart. The length of the axial actuators is fixed with respect to the reacting platen so that test models that are 139.7 cm can be accommodated. A photograph of the Mini-COLTS is shown in Fig. 21.2. Using the COLTS control system, Mini-COLTS can be used to conduct operational checkout of the mechanical loading parameters before loading the actual test article. Once the Mini-COLTS loading has been verified, the controls can be switched back to the COLTS facility.

A 2048-channel data system is used to record data from the strain gages, thermocouples, transducers, and load cells during each test. Additional data systems to supplement this system can also be used. For example, fiber optic, digital video image correlations systems and acoustic emission systems have independent personal computer data systems to record their data but can be linked to the primary data system to acquire pressure and load from one load cell so that readings from the independent systems can be correlated.

21.3 Goals of the Multi-bay Test Article Experiment

NASA has created the Environmentally Responsible Aviation (ERA) Project to explore and document the feasibility, benefits and technical risk of advanced vehicle configurations and enabling technologies that will reduce the impact of aviation on the environment. A critical aspect of this pursuit is the development of a lighter, more robust airframe that will enable the introduction of unconventional aircraft configurations that have higher lift-to-drag ratios, reduced drag, and lower community noise levels. The Hybrid Wing Body (HWB) arrangement offers a significant improvement in aerodynamic performance compared to a traditional tube-and-wing aircraft. However, the HWB design poses challenges in the design of a non-circular pressure cabin that is not only lightweight, but also economical to produce. Developing a viable structural concept is the primary technical challenge to the implementation of a large lifting-body design like HWB [2].

21.3.1 Damage Arresting Composites

To address this challenge, researchers at NASA and The Boeing Company (Boeing) are working together to develop a new structural concept called the Pultruded Rod Stitched Efficient Unitized Structure (PRSEUS) [2–5]. NASA and Boeing are exploring fundamental PRSEUS technologies that could someday be implemented on a transport-size airplane design. In ERA and previous programs, the PRSEUS concept was evaluated analytically and experimentally using a building-block approach [5, 6] that culminates in the test of a large 9.1-meter-long multi-bay pressure box that will be subjected to combined bending and internal pressure loadings.

The PRSEUS design and fabrication approach incorporates damage arrestment, improved load paths, and other weight reducing design features, which result in a highly effective stiffened panel concept. It is a conscious progression away from conventional laminated and bonded methods of assembly, and has evolved to become a one-piece cocured panel design with seamless transitions and damage-arrest interfaces. The highly integrated nature of the PRSEUS stiffened panel design is enabled by the use of through-thickness stitching, which ultimately leads to unprecedented levels of fiber tailoring and load path continuity between the individual structural elements.

A PRSEUS panel geometry consists of dry warp-knit fabric, pre-cured rods, and foam-core materials that are assembled and then stitched together as shown in Fig. 21.3. Load path continuity at the stringer frame intersection is maintained in both

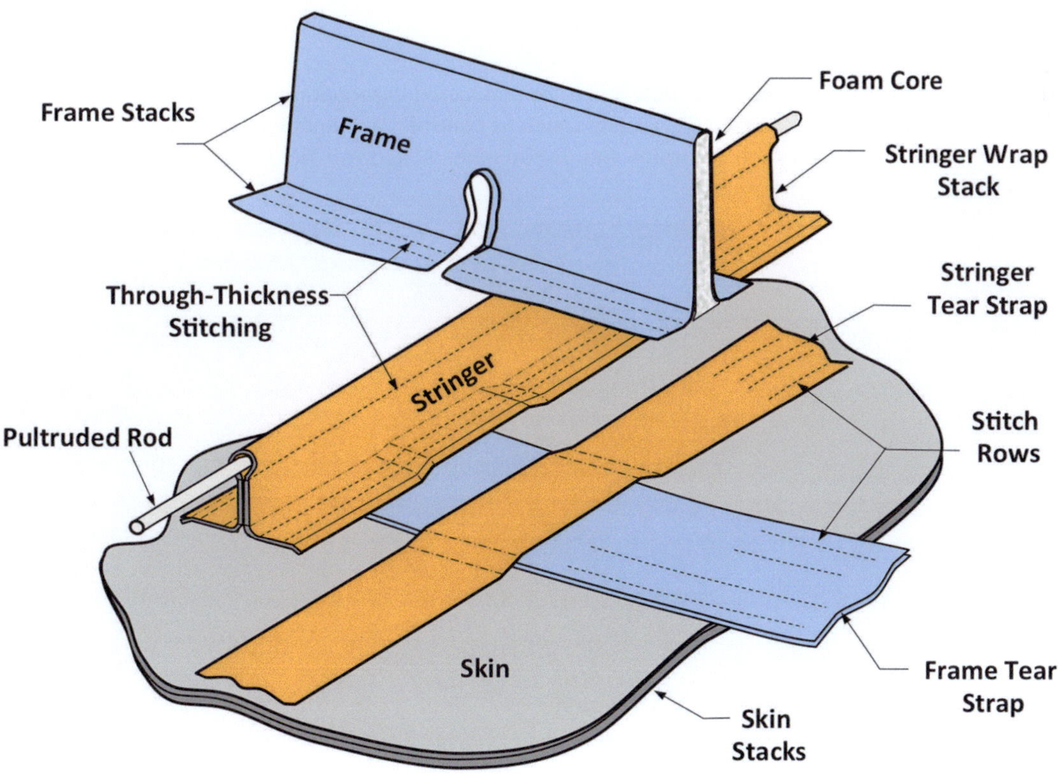

Fig. 21.3 Exploded view of Pultruded Rod Stitched Efficient Unitized Structure (PRSEUS) concept

directions by passing the rod-stringer through a small keyhole in the frame web. The 0° fiber-dominated pultruded rod increases local strength and stability of the stringer section while simultaneously shifting the neutral axis away from the skin to further enhance the overall panel bending capability. Frame elements are placed directly on the inner mold line skin surface and are designed to take advantage of carbon fiber tailoring by placing bending- and shear-conducive lay ups where they are most effective. The stitching is used to suppress out of plane failure modes. Suppressing these modes enables a higher degree of tailoring than would be possible using conventional laminated materials. The resulting bi-directionally stiffened panel design is ideal for the HWB pressure cabin because it is not only highly efficient in all three loading directions, but also stitched to react pull-off loading and increase panel survivability.

21.3.2 Multi-bay Pressure Box

The final step in the building block ground testing in applying the PRSEUS concept to the HWB centerbody is a 9.1-m-long multi-bay pressure box test article [6]. The test article is representative of a portion of the center section of a HWB vehicle and addresses the same construction, analysis, and pressurization issues that would be encountered on an airframe that has flat pressurized surfaces. These technologies include minimum-gauge design, post-buckled skins, crack arrestment, impact damage, manufacturing scale-up to larger structures, and joining technologies. The multi-bay test article contains 11 PRSEUS panels (crown, floor, two upper bulkheads, two lower bulkheads, two side ribs, two side keels, one center keel), four honeycomb inner rib panels, aluminum fittings, access doors, and titanium bolts and load introduction hardware. A photograph of a 9.1-m-long panel used to assemble the double deck closed box multi-bay pressure box test article is shown in Fig. 21.4. A sketch of the multi-bay pressure box is shown in Fig. 21.5 and a photograph of the multi-bay pressure box in its holding fixture while being prepared for testing is shown in Fig. 21.6.

The multi-bay box test article will be subjected to mechanical loads, but it will also be subjected to internal pressure loads. Mechanical and internal pressure loads will be applied simultaneously. The loading conditions will simulate the critical load cases of 2.5 g up-bending, 2.5 g up-bending plus 63.4 kPa pressure, −1.0 g down-bending, and −1.0 g down-bending plus 63.4 kPa internal pressure maneuver conditions, as well as the static internal pressure loads of 126.9 kPa. These loading conditions were used to size the overall test article structure.

21.4 Preparation for the Multi-bay Box Test

Preparation of the COLTS facility and the test article had to be conducted in preparation for the tests described above. These preparations included examination of the COLTS systems such as controls, pneumatics, and hydraulics and instrumentation of the test article to record point-data such as strains and displacements, and full-field images from cameras.

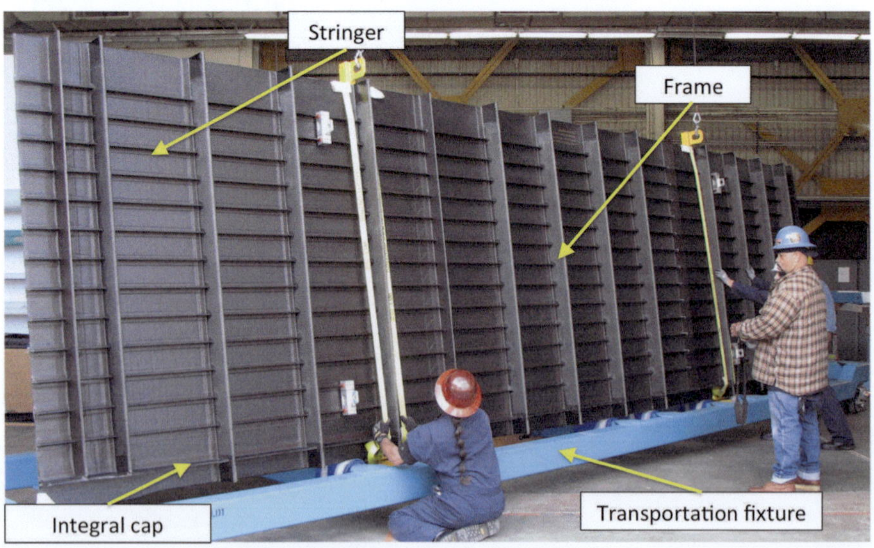

Fig. 21.4 PRSEUS 9.1-m-long bulkhead panel

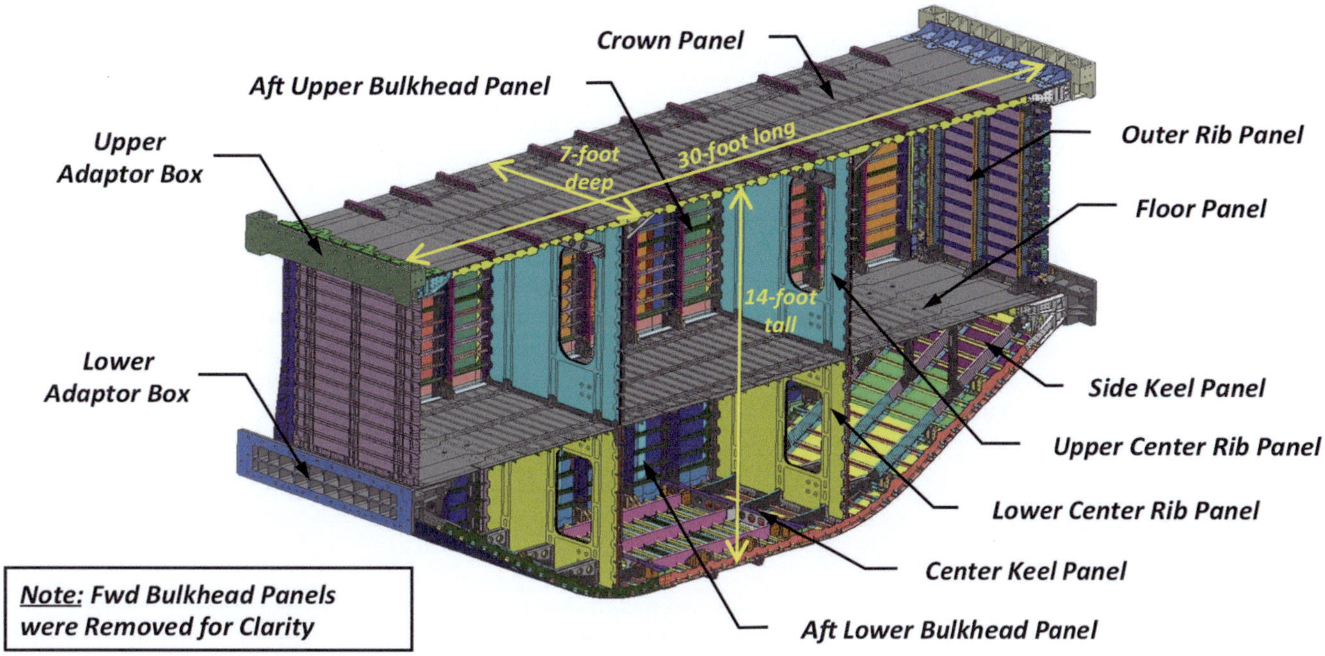

Fig. 21.5 Multi-bay pressure box

Fig. 21.6 Multi-bay pressure box in its holding fixture before moving into test chamber

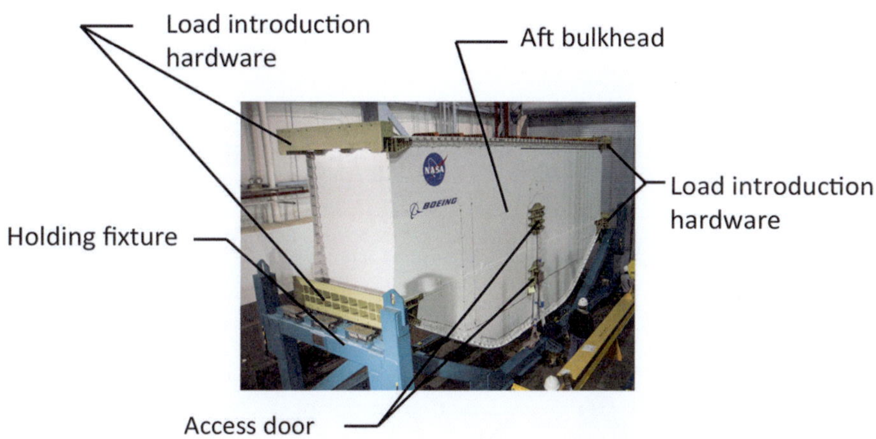

21.4.1 Control System

An MTS control system is used to control the ten COLTS actuators. In the multi-bay pressure box test, only four actuators are used. A combination of load-control, stroke control, and a mix mode will be used during the sequence of tests. The pressure loading commands can be input directly to the pressure controller or by discrete input from the MTS Control system. Air pressure loading is controlled by a separate Fisher controller that receives a discrete digital command from the MTS Control system. Internal pressure loading of the multi-bay box is achieved varying the opening of an inlet valve and an outlet valve.

21.4.2 *Control System Checkout*

A 9.1-m-long checkout beam was used to develop the test operations and procedures for the multi-bay test article evaluation. The multi-bay test article will be subjected to mechanical loading conditions to simulate flight conditions of the 2.5 g and −1.0 g bending maneuver conditions by applying loads on the platens of combined loads test machine. In order to achieve the critical 2.5 g up-bending maneuver load, the center bay of the crown panel will be loaded to an average compressive running load of 875,634 kN/m across its width. The loading methodology to be used for the multi-bay box test was verified using the 9.1-m checkout beam.

The checkout beam is a W36 × 256, A992 steel I-beam structure reinforced at each end with steel plates to allow load introduction. The checkout beam was subject to a maximum applied bending moment 1,346,594 m-N in COLTS in order to verify loading methodologies for the multi-bay box test. A photograph of checkout beam installed in COLTS is presented in Fig. 21.7. The series of tests on this checkout structure placed the beam in pure bending. Strain gages on the I-beam flanges were monitored to develop confidence that the MBB would be loaded in pure bending as planned [7].

21.4.3 *Instrumentation*

Several types of instrumentation will be used to monitor and record data during each test. 262 linear and 36 conventional strain gages, 15 linear variable displacement transducers (LVDT), two pressure transducers, three fiber optic wires, up to 5 video digital image correlation systems, 26 acoustic emission sensors and 14 video cameras will be used to record the behavior of the test article and the COLTS system. Strain gages are located on each PRSEUS panel and several metal fittings. The location of the LVDT's and fiber optic wires are shown in Fig. 21.8.

Fig. 21.7 Checkout beam between the COLTS plates

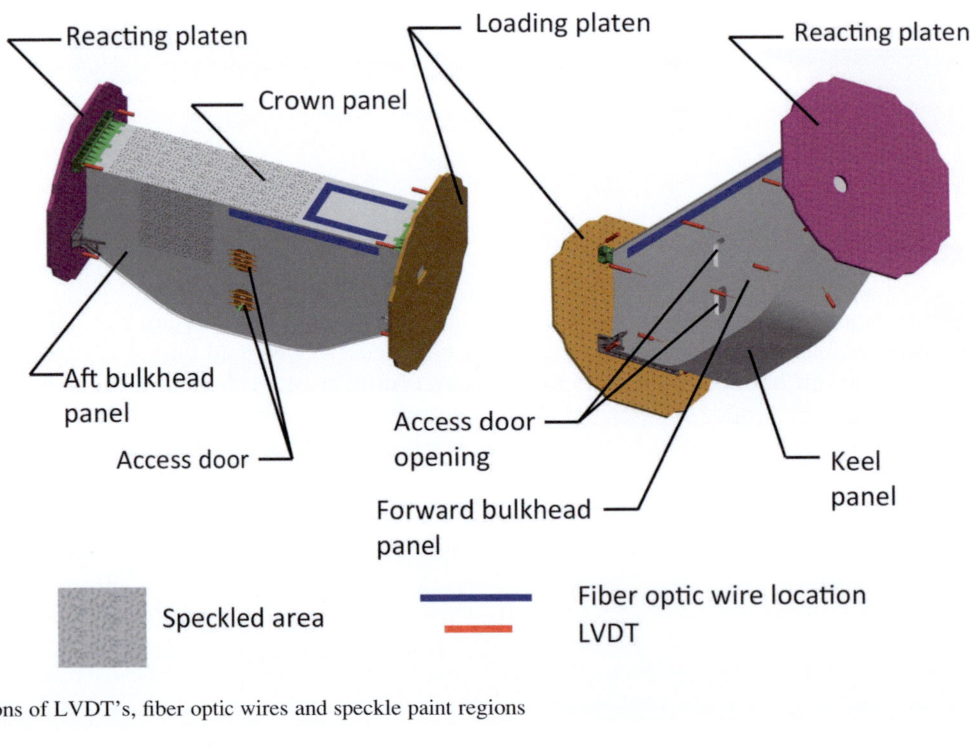

Fig. 21.8 Locations of LVDT's, fiber optic wires and speckle paint regions

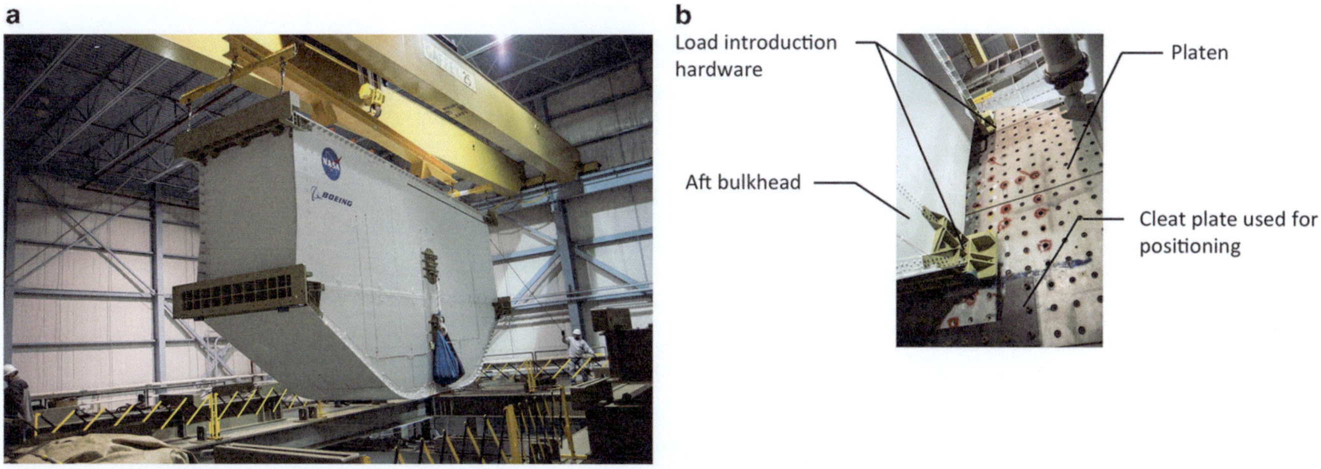

Fig. 21.9 Moving multi-bay box test article into test chamber. (**a**) Multi-bay pressure box suspended from crane above test chamber. (**b**) Multi-bay box load introduction fitting being placed against platen

A speckle pattern, consisting of black paint dots on a white paint background was applied to a portion of the aft bulkhead, the crown and the center keel. Approximate locations of areas to be monitored using the video digital image correlation systems is shown in Fig. 21.8.

Video cameras will be placed inside each of the six bays of the test article to record cracks and deformations in the bulkheads, floor, crown and keel. Additional cameras will be placed outside the test article to see a more global view of the structure. These video images will be monitored during each test. Time will be used to correlate video images with strain gage and other data post-test.

21.4.4 Installation Between Platens

To ensure the fit of proper fit and alignment of the multi-bay pressure box test article between the platens in the COLTS facility, the bolt holes in the load introduction fixtures were not drilled during the assembly of the test article. Instead, these four fittings were attached to the test article and milled to be flat and parallel to each other. These fittings in the undrilled state are shown in Fig. 21.6.

The multi-bay pressure box was initially lowered into the test chamber and positioned between the platens so that the bolt hole locations where the COLTS platens would interface with the test article load introduction fixtures could be marked. Using the 222 kN capacity crane in the COLTS facility, the test article was lifted out of its holding fixture, lifted over the wall between the test preparation area, and lowered between the platens in the test chamber. The test article was placed on the cleat plates on each platen and held in place by the crane for marking. Moving the test article from its holding fixture to place between the platens for marking is shown in Fig. 21.9. The test article was then lifted back out of the test chamber and placed back in its holding fixture so that the holes could be drilled in the load introduction fixtures. After all drilling was completed, the test article was again placed between the platens. Bolts were installed to hold he test article against the platens. A photograph of the test article fully attached to the platens is shown in Fig. 21.10.

21.5 Testing

Data from the strain gages, transducers, and load cells will be recorded at a rate of 1 scans per second. Plots of strain gages and LVDT's will be monitored during each test to track the structural behavior real time to compare to predictions and evaluate the operation of the loading system. Two still cameras will be positioned to view each speckled region to simultaneously photograph the pattern every 2 s during each test. These images will be compared using the Correlated Solutions software [8] to determine the displacements in the x, y and z directions and the in-plane strains. These systems will

Fig. 21.10 Multi-bay box located between platens

Table 21.1 Load at DUL

Load condition	Pressure (kPa)	Actuator load (kN)
Pressure only	127	0
Down-bending	0	424
Up-bending	0	1059
Down-bending with pressure	95	424
Up-bending with pressure	95	1059

provide real-time full-field imaging of displacements and strains. Data from the fiber optic and acoustic emission systems will be used in post-test test-analysis correlation.

Video cameras will be placed inside each bay of the test article to record cracks and deformations in the bulkheads, floor, crown and keel panels. Additional cameras will be placed outside the test article to see a more global view of the structure. These video images will be monitored during each test. Time will be used to correlate video images with strain gage and other data post-test.

21.5.1 Pristine Structure Testing

After all instrumentation and other systems are in place, a pressure test to 13.8 kPa and a checkout mechanical loading test to 50 % of design limit load (DLL) will be conducted to verify that all systems are functioning as expected. A series of ten tests are planned to evaluate the pristine test article's response to critical load conditions. Loading will be conducting starting with the least critical load condition to minimize the possibility of premature damage to the test article. The test article will be loaded to DLL in down-bending alone, down-bending with pressure, pressure only, up-bending alone, and up-bending with pressure conditions. In each case, load will be quasistatic with pause points in the loading to make sure the actuators are staying in sync with each other and with the pressure load. The four active actuators will be slaved together so that all ramp to the maximum load at the same time. When pressure is simultaneously applied, mechanical load and pressure will ramp such that the maximum pressure is achieved at the same time as the maximum mechanical load. The loading will be paused briefly at the maximum load then the load will be slowly reduced to zero. If anomalies are seen in the test data, ultrasonic scans may be conducted between tests to quantify possible damage. After all DLL tests are complete, the series of loadings will be repeated with load to design ultimate load (DUL). DUL for each load condition is given in Table 21.1.

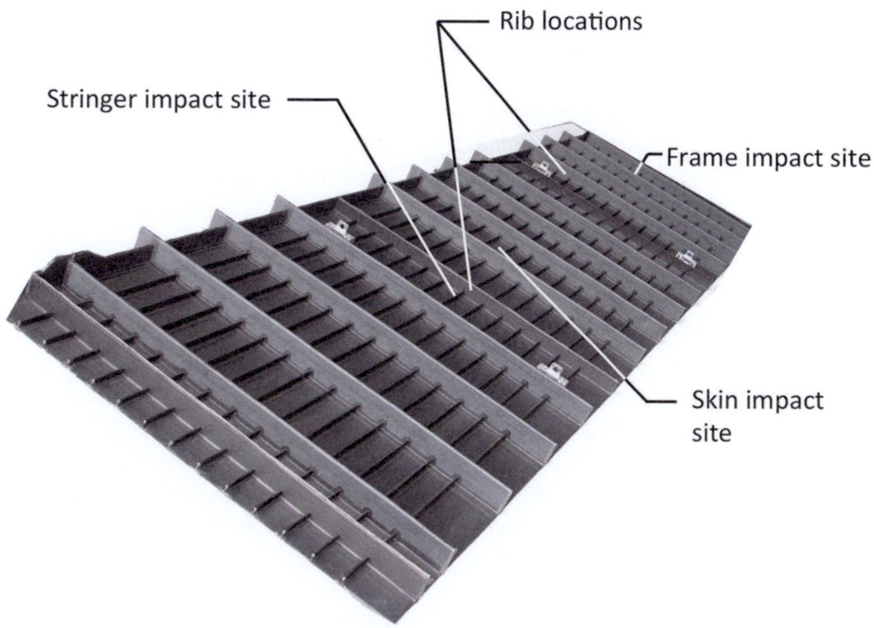

Fig. 21.11 Planned interior impact damage sites shown on bulkhead panel prior to test article assembly

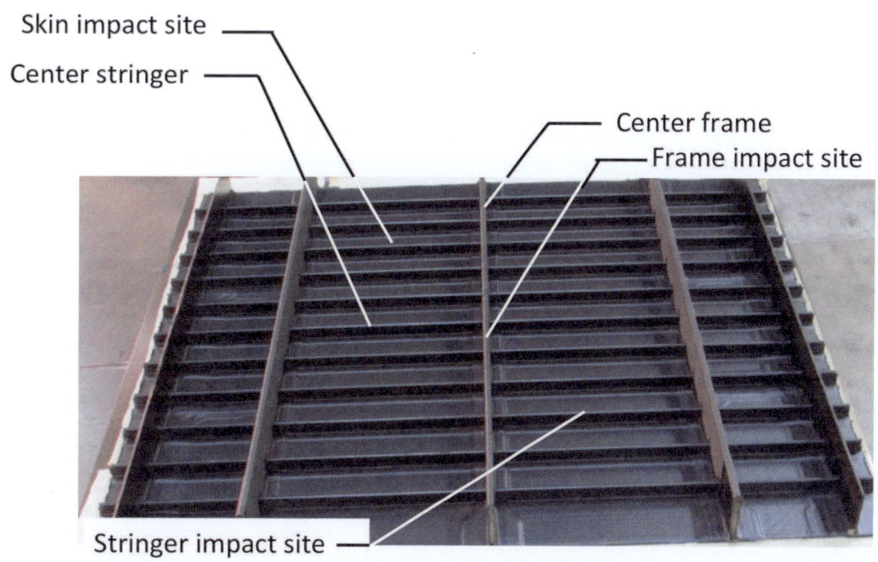

Fig. 21.12 Planned exterior impact damage sites shown on keel panel prior to test article assembly

21.5.2 Barely Visible Impact Damage Testing

After the completion of the pristine structure tests, barely visible impact damage (BVID) will be inflicted to the forward upper bulkhead panel. Three impacts to the interior of the structure and three impacts to the exterior if the structure will be inflicted. Interior impact energies will be 20 J in a skin area, 27 J on the top of a stringer, and 27 J on the top of a frame, as shown in Fig. 21.11. These impacts will be inflicted using a spring-loaded mechanism with an impactor tup diameter of 2.54 cm. Three exterior impacts will be imparted to the center keel panel. Exterior impact will be in a skin area with no interior support with an energy of 20 J, an area where the skin is reinforced by the frame flange with an impact energy of 81 J, and an area where the skin is reinforced by the stringer flange with an impact energy of 68 J, as shown in Fig. 21.12. These impacts will be inflicted using a gravity fed impactor with a 2.54 cm diameter tup. Immediately before and immediately after the impacts, ultrasonic scans will be conducted to so that the extent of damage caused by the impacts can be quantified. After all NDI is complete and verification that all instrumentation is fully functional, the series of tests conducted for the pristine structure will be repeated for the damaged structure. The final test is in the load condition of 2.5 g up-bending with pressure. However, if catastrophic failure has not resulted by DUL, this test will be modified such that after achieving DUL, pressure will be held constant and mechanical load will increase by an additional 10 %. If catastrophic failure has still not occurred, mechanical loading will be reduced back to DUL, then pressure will be slowly removed while mechanical loading is held constant at DUL and then mechanical load will be increased to 200 % of DLL. If catastrophic failure has still not occurred, loading will be removed, and NDI will be conducted. An additional loading with more severe damage to the structure is being considered.

21.6 Concluding Remarks

The COmbined Loads Test System facility at NASA Langley Research Center provides a test capability to help develop validated structures technologies. Preparations are almost complete for the testing of a large-scale composite multi-bay pressure box test article in the COLTS facility. The 9.1-m-long test article will be subjected to a series of critical loadings in pristine and damaged conditions. Combined mechanical and pressure loading applied to the multi-bay pressure box will be used to experimentally verify the structural performance of a composite structure which is representative of a section of a hybrid wing body fuselage section.

Marshall Rouse is a senior research engineer in the Structures Testing Branch (STB) of the Research Directorate at NASA Langley Research Center who is studying advanced-composite structural components for transport aircraft and space applications. Mr. Rouse conducts experimental and analytical research on the response, damage tolerance and failure response of undamaged and damaged advanced-composite structural panels. He is responsible for designing specimens representing advanced structural concepts and testing them in the laboratory using test procedures suitable for composite structural components, or when required, developing new test procedures or fixtures. Mr. Rouse also correlates the experimental results with linear and nonlinear analysis results as appropriate. Mr. Rouse received a M.S., Engineering Science, George Washington University in 1986 and a Bachelor of Science degree in Mechanical Engineering, North Carolina Agricultural and Technical State University in 1977.

References

1. D.R. Ambur, M. Rouse, J.H. Starnes, M.J. Shuart, Facilities for combined loads testing of aircraft structures to satisfy structural technology development requirements. in Presented at the 5th Annual Advanced Composites Technology Conference (Seattle, WA, August 22–36, 1994)
2. A. Velicki, Damage arresting composites for shaped vehicles, Phase I Final Report. NASA CR-2009-215932 (NASA Langley Research Center, Hampton, VA, September 2009)
3. A. Velicki, N.P. Yovanof, J. Baraja, K. Linton, V. Li, A. Hawley, P. Thrash, S. DeCoux, R. Pickell, Damage arresting composites for shaped vehicles—phase II Final Report. NASA CR-2011-216880 (NASA Langley Research Center, Hampton, VA, January 2011)
4. P. Thrash, Manufacturing of a stitched resin infused fuselage test article. in Presented at SAMPE Fall Technical Conference (Orlando, FL, October 2014)
5. A. Velicki, D.C. Jegley, PRSEUS structural concept development. in Presented at the 52nd AIAA Aerospace Sciences Meeting (National Harbor, MD, January 2014)
6. D.C. Jegley, A. Velicki, Development of the PRSEUS multi-bay pressure box for a hybrid wing body vehicle. in Presented at the 53rd AIAA Aerospace Sciences Meeting (Orlando, FL, January 2015)
7. M. Rouse, Methodologies for combined loads tests using a multi-actuator test machine. in Presented at the 2013 Annual Society for Experimental Mechanics Meeting (Chicago, IL, June 2013)
8. VIC-3D Reference Manual, Correlated Solutions, Inc., www.correlatedsolutions.com (2014)

Chapter 22
A Study on the Generation of a Fine-Grained Layer in Upsetting Between Flat and Conical Dies

Robert Goldstein, Sergei Alexandrov, Dragisa Vilotic, and Leposava Sidjanin

Abstract A narrow fine-grained layer is generated in the vicinity of frictional interfaces in many manufacturing processes. The majority of previous studies on this topic is devoted to machining. The present is paper concerned with the generation of a fine-grained layer in upsetting of hollow cylinders between flat and conical dies. The layer occurs in the vicinity of the friction surface between the specimen and the conical die. The die angle varies to change the state of stress and strain near the friction surface. No lubricant is used to increase the friction stress. Based on a proposed criterion that defines the fine-grained layer, it is demonstrated that its thickness depends on the die angle. Moreover, the thickness varies along the friction surface. A general continuum mechanics model is proposed to connect the thickness of the fine-grained layer and the strain rate intensity factor. The model is based on dimensional analysis and a number of general assumptions. The strain rate intensity factor in the upsetting process is determined by an approximate method. Then, the experimental results are used to find some input parameters of the theoretical model.

Keywords Frictional interfaces • Fine-grained layer • Thickness • Strain rate intensity factor

22.1 Introduction

There are a great number of experimental observations that demonstrate that a narrow layer with drastically modified microstructure is generated near frictional interfaces. This layer is usually called white layer in papers devoted to machining processes and fine grain layer in papers devoted to metal forming processes. A complete review of results on white layer/fine grain layer generation published before 1987 has been presented in [1]. According to this review article there are three main contributory mechanisms responsible for white layer generation. Two of these mechanisms are: (a) the mechanism of rapid heating and quenching and (b) the mechanism of intensive plastic deformation. The present paper is devoted to an experimental/theoretical study of the latter mechanism. Severe plastic deformation in the vicinity of frictional interfaces results in mechanical behavior that cannot be represented by conventional material tests. Therefore, numerical results based on conventional material models are not capable to adequately describe material flow in the vicinity of surfaces with high friction. The present paper combines the strain rate intensity factor and special purpose tests to develop a novel approach for predicting the generation of fine grain layers in deformation processes. The strain rate intensity factor has been introduced in [2]. This factor is the coefficient of the leading singular term in a series expansion of the equivalent strain rate near maximum friction surfaces. This expansion shows that the equivalent strain rate approaches infinity in the vicinity of the friction surface. It is therefore evident that the strain rate intensity factor controls the magnitude of the equivalent strain rate in a narrow layer near the surface. A necessary condition for the existence of the strain rate intensity factor is the friction boundary condition that postulates that the friction stress at sliding is equal to the shear yield stress (maximum friction law). This boundary condition is often adopted at the tool-chip interface (at least, over a portion of this interface) in machining processes [3–10]. This zone is usually called the sticking zone. The existence of such zones has been reported in deformation processes as well [11–14].

R. Goldstein (✉) • S. Alexandrov
Laboratory on Mechanics of Strength and Fracture of Materials and Structures, The Institute for Problems in Mechanics of the Russian Academy of Sciences, Prospect Vernadskogo, 101-1, Moscow 119526, Russia
e-mail: goldst@ipmnet.ru

D. Vilotic • L. Sidjanin
University of Novi Sad, Obradovich Street, 6 Novi Sad 21000, Serbia

C. Sciammarella et al. (eds.), *Experimental and Applied Mechanics, Volume 4*,
Conference Proceedings of the Society for Experimental Mechanics Series, DOI 10.1007/978-3-319-22449-7_22

The experimental program includes upsetting of specimens between conical and flat dies. The surface of the conical die is very rough. By assumption, this is a maximum friction surface. It is reasonable to expect that both the thickness of the fine grain layer found experimentally and the strain rate intensity factor found theoretically depend on the die angle. Therefore, using dies with different angles it is in general possible to establish a relation between the strain rate intensity factor and the thickness of the layer (or any other parameters that characterize the fine grain layer). A difficulty is that commercial finite element packages are not capable to calculate the strain rate intensity factor. Probably, the generalized finite element method [15] can be used for this purpose. However, no specific code is currently available. Therefore, the only possibility to determine the strain rate intensity factor is to use approximate solutions. The upper bound theorem [16] is used in the present paper.

22.2 Conceptual Approach

For a sake of simplicity, it is assumed that the material is rigid perfectly plastic. In this case, the maximum friction law postulates that

$$\tau_f = k \tag{22.1}$$

at sliding. Here τ_f is the friction stress and *k* is the shear yield stress, a material constant. The asymptotic behavior of the equivalent strain rate in the vicinity of friction surfaces where (22.1) is valid is given by [2]

$$\xi_{eq} = \frac{D}{\sqrt{x}} + o\left(\frac{1}{\sqrt{x}}\right) \tag{22.2}$$

as $x \to 0$. Here ξ_{eq} is the equivalent strain rate (the quadratic invariant of the strain rate tensor), *D* is the strain rate intensity factor and *x* is the normal distance to the maximum friction surface.

It is evident from (22.2) that $\xi_{eq} \to \infty$ as $x \to 0$ and that the strain rate intensity factor controls the magnitude of the equivalent strain rate in the vicinity of the maximum friction surface. Equation (22.2) predicts severe plastic deformation in the vicinity of frictional interfaces. This is in qualitative agreement with the occurrence of narrow fine grain layers near frictional interfaces in experiment since severe plastic deformation is one of the main contributory mechanisms responsible for the generation of such layers [1]. However, the equivalent strain rate cannot be used in constitutive equations since it approaches infinity as $x \to 0$. Therefore, a reasonable assumption is that the magnitude of the strain rate intensity factor controls the generation of fine grain layers in the vicinity of frictional interfaces. Such an approach is similar to that in the mechanics of cracks where the stress intensity factor is used instead of stress to predict the propagation of cracks [17].

Dimensional analysis is a powerful means which can be used without sufficient preliminary study of the problem. A great account on this technique is given in [18]. Without having any physical theory for predicting the thickness of the fine grain layer near friction surfaces it is reasonable to guess that this thickness depends on the strain rate intensity factor and the distance travelled by a material particle along the friction surface, S. The current thickness of the layer is denoted by h. Let D_{cr} be a threshold magnitude of the strain rate intensity factor such that the intensive change in material microstructure near the friction surface occurs only if $D \geq D_{cr}$. Material flow near maximum friction surfaces is very similar to that in conventional processes of severe plastic deformation [19]. In the latter processes, fine microstructures become stable after a certain amount of plastic deformation [20]. Therefore, it is reasonable to assume that the thickness of the fine grain layer increases over an initial segment of the total distance travelled by the particle along the friction surface and then does not change. Let S_{cr} be the value of S that is necessary to attain the stable thickness of the layer if $D = D_{cr}$ throughout the process of deformation. For the theory developed in the present paper it is essential that $D_{cr} \neq 0$ and $S_{cr} < \infty$. In general, D is not a monotonic function of S. On the other hand, it is natural to assume that h is a non-decreasing function of S. Therefore, it is reasonable to introduce the following function δ

$$\delta = \int_0^S D_s ds, \quad D_s = \begin{cases} D & \text{if} \quad D \geq D_{cr} \\ 0 & \text{if} \quad D < D_{cr} \end{cases} \tag{22.3}$$

It is now hypothesized that the equation for h is

$$h = S_{cr}\Phi_0\left(\frac{\delta}{D_{cr}S_{cr}}, \frac{S}{S_{cr}}\right) \tag{22.4}$$

where Φ_0 is an arbitrary function of its arguments. It is worthy of note that it follows from (22.2) that $D \underline{\Delta} L^{1/2}T^{-1}$ and from (22.3) that $\delta \underline{\Delta} L^{3/2}T^{-1}$. Here the symbol $\underline{\Delta}$ means 'is dimensionally equal to', L is the dimension symbol for length and T is the dimension symbol for time. Therefore, the arguments of the function Φ_0 are dimensionless. An additional condition imposed of the function Φ_0 is

$$h = 0 \tag{22.5}$$

for $S = 0$. Assume that $D < D_{cr}$ throughout the entire process of deformation. Then, $\Phi_0(0, \ S/S_{cr}) = 0$. Thus the right hand side of (22.4) may only depend on S/S_{cr} in a combination with the first argument of Φ_0. In this case (22.4) becomes

$$h = S_{cr}\Phi_1\left[\frac{\delta}{D_{cr}S_{cr}}, f\left(\frac{\delta}{D_{cr}S_{cr}}, \frac{S}{S_{cr}}\right)\right] \tag{22.6}$$

where Φ_1 and f are new arbitrary functions. Assume that $D \geq D_{cr}$ in the range $0 \leq S \leq S_1$ and $D < D_{cr}$ in the range $S > S_1$. By the constitutive equation adopted, $dh/dS = 0$ in the range $S > S_1$. Moreover, $d\delta/dS = 0$ but $\delta \neq 0$. Differentiating (22.6) and using these conditions give $\frac{dh}{dS} = S_{cr}\frac{\partial\Phi_1}{\partial f}\frac{\partial f}{\partial S} = 0$.

Thus the function f is independent of S. Then, (22.6) becomes

$$h = S_{cr}\Phi\left(\frac{\delta}{D_{cr}S_{cr}}\right) \tag{22.7}$$

where Φ is a new arbitrary function of its argument. It follows from (22.5) that this function should satisfy the condition $\Phi(0) = 0$. In order to specify the function Φ for this or that material it is necessary to (*i*) design and carry out an experimental program and (*ii*) calculate the strain rate intensity factor.

22.3 Experiment

In order to promote the generation of fine grain layer in experiment, it is necessary to create intensive shearing deformation in the vicinity of a friction surface. In the present paper, upsetting of hollow cylindrical specimens between conical and flat dies is used (Fig. 22.1). The conical die moves down with velocity V. The flat die is motionless. No lubricant is used between the specimen and the conical die to approach the condition (22.1). On the other hand, the Rastegaev method [21] is used to significantly reduce friction between the specimen and the flat die. The initial outer, R_0, and inner, r_0, radii of the specimen are chosen to be the same in all tests. In particular, $R_0 = 10$ mm and $r_0 = 2.5$ mm. The specimens were machined such that the initial shape of the friction surface is the same as the shape of the die surface. Two conical dies with $\alpha = 60^0$ and $\alpha = 45^0$ were used to upset specimens. All the specimens were made of the steel C45E (European standard, the nominal composition of 0.46 % C and 0.65 % Mn). Typical microstructures generated at approximately the same distance from the outer and inner surfaces of the specimens are shown in Fig. 22.2 after upsetting with the $\alpha = 60^0$ die and in Fig. 22.3 after upsetting with the $\alpha = 45^0$ die. The layer is clearly seen in these figures. Also seen is that the angle α affects its thickness.

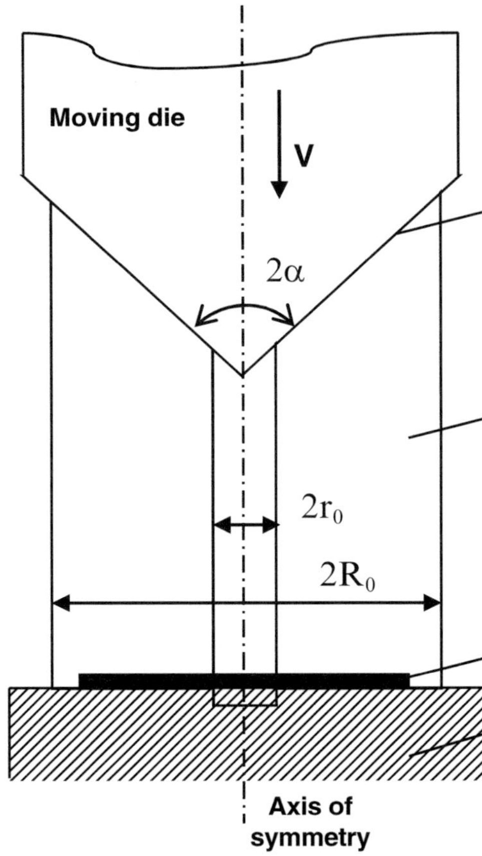

Fig. 22.1 Scheme of the upsetting

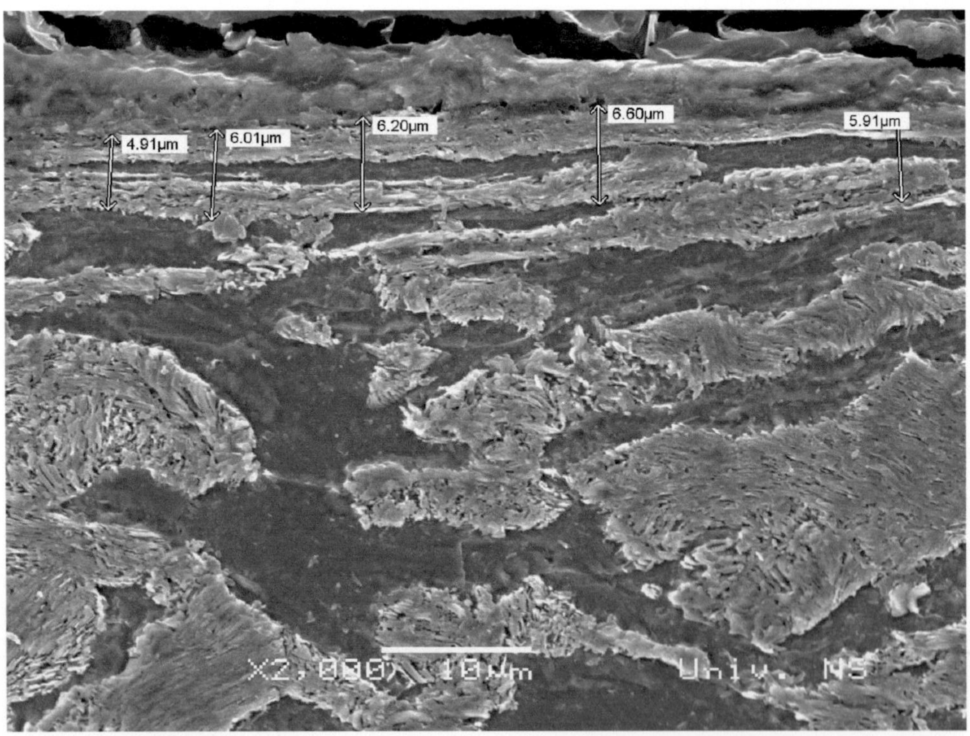

Fig. 22.2 Distribution of microstructure after upsetting with the α=60° die

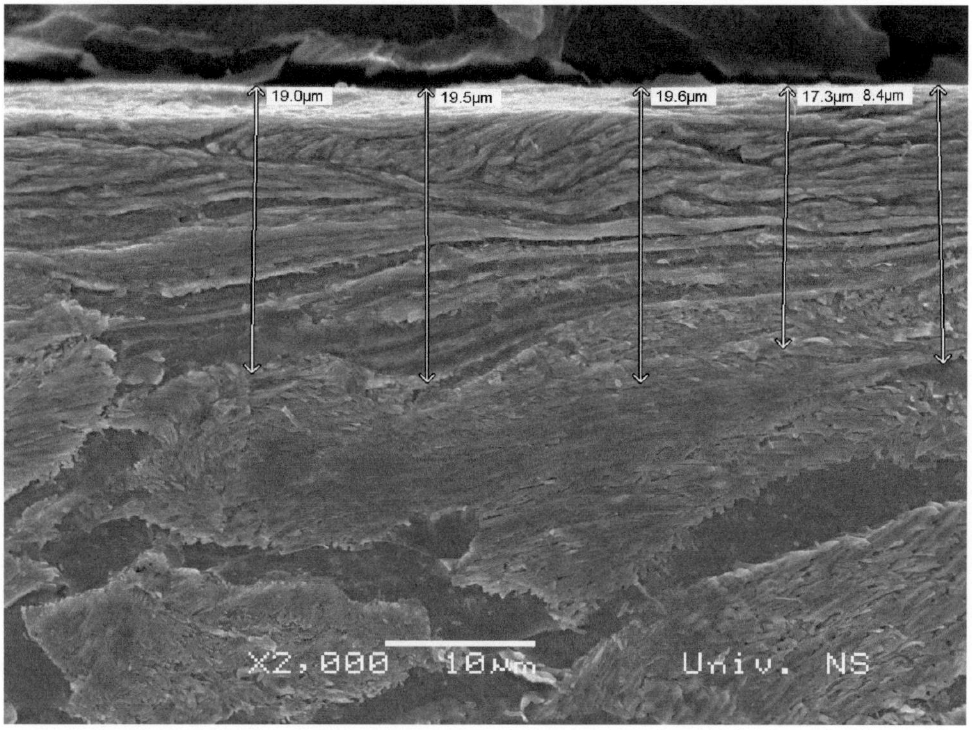

Fig. 22.3 Distribution of microstructure after upsetting with the α=45° die

22.4 Strain Rate Intensity Factor in Upsetting

The upper bound theorem is a convenient tool to obtain approximate solutions in metal forming. An approximate magnitude of the strain rate intensity factor can be found if the cinematically admissible velocity field accounts for the asymptotic expansion of the real velocity field that leads to (22.2). Introduce a cylindrical coordinate system (r, φ, z) such that its z-axis coincides with the axis of symmetry of the process and the plane $z = 0$ coincides with the interface between the specimen and the flat die. In addition, introduce a new orthogonal coordinate system (ρ, φ', θ) by the following transformation equations

$$r = \rho \cos\theta - l, \quad z = \rho \sin\theta, \quad \varphi = \varphi' \tag{22.8}$$

where $l = L \tan\alpha - R_0$. Here L is the height of the specimen at $r = R_0$. In the new coordinate system, the interface between the specimen and the conical die is determined by the equation $\theta = \pi/2 - \alpha$ and the interface between the specimen and the conical die by the equation $\theta = 0$. Therefore, the essential boundary conditions are

$$u_\theta = 0 \tag{22.9}$$

for $\theta = 0$ and

$$u_\theta = -V \sin\alpha \tag{22.10}$$

for $\theta = \pi/2 - \alpha$. Using (22.8) the scale factors of the coordinate system (ρ, φ', θ) are found as

$$h_\rho = 1, \quad h_\theta = \rho, \quad h_\varphi = \rho \cos\theta - 1. \tag{22.11}$$

According to the upper bound theorem it is necessary to minimize the following functional

$$\sqrt{3}\int_0^{\pi/2-\alpha}\int_{\rho_1(\theta)}^{\rho_2(\theta)} \xi_{eq} h_\rho h_\theta h_\varphi d\rho d\theta + \int_{\rho_{min}}^{\rho_{max}} \left| V\cos\alpha + u_\rho\big|_{\theta=\pi/2-\alpha}\right| d\rho \to \min,$$
$$\rho_1(\theta) = \frac{r_0+1}{\cos\theta}, \quad \rho_2(\theta) = \frac{R_0+1}{\cos\theta}, \quad \rho_{min} = \frac{r_0+1}{\sin\alpha}, \quad \rho_{max} = \frac{R_0+1}{\sin\alpha} \tag{22.12}$$

It has been taken into account here that the condition (22.1) is satisfied. In the case under consideration the equivalent strain rate is defined as

$$\xi_{eq} = \sqrt{\frac{2}{3}\left(\xi_{\rho\rho}^2 + \xi_{\theta\theta}^2 + \xi_{\varphi\varphi}^2 + 2\xi_{\rho\theta}^2\right)} \tag{22.13}$$

The strain rate components involved in (22.13) are given by

$$\xi_{\rho\rho} = \frac{\partial u_\rho}{\partial \rho}, \qquad \xi_{\varphi\varphi} = \frac{u_\rho\cos\theta - u_\theta\sin\theta}{\rho\cos\theta - 1}, \qquad \xi_{\theta\theta} = \frac{1}{\rho}\left(\frac{\partial u_\theta}{\partial\theta} + u_\rho\right),$$
$$\xi_{\rho\theta} = \frac{1}{2}\left(\frac{\partial u_\theta}{\partial\rho} + \frac{1}{\rho}\frac{\partial u_\rho}{\partial\theta} - \frac{u_\theta}{\rho}\right) \tag{22.14}$$

The simplest cinematically velocity field is obtained by assuming that

$$\frac{u_\theta}{V} = -\tan\alpha\sin\theta. \tag{22.15}$$

It is possible to verify by inspection that (22.15) satisfies (22.9) and (22.10). The incompressibility equation is $\xi_{\rho\rho} + \xi_{\theta\theta} + \xi_{\varphi\varphi} = 0$. Substituting (22.14) and (22.15) into this equation results in the following equation for u_ρ

$$\frac{\partial u_\rho}{\partial\rho} + \frac{u_\rho\cos\theta + V\tan\alpha\sin^2\theta}{\rho\cos\theta - 1} + \frac{u_\rho - V\tan\alpha\cos\theta}{\rho} = 0. \tag{22.16}$$

The general solution to this equation is

$$\frac{u_\rho}{V} = \left(\frac{\rho^2}{2}\cos 2\theta - l\rho\cos\theta + \omega R_0^2\right)\frac{\tan\alpha}{(\rho\cos\theta - 1)\rho} \tag{22.17}$$

where ω is an arbitrary function of θ. It is seen from (22.14) and (22.17) that the condition $|\xi_{\rho\theta}| \to \infty$ as $\theta \to \pi/2 - \alpha$ is equivalent to the condition $|d\omega/d\theta| \to \infty$ as $\theta \to \pi/2 - \alpha$. The asymptotic expansion (22.2) is obtained if

$$\omega = \omega_0 + \omega_1\sqrt{\frac{\pi}{2} - \alpha - \theta} + o\left(\sqrt{\frac{\pi}{2} - \alpha - \theta}\right) \tag{22.18}$$

as $\theta \to \pi/2 - \alpha$. In the simplest case (22.18) reduces to

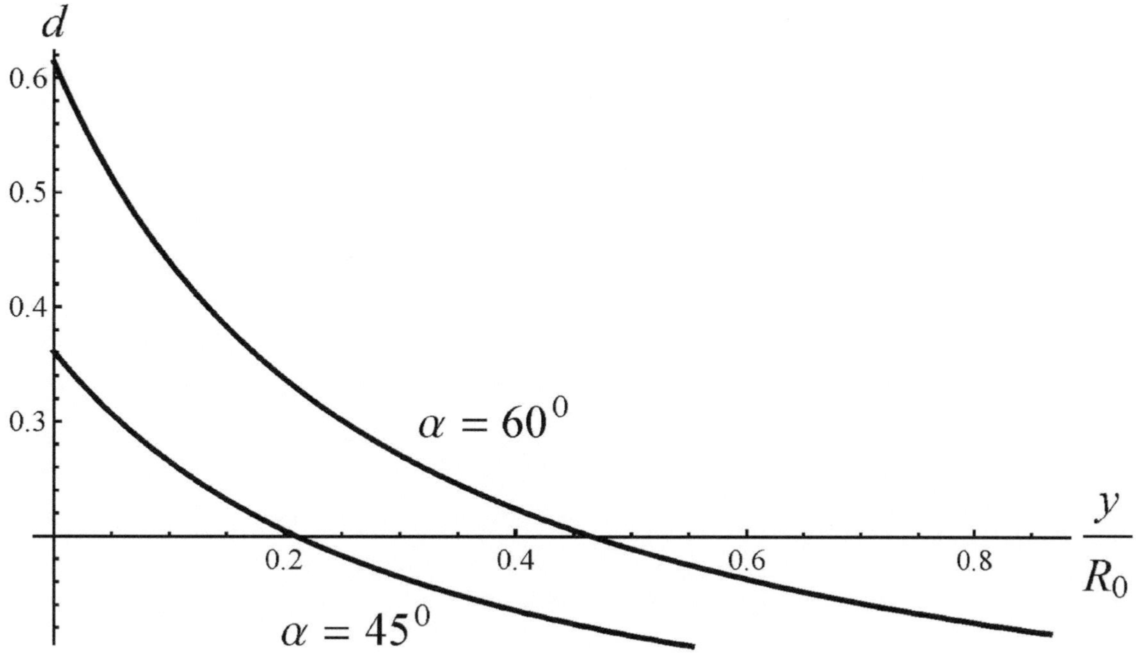

Fig. 22.4 Variation of the dimensionless strain rate intensity factor along the friction surface

$$\omega = \omega_0 + \omega_1 \sqrt{\frac{\pi}{2} - \alpha - \theta} \tag{22.19}$$

where ω_0 and ω_1 are arbitrary parameters. These parameters should be found numerically from (22.12). Substituting (22.19) into (22.17) and the resulting expression into (22.14) yields

$$\xi_{\rho\theta} = -\frac{R_0^2 V \omega_1 \tan\alpha}{4\rho^2(\rho \sin\alpha - 1)}\left(\frac{\pi}{2} - \alpha - \theta\right)^{-1/2} + o\left[\left(\frac{\pi}{2} - \alpha - \theta\right)^{-1/2}\right] \tag{22.20}$$

as $\theta \to \pi/2 - \alpha$. The strain rate components are finite in the vicinity of the surface $\theta = \pi/2 - \alpha$. Therefore, substituting (22.20) into (22.13) and comparing the resulting expression with (22.2) show that

$$D = \frac{R_0^2 V \tan\alpha |\omega_1|}{2\sqrt{3}(\rho \sin\alpha - 1)\rho^{3/2}} \tag{22.21}$$

Numerical minimization in (22.12) supplies $\omega_1 = -2.579$ for the $\alpha = 60°$ die and $\omega_1 = -1.217$ for the $\alpha = 45°$ die. Substituting these values into (22.21) determines the strain rate intensity factors. The variation of the dimensionless strain rate intensity factors, $d = D\sqrt{R_0}/V$, along the friction surface is depicted in Fig. 22.4 where $y = \rho - (l + r_0)/\sin\alpha$. Thus $y = 0$ corresponds to the inner radius of the specimen.

22.5 Conclusions

A novel approach to predict the thickness of fine grain layers in the vicinity of frictional interfaces in metal forming processes has been proposed. The approach is based on the strain rate intensity factor and dimensional analysis. An experimental program to determine the function Φ involved in (22.7) has been designed and preliminary experimental results have been obtained. In particular, it has been shown that the angle α (see Fig. 22.1) has a significant effect on the thickness of the fine grain layer (Figs. 22.2 and 22.3). The upper bound method has been adopted to roughly evaluate the effect of the angle α on the strain rate intensity factor at the initial instant. It has been demonstrated that this effect is also significant (Fig. 22.4). A comprehensive experimental program is to be carried out and an efficient numerical method for calculating the strain rate intensity factor is to be developed to connect theoretical and experimental data by (22.7).

Acknowledgment Robert Goldstein and Sergei Alexandrov gratefully acknowledge the support of the Russian Science Foundation through RSF-14-11-00844 Project.

References

1. B.J. Griffiths, Mechanisms of white layer generation with reference to machining and deformation processes. Trans. ASME J. Tribol. **109**, 525–530 (1987)
2. S. Alexandrov, O. Richmond, Singular plastic flow fields near surfaces of maximum friction stress. Int. J. Non-Linear Mech. **36**, 1–11 (2001)
3. E.-G. Ng, D.K. Aspinwall, D. Brazil, J. Monaghan, Modeling of temperature and forces when orthogonally machining hardened steel. Int. J. Mach. Tool Manu. **39**, 885–903 (1999)
4. K. Yiğit, Ö. Tuğrul, Predictive analytical and thermal modeling of orthogonal cutting process—part I: predictions of tool forces, stresses, and temperature distributions. J. Manu. Sci. Eng. **128**, 435–444 (2006)
5. A. Ramesh, S.N. Melkote, Modeling of white layer formation under thermally dominate condition in orthogonal machining of hardened AISI 52100 steel, international. Int. J. Mach. Tool. Manu. **48**, 402–414 (2008)
6. D.I. Lalwani, N.K. Mehta, P.K. Jain, Extension of Oxley's predictive machining theory for Johnson and Cook flow stress model. J. Mater. Process. Tech. **209**, 5305–5312 (2009)
7. F. Akbar, P.T. Mativenga, M.A. Sheikh, An experimental and coupled thermo-mechanical finite element study of heat partition effects in machining. Int. J. Adv. Manu. Technol. **46**, 491–507 (2010)
8. A. Molinari, R. Cheriguene, H. Miguelez, Contact variables and thermal effects at the tool-chip interface in orthogonal cutting. Int. J. Solids Struct. **49**, 3774–3796 (2012)
9. G. Chen, J. Li, Y. He, C.A. Ren, New approach to the determination of plastic flow stress and failure initiation strain for aluminium alloys cutting process. Comp. Mater. Sci. **95**, 568–578 (2014)
10. M. Agmell, A. Ahadi, J.-E. Stahl, Identification of plastic constants from orthogonal cutting and inverse analysis. Mech. Mater. **77**, 43–51 (2014)
11. T.-K. Kim, K. Ikeda, Flow behavior of the Billet surface layer in porthole die extrusion of aluminium. Metall. Mater. Trans. A **31**, 1635–1643 (2000)
12. F. Widerøe, T. Welo, Conditions for sticking friction between aluminium alloy AA6060 and tool steel in hot forming. Key Eng. Mat. **491**, 121–128 (2012)
13. V. Sanabria, S. Mueller, S. Gall, W. Reimers, Investigation of friction boundary conditions during extrusion of aluminium and magnesium alloys. Key Eng. Mat. **611–612**, 997–1004 (2014)
14. V. Sanabria, S. Mueller, W. Reimers, Microstructure evolution of friction boundary layer during extrusion of AA 6060. Procedia Eng. **81**, 586–591 (2014)
15. T.-P. Fries, T. Belytschko, The extended/generalized finite element method: an overview of the method and its applications. Int. J. Numer. Meth. Engng. **84**(3), 253–304 (2010)
16. R. Hill, *The mathematical theory of plasticity* (Clarendon, Oxford, 1950)
17. M.F. Kanninen, C.H. Popelar, *Advanced fracture mechanics* (Oxford University Press, New York, 1985)
18. J.C. Gibbings, *Dimensional analysis* (Springer, London, 2011)
19. S.E. Alexandrov, R.V. Goldstein, The similarity of grain-size evolution near frictional interfaces and in the process of equal-channel angular pressing. Doklady Phys. **58**, 177–180 (2013)
20. F.J. Humphreys, P.B. Prangnell, J.R. Bowen, A. Gholinia, C. Harris, Developing stable fine-grain microstructures by large strain deformation. Phil. Trans. R. Soc. Lond. Ser. A **357**, 1663–1681 (1999)
21. W. Reiss, K. Pohlandt, The Rastegaev upset test—a method to compress large material volumes homogeneously. Exp. Technique **10**, 20–24 (1986)

Chapter 23
Interface Microstructure and Strength of TLP Bonded Iron and Steel

A.H.M.E. Rahman and M.N. Cavalli

Abstract Commercially pure iron (Fe) and stainless steel 321 (SS 321) were transient liquid phase (TLP) diffusion bonded using Cu and/or Au-12Ge interlayers. The joining pressures, temperatures and times were investigated. Au atoms appeared to diffuse back at the joining interface during slow cooling from the joining temperature. A faster cooling method prevented the back-diffusion of Au atoms from interior to the joining interface. Residual interlayer disappeared when faster cooling methods such as air cooling, water cooling or water quenching were applied. However, microcracks appeared in the joint centerline if water cooling or quenching was applied. Au-rich particles also appeared in the joint microstructure. No microcracks appeared for the samples cooled in air from the bonding temperature. The highest strength of the TLP diffusion bonded Fe reached almost same as the normalized Fe. The maximum tensile strengths (UTS) obtained for the bonded Fe were 291 ± 2 MPa using a Cu interlayer and 315 ± 4 MPa. The maximum strength found was 387 ± 4 MPa for TLP diffusion bonded steel using an Au-12Ge interlayer. The maximum strength for steel was obtained for the samples bonded at 1050 °C for 20 h in vacuum and cooled in air. The maximum strength found was 387 ± 4 MPa which is 70.4 % of the base alloy strength.

Keywords Diffusion bonding • TLP bonding • Microstructure • Tensile strength • Au-12Ge interlayer

23.1 Introduction

The most common metal joining processes are fusion welding processes. However, these joining processes have some detrimental effects. Fusion welded T91 martensitic steels are prone to hot cracking [1]. Duplex stainless steels containing ferrite and austenite possess superior mechanical and corrosion-resistant properties. Fusion welding of these alloys disrupts the austenite-ferrite ratios, thereby affecting the mechanical and corrosion-resistant properties. According to Floreen and Hayden [2], cleavage fracture in the ferrite phase is hindered by the more ductile austenite phase in duplex stainless steel. Rapid cooling during fusion welding produces a change in microstructure (additional ferrite is formed) causing poor mechanical properties, less ductility and poor corrosion resistance [3]. Among the austenitic alloys, SS 321 is utilized for many purposes mentioned above as well as nuclear facilities. As it is stabilized, it provides excellent intergranular corrosion-resistance when exposed to high temperatures. This alloy is stabilized with titanium to prevent the formation of chromium carbide during fusion welding. However, fusion welding still causes segregation, formation of intermetallics, loss of titanium and stress concentration sites. Fusion welded SS 321 is susceptible to knife-line attack if used in highly oxidizing environments [4].

Engine efficiency is a prime issue for engineers because it causes reduced fuel consumption and engine emissions. However, higher efficiency requires higher operating temperatures. The use of single crystal or columnar-grained blades permits turbine engine operation at higher temperatures and significantly increases the creep resistance [5–8]. For example, single crystal blades withstand up to 50 °C higher temperatures compared to their polycrystalline counterparts [9]. This is

A.H.M.E. Rahman (✉)
Department of Mechanical Engineering, Penn State Harrisburg, 777 West Harrisburg Pike,
Middletown, PA 17057, USA
e-mail: aer15@psu.edu

M.N. Cavalli
Department of Mechanical Engineering, University of North Dakota, 243 Centennial Drive Stop 8359,
Grand Forks, ND 58202, USA

C. Sciammarella et al. (eds.), *Experimental and Applied Mechanics, Volume 4*,
Conference Proceedings of the Society for Experimental Mechanics Series, DOI 10.1007/978-3-319-22449-7_23

due to the lesser amount of grain boundary strengthening elements which are known to be responsible for reducing creep strength at high temperatures. However, the casting of single crystal blade is a very challenging process. The yield of this process is comparatively low and gets lower as the blade size increases. A fabrication process for larger single crystal blades by joining smaller parts might overcome this size limitation. To combat both problems discussed above, solid/liquid phase diffusion bonding is considered as a preferred joining method.

In the current work, commercially pure iron and SS 321 was transient liquid phase bonded using a Cu and Au-12Ge eutectic interlayers. The effects of bonding parameters such as bonding environment, bonding temperature, bonding time and cooling method from the bonding temperature were studied. The microstructure of joint interface and the bond strength were determined.

23.2 Experimental Procedure

Commercially pure Fe and Stainless steel 321 (SS 321) rods 6.35 mm in diameter and ~8 mm in length were used for transient liquid phase diffusion bonding and subsequent microstructural examination. The samples were polished to a 1200 grit finish and the surfaces were cleaned in an ultrasonic bath using isopropyl alcohol and stored in alcohol before bonding. The samples for tensile testing had a gauge length of 59 $\pm$ 0.5 mm and a diameter of 9.0 $\pm$ 0.1 mm in the gauge section. The diffusion bonding was done with Au-12Ge (Alfa Aesar, USA) interlayers. The samples to be bonded were held in a jig made of Kovar under static pressure and placed in a tube furnace. Schematics of the diffusion bond geometry and diffusion bonded samples are shown elsewhere [10]. The heating chamber was repeatedly evacuated and filled with argon gas ten times to make it oxygen-free. The bonding was performed under either vacuum or argon atmosphere. An oxygen trap was used in the line of argon flow into the heating furnace. Fe joints with Au-12Ge interlayer were performed from 900 to 990 °C for 1–20 h. SS 321 joints with Au-12Ge interlayers (100 μm) were bonded at 900–1050 °C for 1–20 h. The joining temperatures were chosen based on the interlayer melting temperature and binary phase diagrams between major alloying elements and interlayer elements. Metallographic samples were mounted in epoxy resin. Longitudinal sections were polished down to approximately half of the diameter for microstructural observation.

Scanning electron microscopy (SEM) observations of bonded samples were carried out using a Hitachi S-3400 N scanning electron microscope (SEM) and the compositions and different phases in the bonded zone were analysed by energy dispersive spectrometry (EDS). To determine the mechanical properties of the bonded specimen, uniaxial tensile tests were performed using a screw-driven AG-IS 50 kN universal testing machine (Shimadzu) with a crosshead speed of 1 mm/min. Cylindrical tensile specimens were prepared according to ASTM standard E 8 M-99. For each data point the average of three specimens was used.

23.3 Results and Discussion

23.3.1 Microstructure of Diffusion Bonded Fe

Figure 23.1 shows SEM images of TLP diffusion bonded Fe using Au-12Ge interlayer (100 μm) at (a) 900 °C for 1 h. The thickness of the remaining interlayer found was ~12.6 μm. The composition of the residual interlayer was 15.5–28.9 wt% Fe, 61–77 wt% Au and no Ge. The composition of the parent metal at the joining interface was 78–79 wt% Fe, 11–11.4 wt% Au and 4.4 wt% Ge. The content of Ge and Au decreased gradually away from the interface. The microstructure of the residual interlayer appeared to be an irregular fingerprint structure. The melting point of Au-12Ge eutectic interlayer is 361 °C but it increases as the percentage of Ge decreases. The interlayer did not diffuse completely due to the faster diffusion of Ge in the base metal as no Ge was found in the residual interlayer. According to the Au-Fe phase diagram [11], Au can dissolve 26 wt% Fe at 900 °C. However, the percentage of Fe in the residual interlayer was not uniform. According to the element mapping, the black lines in the residual interlayer contained more Fe than the white area did. The evolution of this microstructure can be explained by referring to the Au-Fe equilibrium diagram. The solubility of Fe in Au decreases as the temperature decreases. As a result, a solid solution of Fe in Au becomes supersaturated and the excess Fe comes out as another solid solution of Au in Fe (α-Fe).

Figure 23.2 shows SEM images of TKP diffusion bonded Fe using Au-12Ge interlayer (100 μm) at 900 °C for 10 h. The interlayer completely diffused into the base metal in certain areas. In most areas the interlayer did not diffuse completely and

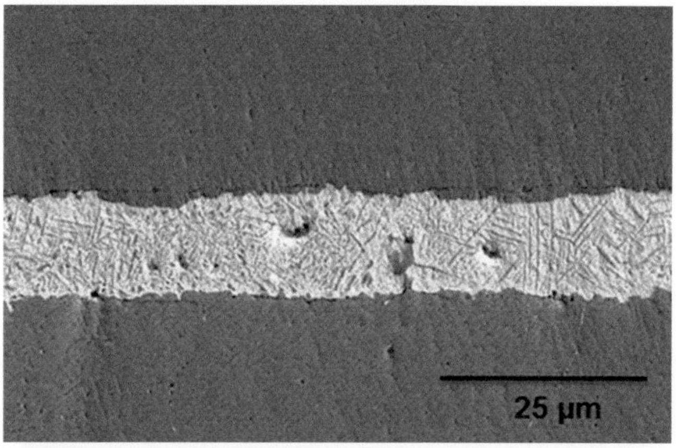

Fig. 23.1 SEM image of TLP diffusion bonded Fe using Au-12Ge interlayer (100 μm) at 900 °C for 1 h

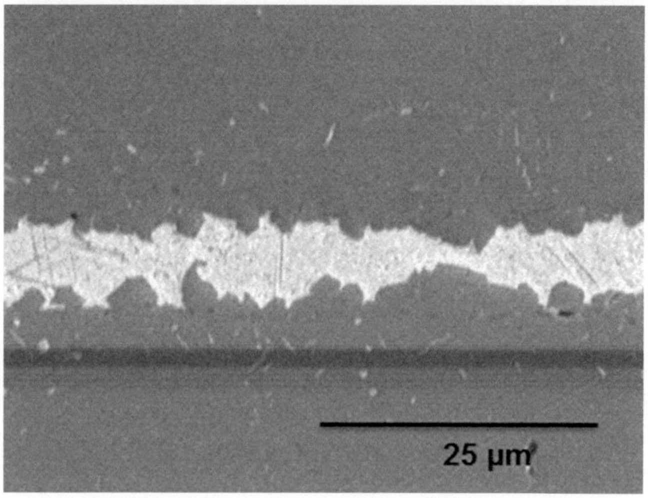

Fig. 23.2 SEM image of diffusion bonded Fe using Au-12Ge interlayer (100 μm) at 900 °C for 10 h

the average thickness of the residual interlayer found was ~6.3 μm. The microstructure of the residual interlayer was similar to that of the previous conditions—irregular fingerprint type. The residual interlayer contained ~18.5 wt% Fe, ~75.6 wt% Au and no Ge. Base metal at the interface contained 85 wt% Fe, 7.6 wt% Au and 2.1 wt% Ge. White particles near the interface were formed possibly during the cooling process. EDS analysis confirmed these particles were Au-rich particles. The solubilities of Au in Fe and Fe in Au decrease as the temperature decreases; the solubilities are almost zero at room temperature. During the cooling process, solid solutions of Au in Fe became supersaturated and excess Au comes out of the solution and forms solid solutions of Fe in Au.

Figure 23.3 shows SEM images of TLP diffusion bonded Fe using Au-12Ge interlayer (100 μm) at (a) 950 °C and (b) 990 °C for 15 h. At 950 °C, the composition of the residual interlayer and base metal near interface obtained were similar to those at 920 °C. The thickness of the residual interlayer was also in the similar range. The amount of residual interlayer found was less than at previous conditions. At 990 °C for 15 h, the interlayer completely diffused into the metal in most areas while it retained in few area. Au-rich particles were found with some pores in the bond centerline where interlayer diffused completely.

23.3.2 Microstructure of Diffusion Bonded SS321

The composition of the base metal alloy (SS 321) is shown in Table 23.1 [12]. Figure 23.4 shows SEM image of TLP diffusion bonded SS 321 using Au-12Ge interlayer (100 μm) at 900 °C for 1 h with slow cooling. Most part of the interlayer diffused into the base material at these joining conditions. The average thickness of the residual interlayer obtained was ~5.8 μm. The compositions found in the residual interlayer were 78.4 wt% Au, 6.4 wt% Fe, 1.6 wt% Ni, 6 wt% Cr and no Ge with small amounts of other alloying elements. The composition found at the interface on the base metal side was 25.3 wt%

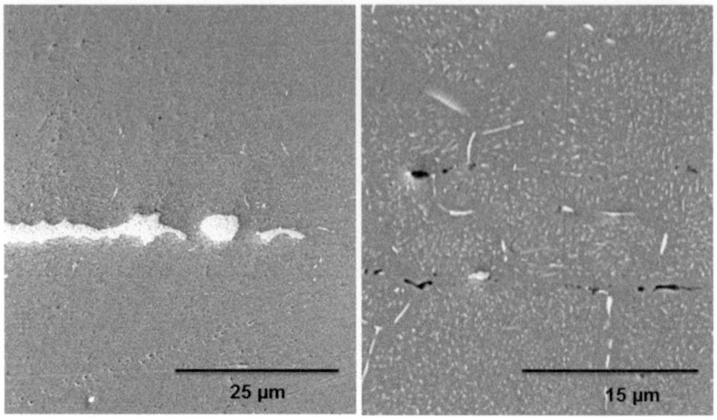

Fig. 23.3 SEM images of TLP diffusion bonded Fe using Au-12Ge interlayer (100 μm) at (**a**) 950 °C and (**b**) 990 °C for 15 h

Table 23.1 Composition (wt%) of AISI type stainless steel 321 (SS 321) [12]

Fe	Cr	Ni	Mn	Si	Ti	C	S	P
Bal.	17–19	9–12	2	1	5× %(C+N)−0.7	0.08	0.03	0.045

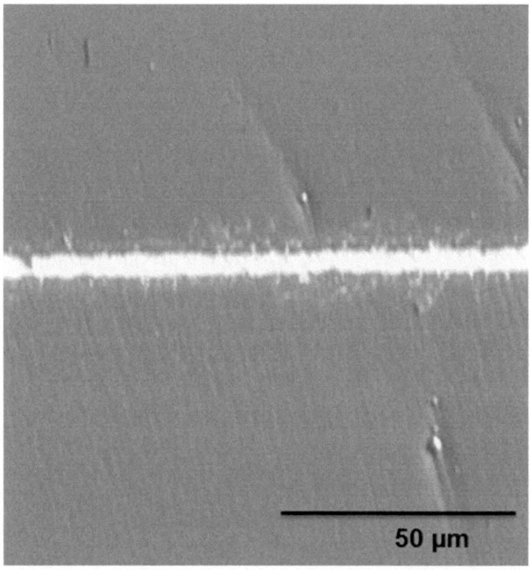

Fig. 23.4 SEM image of TLP diffusion bonded SS 321 using Au-12Ge interlayer (100 μm) at 900 °C for 1 h, Thickness of residual interlayer ~5.8 μm

Au, 4.1 wt% Ge, 44.7 % Fe, 7.4 % Ni and 14.7 wt% Cr. It is expected that the microstructure of the joint would be free of residual interlayer or any intermetallics. The reason is that the residual interlayer and intermetallics will reduce the strength of the joint significantly.

Figure 23.5 shows SEM image of TLP diffusion bonded SS 321 using Au-12Ge interlayer (100 μm) at 950 °C for 10 h with slow cooling. The joint centerline microstructure consisted of residual interlayer and Au-rich particles near the interlayer/base metal interface. The average thickness of the residual interlayer found was ~5.5 μm with a composition of 79.3 wt% Au, 5.3 wt% Fe, 1.5 wt% Ni, 4.2 wt% Cr, 3 wt% Mn and no Ge. It is noticeable that even though the bonding temperature and time were increased, the overall interdiffusion was not increased appreciably. The residual interlayer (~5.5 μm) obtained was little thinner than obtained with previous bonding conditions.

Since the residual interlayer is generally not desirable, the joining temperature and time were increased to enhance interdiffusion. Figure 23.6 shows SEM image of TLP diffusion bonded SS 321 using Au-12Ge interlayer at 1050 °C for 15 h with slow cooling. The bond centerline consisted of residual interlayer and Au-rich particles. The residual interlayer had a composition of 78.3 wt% Au, 3.1 wt% Fe, 1.7 wt% Ni, 3.3 wt% Cr, 6.0 wt% Mn and no Ge. The average thickness of the residual interlayer found was ~3.8 μm which was 8.6 % thicker than for previous bonding conditions. The precipitated Au-rich particles were elongated in shape and precipitated in a region extending ~45.5 μm from the bond centerline. The number of Au-rich particles decreased but their average size increased compared to previous conditions. Changing the

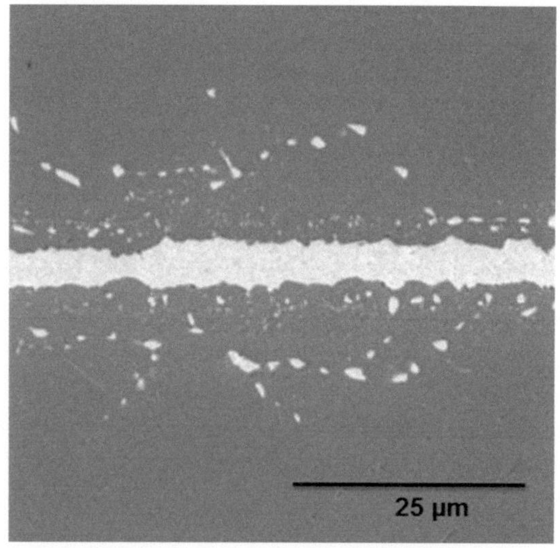

Fig. 23.5 SEM image of TLP diffusion bonded SS 321 using Au-12Ge interlayer (100 μm) at 950 °C for 10 h with slow cooling, Thickness of residual interlayer ~5.5 μm

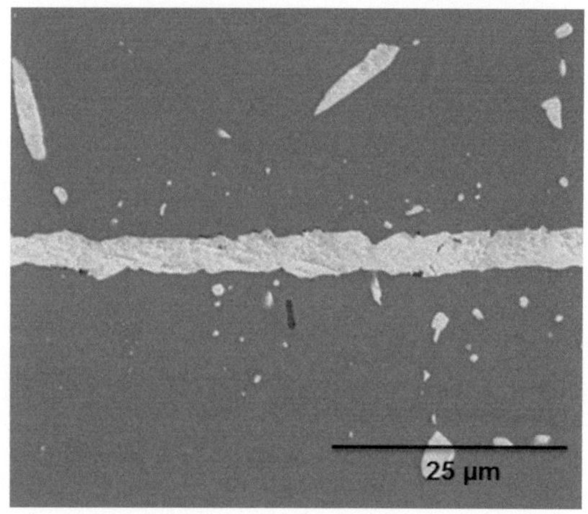

Fig. 23.6 SEM image of TLP diffusion bonded SS 321 using Au-12Ge interlayer at 1050 °C for 15 h with slow cooling. Residual interlayer ~3.8 μm

bonding temperature and time from 900 to 1050 °C and 1 to 15 h, respectively, the average residual interlayer thickness decreased from 5.8 to 3.8 μm. Two possible explanations for the similar results across a range of temperatures and times are (1) interdiffusion was not increased appreciably or (2) the diffusing elements (primarily Au) precipitated during cooling process preferentially on the joint centerline. The later phenomenon can be explained referring the Au-Fe phase diagram [11]. The solubility of Fe in Au and Au in Fe is very small at room temperature compared to the bonding temperature. During cooling, both Fe and Au solid solutions become supersaturated. As a result, excess Au comes out of the solution and precipitates. A faster cooling rate might reduce or eliminate this phenomenon because precipitation requires diffusion. At lower temperatures the atoms might not get enough time to diffuse. It should be noted that the presence of additional alloying elements in the system will also affect the interdiffusion and precipitation.

To inhibit precipitation, a faster cooling process was applied. Figure 23.7 shows SEM image of TLP diffusion bonded SS 321 using Au-12Ge interlayer at 1050 °C for 15 h in vacuum with cooling in air. The interlayer was diffused completely in some areas and it was retained in some areas. The faster cooling seemed to prevent the precipitation of diffusing elements, primarily Au, out of the supersaturated base alloy matrix. Spherical Au-rich particles were observed near the bond centerline along the whole cross-section. The size of these particles was much smaller than was observed in previous conditions where a slower cooling process was applied. The coarsening of these particles was prevented by fast cooling. The composition of the residual interlayer was 75 wt% Au, 5.4 wt% Fe, 2.0 % wt. % Ni, 3.9 wt% Cr, 7.4 wt% Mn and no Ge. The composition in the bond centerline, where no residual interlayer was found, was ~10 wt% Au, 62 %wt. Fe, 15.6 wt% Cr, 8 wt% Ni, 0.28 % wt. Mn and 1.9 wt% Ge.

The joining time was then increased to 20 h and the cooling rate was increased by quenching in water. Figure 23.8 shows SEM image of TLP diffusion bonded SS 321 at 1050 °C for 20 h using Au-12Ge interlayer in vacuum and quenched in water.

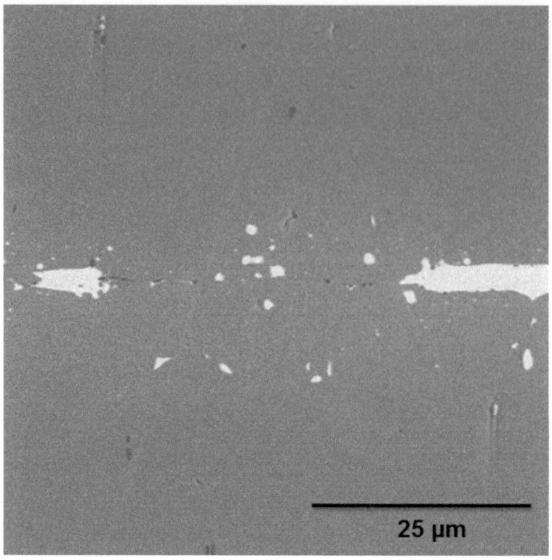

Fig. 23.7 SEM image of TLP diffusion bonded SS 321 using Au-12Ge interlayer at 1050 °C for 15 h in vacuum with cooling in air

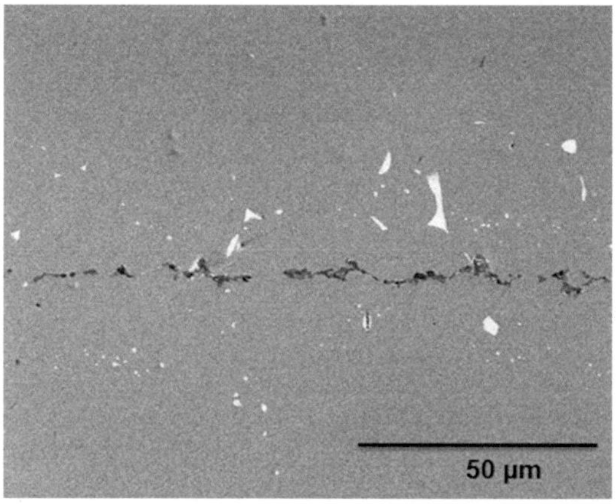

Fig. 23.8 (**a**) SEM image of TLP diffusion bonded SS 321 at 1050 °C for 20 h using Au-12Ge interlayer in vacuum and quenched in water

The bond centerline microstructure consisted of Au-rich particles, precipitates (black particles on the image) and microcracks which ran through the entire cross section. EDS analysis near the microcracks at the joint centerline showed 8.7 wt% C, 28.2 wt% O, 43.1 wt% Al, 8.9 wt% Ti and small amounts of other alloying and interlayer elements while the base alloy had only 0.08 wt% C, .4–0.7 wt% Ti and no Al. This analysis indicated that high percentages of Al, O, Ti and C appeared from polishing media (diamond paste and γ-Al_2O_3), TiC and possibly TiO_2. The microstructures in the joint centerline other than precipitates contained 59.5 wt% Fe, 15.3 wt% Cr, 8 wt% Ni, 10 wt% Au, 3.1 wt% Ge and .3 wt% Mn with trace amounts of other alloying elements.

The polishing media might have trapped in the microcracks during polishing. TiC and TiO_2 formed possibly at the bonding temperature and during cooling process, respectively. The stabilising anneal temperature for SS 321 is 845–900 °C and TiC forms during this annealing process. Ti is added to SS 321 to prevent the formation of chromium carbide at elevated temperature. Formation of chromium carbide depletes Cr from the grain boundary, which in turn, leads to grain boundary corrosion.

Figure 23.9 shows (a) SEM image of TLP diffusion bonded SS 321 using Au-12Ge interlayer at 1050 °C for 20 h in vacuum and air-cooled and (b) magnified image of (a). At these joining conditions no residual interlayer was found in the joint area. The microstructure of the joint centerline consisted of a matrix of base alloy with Au-rich particles along the grain boundaries as well as in the grain areas. No microcracks were found in the bonded area. The size of the Au-rich particles along the grain boundary was much bigger than the ones in the grain areas. The bond centerline microstructure consisted of 17 wt% Au, 47 wt% Fe, 16 wt% Cr, 7 wt% Ni, 3.8 wt% Ge and 1.8 wt% Mn. The Au-rich particles contained 70–72 wt% Au, 8–11 wt% Mn, 5 wt% Fe, 2.5 wt% Cr and 1.1 wt% Ni. There were only a few black particles observed in the joint centerline microstructure. These particles are possibly a mixture of Al_2O_3, C, TiO_2 or TiC.

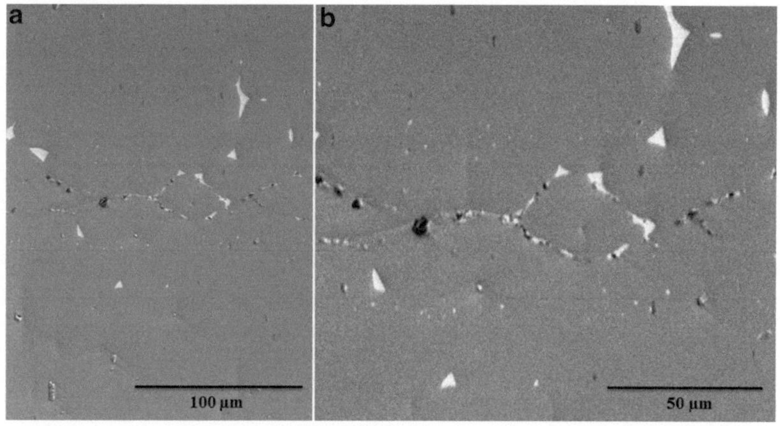

Fig. 23.9 (**a**) SEM image of TLP diffusion bonded SS 321 using Au-12Ge interlayer at 1050 °C for 20 h in vacuum and air-cooled and (**b**) magnified image of (**a**)

Table 23.2 Summary of the microstructure of TLP bonded SS 321 using Au-12Ge, bonding environment and cooling method

	Bonding atmosphere	Cooling method	Microstructure
1	Vacuum	Quench in water	Almost continuous line of precipitates of Al_2O_3, TiO_2, TiC, microcracks and Au-rich particles in the bond centerline
2	Vacuum	Flow of water	Reduced microcracks, precipitates and Au-rich particles in the joint centerline. The size and amount of the Au-rich particles increased
3	Vacuum	Air	No microcracks. Au-rich particles seemed to precipitate along the grain boundary in the joint centerline. The size and amount of the Au-rich particles increased more than previous two cases. There are only a few precipitates in the joint centerline
4	Argon	Air	No microcracks with Au-rich particles and precipitates in the joint centerline
5	Argon	Flow of water	Microcracks and precipitates along the joint centerline. Au-rich particles in the joint centerline area

Table 23.2 shows the summary of the microstructure of TLP bonded SS 321 using Au-12Ge, bonding environment and cooling method. For all joining environments and cooling methods Au-rich particles appeared in the joint centerline. The size and amount of Au-rich particles seemed to increase as the cooling method changed from quenching in water to air-cooling. The faster the cooling method, the less time Au atoms have to precipitate. However, with quenching in water and cooling with water-flow microcracks appeared in the joint centerline. Some precipitates such as Al_2O_3, TiC and possibly TiO_2 appeared in the joint centerline. Al_2O_3 is a contamination from polishing media and TiC formed during joining process. TiO_2 might form during cooling process. The formation of this type of oxide can be prevented by cooling in a controlled environment. However, the amount and the effect of the TiO_2 in the microstructure on the strength need to be determined. The Au-12Ge interlayer had a thickness of 100 µm. A thinner foil may have eliminated or significantly reduced the precipitation of Au-rich particles in the joint centerline.

23.3.3 Mechanical Properties

Table 23.3 shows tensile strengths of diffusion bonded Fe using Au-12Ge interlayers. The strength of bonded Fe using Au-12Ge at 920 °C for 15 h was 284 ± 1 MPa. These samples failed away from the joint area. The strengths obtained for annealed and normalized cp-Fe were 282 ± 2 and 316 ± 2 MPa, respectively. The strength of the diffusion bonded Fe at 920 °C was almost same as the annealed cp-Fe. To determine the strength of the joint diffusion bonded Fe was normalized (air cooled from the bonding temperature). The normalized diffusion bonded Fe had maximum ultimate tensile strengths of 315 ± 4 MPa at 950 °C for 15 h and they failed in the bonded area. Normalized diffusion bonded Fe did not fail in the base metal area because the grains of base metal were refined by normalizing. The strength of metals is inversely related to grain size. The tensile strengths for the samples bonded at 970 °C for 15 h were almost same as for samples bonded at 950 °C for 15 h. However, strengths decreased to 302 ± 3 MPa for the samples bonded at 990 °C for 15 h. The lower strength at this bonding condition was due to the porous microstructure in the joint centerline.

Table 23.3 Ultimate tensile strengths (UTS) of diffusion bonded Fe using Au-12Ge (100 μm) interlayer

Bonding temp. (°C)	UTS (MPa)
cp-Fe (Normalized, 965° C, 1 h)	316 ± 2
cp-Fe (Annealed, 920° C, 15 h)	282 ± 2
920	284 ± 1[a]
890 (Normalized)	287.28
920 (Normalized)	302 ± 1
950 (Normalized)	315 ± 4
970 (Normalized)	314 ± 4
990 (Normalized)	302±3

[a]Specimen failed away from bondline

Table 23.4 Tensile strengths of TLP bonded SS 321 using Au-12Ge interlayer

Interlayer	Temp. (°C)	Time (h)	Clamping torque	UTS (MPa)	Corrected UTS (MPa)	Environment	Cooling
Base material (as machined specimen)				672 ± 3	–	–	–
Au-12Ge	950	10	40 in-lb	272 ± 3	336 ± 4	Vacuum	Slow
Au-12Ge	1050	15	50 in-lb	254 ± 5	309 ± 5	Vacuum	Slow
Au-12Ge	1050	15	35 in-lb	278 ± 4	352 ± 5	Vacuum	Slow
Au-12Ge	1050	20	40 in-lb	222 ± 3	280 ± 4	Argon	Air
Au-12Ge	1050	20	35 in-lb	371 ± 4	466 ± 5	Argon	Air
Au-12Ge	1050	20	35 in-lb	354 ± 2	448 ± 2	Argon	Water flow
Au-12Ge	1050	20	35 in-lb	387 ± 4	490 ± 5	Vacuum	Air
Base alloy	1050	20		550		Vacuum	Air

Table 23.4 shows tensile strengths of TLP bonded SS 321 using Au-12Ge interlayers. The maximum strength obtained for bonded SS 321 was 387 ± 4 MPa with Au-12Ge interlayer. This sample was bonded at 1050 °C for 20 h in vacuum and cooled in air. This maximum strength is 57.6 % of the tensile strength of as-machined SS 321 and 70.4 % of the tensile strength of the base alloy that went through similar heat treatment process as diffusion bonded sample. There are several factors that are believed to affect the strength of the joint including the microstructure of the joint, the clamping load applied to the sample during bonding, the bonding temperature, the bonding time and the cooling rate. The coefficients of thermal expansions of the jig, base alloy and mild steel bolts at the bonding temperatures were not available; thus it was not possible to calculate the exact clamping load. As a result, samples bonded at different temperatures likely experienced different clamping loads (i.e., bonding pressure). The 35 in-lb clamping torque provided the highest strength of the joint compared to other clamping torque. Although the argon bonding environment provided the best microstructure, the maximum strength was obtained for the sample bonded in vacuum. Among the three different cooling methods, air cooling provided the best strength of the bonded sample. A factor that could affect the joint strength is the foil thickness. As discussed earlier, there were Au-rich particles in the joint centerline for all joining conditions. The precipitation of these particles disrupts the joint microstructures. Using a thinner foil would potentially decrease the Au-rich particles in the joint centerline. This, in turn, may increase the strength of the joint. A thinner foil with lower percentage of Au may also provide a positive effect on the joint strength.

23.4 Conclusion

Fingerprint-like microstructure appeared in the residual interlayer when Fe bonded with a Au-12Ge interlayer. The microstructure of the base metal near bond centerline contained dispersed Au-rich particles. The highest strength observed was 315 ± 4 MPa for the sample bonded at 950 °C for 15 h and normalized. This is almost the same strength as normalized Fe. The best microstructures were obtained for diffusion bonded SS 321 using Au-12Ge interlayer at 1050 °C bonding temperature in vacuum followed by cooling in air. Some Au-rich particles observed in the microstructure of bonded area for all joining conditions. Although these particles were smaller for water cooling, microcracks appeared in the joint centerline at faster cooling rates. The maximum strength of the joint for SS 321 obtained was 70.4 % of the strength of base alloy. It is expected that using a thinner Au-12Ge foil and/or a foil with lower percentage of Au would decrease the second phase particles, which in turn, may increase the strength of the joint.

Acknowledgement Financial support from the National Science Foundation (Award #DMR-0907616) and Penn State Capital College during the course of this work is gratefully acknowledged.

References

1. T.I. Khan, E.R. Wallach, Transient liquid-phase bonding of ferritic oxide dispersion strengthened superalloy MA957 using a conventional nickel braze and a novel iron-base foil. J. Mater. Sci. **30**(20), 5151–5160 (1995)
2. S. Floreen, H.W. Hayden, The influence of austenite and ferrite on the mechanical properties of two phase stainless steels having microduplex structures. Trans. Quart. ASM **61**, 489–499 (1968)
3. D. Blumfield, G.A. Clark, P. Guha, Welding duplex austenitic-ferritic stainless steel. Met. Construct. **5**, 269–273 (1981)
4. ASM International, Corrosion of Weldments (#05182G) (2006), http://www.asminternational.org/content/ASM/StoreFiles/ACFAB86.pdf. Accessed 17 Mar 2013
5. R.F. Singer, Advanced materials and processes for land-based gas turbines. Adv. Power Eng. **2**, 1707–1729 (1994)
6. B.B. Seth, Superalloys—the utility gas turbine perspective. in Proceedings of the 9th International Symposium on Superalloys (Champion, PA, 2000)
7. R.C. Reed, *The superalloys: fundamentals and applications* (Cambridge University Press, Cambridge, 2006)
8. C. Rae, Alloys by design: modelling next generation superalloys. Mater. Sci. Technol. **25**(4), 479–487 (2009)
9. P. Caron, T. Khan, Evolution of Ni-based superalloys for single crystal gas turbine blade applications. Aerosp. Sci. Technol. **3**(8), 513–523 (1999)
10. A.H.M.E. Rahman, M.N. Cavalli, Strength and microstructure of diffusion bonded titanium using silver and copper interlayers. Mater. Sci. Eng. A **527**, 5189–5193 (2010)
11. H. Okamoto, T.B. Massalski, L.J. Swartzendruber, P.A. Beck, Alloy phase diagram. in ASM Handbook, vol. 3. 10 edition (1992), p. 2.69
12. Standard Specification for Stainless Steel Bars and Shapes ASTM Standard A276-08a. Annual Book of ASTM Standard, Section One, Iron and Steel Products, vol. 01.03 (2012), pp. 69–75.

ERRATUM

Experimental and Applied Mechanics, Volume 4

Proceedings of the 2015 Annual Conference on Experimental and Applied Mechanics

Cesar Sciammarella, John Considine, and Paul Gloeckner

C. Sciammarella et al. (eds.), *Experimental and Applied Mechanics, Volume 4*,
Conference Proceedings of the Society for Experimental Mechanics Series, DOI 10.1007/978-3-319-22449-7

DOI 10.1007/978-3-319-22449-7_24

The affiliation of second editor Dr. John Considine is incorrect in the Copyright page.
The correct affiliation is as follow:

U.S. Forest Service
Forest Products Laboratory
One Gifford Pinchot Drive
Madison, WI, 53726, USA

The online version of the original book can be found at
http://dx.doi.org/10.1007/978-3-319-22449-7

C. Sciammarella et al. (eds.), *Experimental and Applied Mechanics, Volume 4*,
Conference Proceedings of the Society for Experimental Mechanics Series, DOI 10.1007/978-3-319-22449-7_24

MIX
Papier aus verantwortungsvollen Quellen
Paper from responsible sources
FSC® C105338

If you have any concerns about our products,
you can contact us on
ProductSafety@springernature.com

In case Publisher is established outside the EU,
the EU authorized representative is:
Springer Nature Customer Service Center GmbH
Europaplatz 3, 69115 Heidelberg, Germany

Printed by Libri Plureos GmbH
in Hamburg, Germany